POWER SYSTEM HARMONIC ANALYSIS

POWER SYSTEM HARMONIC ANALYSIS

Jos Arrillaga, Bruce C Smith
Neville R Watson, Alan R Wood
University of Canterbury, Christchurch, New Zealand

JOHN WILEY & SONS
Chichester · New York · Weinheim · Brisbane · Singapore · Toronto

Baffins Lane, Chichester,
West Sussex PO19 1UD, England

National 01243 779777
International (+44) 1243 779777

e-mail (for orders and customer service enquiries): cs-books@wiley.co.uk

Visit our Home Page on http://www.wiley.co.uk
or
http://www.wiley.com

Other Wiley Editorial Offices

John Wiley & Sons, Inc., 605 Third Avenue,
New York, NY 10158-0012, USA

WILEY-VCH Verlagsgesellschaft GmbH, Pappelallee 3,
D-69469 Weinheim, Germany

Jacaranda Wiley Ltd, 33 Park Road, Milton,
Queensland 4064, Australia

John Wiley & Sons (Asia) Pte Ltd, 2 Clementi Loop #02-01,
Jin Xing Distripark, Singapore 129809

John Wiley & Sons (Canada) Ltd, 22 Worcester Road,
Rexdale, Ontario M9W 1L1, Canada

Library of Congress Cataloguing in Publication Data

Power system harmonic analysis / Jos Arrillaga . . . [et al.].
p. cm.
Includes bibliographical references and index.
ISBN 0 471 97548 6
1. Electric power systems — Mathematical models. 2. Harmonics (Electric waves) — Mathematics. I. Arrillaga, J.
TK3226.P378 1997 97-309
621.319'1 — dc21 CIP

British Library Cataloguing in Publication Data

A catalogue record for this book is available from the British Library

ISBN 0 471 97548 6

Cover design by J. N. Arrillaga
Typeset in 10/12pt Times by Dobbie Typesetting Limited
Printed and bound in Great Britain by Bookcraft (Bath) Ltd
This book is printed on acid-free paper responsibly manufactured from sustainable forestation, for which at least two trees are planted for each one used for paper production

CONTENTS

PREFACE

The subject of *Power System Harmonics* was first discussed in a book published by J. Wiley & Sons in 1985 which collected the state of the art, explaining the presence of voltage and current harmonics with their causes, effects, standards, measurement, penetration and elimination. Since then, the increased use of power electronic devices in the generation, transmission and utilisation of systems has been accompanied by a corresponding growth in power system harmonic problems.

Thus, *Power System Harmonic Analysis* has become an essential part of system planning and design. Many commercial programmes are becoming available, and CIGRE and IEEE committees are actively engaged in producing guidelines to facilitate the task of assessing the levels of harmonic distortion.

This book describes the analytical techniques, currently used by the power industry for the prediction of harmonic content, and the more advanced algorithms developed in recent years.

A brief description of the main harmonic modelling philosophies is made in Chapter 1 and a thorough description of the Fourier techniques in Chapter 2.

Models of the linear system components, and their incorporation in harmonic flow analysis, are considered in Chapters 3 and 4. Chapters 5 and 6 analyse the harmonic behaviour of the static converter in the frequency domain. The remaining chapters describe the modelling of non-linearities in the harmonic domain and their use in advanced harmonic flow studies.

The authors would like to acknowledge the assistance received directly or indirectly from their present and previous colleagues, in particular from E. Acha, G. Bathurst, P. S. Bodger, S. Chen, T. J. Densem, J. F. Eggleston, B. J. Harker, M. L. V. Lisboa and A. Medina. They are also grateful for the advice received from J. D. Ainsworth, H. Dommel, A. Semylen and R. Yacamini. Finally, they wish to thank Mrs G. M. Arrillaga for her active participation in the preparation of the manuscript.

1

INTRODUCTION

1.1 Power System Harmonics

The presence of voltage and current waveform distortion is generally expressed in terms of harmonic frequencies which are integer multiples of the generated frequency [1].

Power system harmonics were first described in book form in 1985 (Arrillaga) [2]. The book collected together the experience of previous decades, explaining the reasons for the presence of voltage and current harmonics as well as their causes, effects, standards, measurement, simulation and elimination.

Since then the projected increase in the use and rating of solid state devices for the control of power apparatus and systems has exceeded expectations and accentuated the harmonic problems within and outside the power system. Corrective action is always an expensive and unpopular solution, and more thought and investment are devoted at the design stage on the basis that prevention is better than cure. However, preventative measures are also costly and their minimisation is becoming an important part of power system design, relying heavily on theoretical predictions.

Good harmonic prediction requires clear understanding of two different but closely related topics. One is the non-linear voltage/current characteristics of some power system components and its related effect, the presense of harmonic sources. The main problem in this respect is the difficulty in specifying these sources accurately. The second topic is the derivation of suitable harmonic models of the predominantly linear network components, and of the harmonic flows resulting from their interconnection. This task is made difficult by insufficient information on the composition of the system loads and their damping to harmonic frequencies. Further impediments to accurate prediction are the existence of many distributed non-linearities, phase diversity, the varying nature of the load, etc.

1.2 The Main Harmonic Sources

For simulation purposes the harmonic sources can be divided into three categories:

(1) Large numbers of distributed non-linear components of small rating.
(2) Large and continuously randomly varying non-linear loads.

(3) Large static power converters and transmission system level power electronic devices.

The first category consists mainly of single-phase diode bridge rectifiers, the power supply of most low voltage appliances (e.g. personal computers, TV sets, etc.). Gas discharge lamps are also included in this category. Although the individual ratings are insignificant, their accumulated effect can be important, considering their large numbers and lack of phase diversity. However, given the lack of controllability, these appliances present no special simulation problem, provided there is statistical information of their content in the load mix.

The second category refers to the arc furnace, with power ratings in tens of megawatts, connected directly to the high voltage transmission network and normally without adequate filtering. The furnace arc impedance is randomly variable and extremely asymmetrical. The difficulty, therefore, is not in the simulation technique but in the variability of the current harmonic injections to be used in each particular study, which should be based on a stochastic analysis of extensive experimental information obtained from measurements in similar existing installations.

As far as simulation is concerned, it is the third category that causes considerable difficulty. This is partly due to the large size of the converter plant in many applications, and partly to their sophisticated point on wave switching control systems. The operation of the converter is highly dependent on the quality of the power supply, which is itself heavily influenced by the converter plant. Thus the process of static power conversion needs to be given special attention in power system harmonic simulation.

1.3 Modelling Philosophies

A rigorous analysis of the electromagnetic behaviour of power components and systems requires the use of field theory. However, the direct applicability of Maxwell's equations to the solution of practical problems is extremely limited. Instead, the use of simplified circuit equivalents for the main power system components generally leads to acceptable solutions to most practical electromagnetic problems.

Considering the (ideally) single frequency nature of the conventional power system, much of the analytical development in the past has concentrated on the fundamental (or power) frequency.

Although the operation of a power system is by nature dynamic, it is normally subdivided into well-defined quasi steady state regions for simulation purposes. For each of these steady-state regions, the differential equations representing the system dynamics are transformed into algebraic ones by means of the factor ($j\omega$), and the circuit is solved in terms of voltage and current phasors at fundamental frequency ($\omega = 2\pi f$).

By definition, harmonics result from periodic steady state operating conditions and therefore their prediction should also be formulated in terms of (harmonic) phasors, i.e. in the frequency domain.

If the derivation of harmonic sources and harmonic flows could be decoupled, the theoretical prediction would be simplified. Such an approach is often justified in assessing the harmonic effect of industrial plant, where the power ratings are relatively small. However, the complex steady state behaviour of some system components, such as an HVdc converter, require more sophisticated models either in the frequency or time domains.

As with other power system studies, the digital computer has become the only practical tool in harmonic analysis. However, the level of complexity of the computer solution to be used in each case will depend on the economic consequences of the predicted behaviour and on the availability of suitable software.

1.4 Time Domain Simulation

The time domain formulation consists of differential equations representing the dynamic behaviour of the interconnected power system components. The resulting system of equations, generally non-linear, is normally solved using numerical integration.

The two most commonly used methods of time domain simulation are state variable and nodal analysis, the latter using Norton equivalents to represent the dynamic components.

Historically, the state variable solution, extensively used in electronic circuits [3–5], was first applied to ac–dc power systems [6]. However, the nodal approach is more efficient and has become popular in the electromagnetic transient simulation of power system behaviour [7–8].

The derivation of harmonic information from time domain programmes involves solving for the steady state and then applying the Fast Fourier Transform. This requires considerable computation even for relatively small systems and some acceleration techniques have been proposed to speed up the steady state solution [9,10]. Another problem attached to time domain algorithms for harmonic studies is the difficulty of modelling components with distributed or frequency-dependent parameters.

It is not the purpose of this book to discuss transient simulation. However, in several sections use is made of standard EMTP programmes to verify the newly proposed frequency domain algorithms.

1.5 Frequency Domain Simulation

In its simplest form the frequency domain provides a direct solution of the effect of specified individual harmonic (or frequency) injections throughout a linear system, without considering the harmonic interaction between the network and the non-linear component(s).

The simplest and most commonly used model involves the use of single phase analysis, a single harmonic source and a direct solution.

The supply of three-phase fundamental voltage at points of common coupling is within strict limits well balanced, and under these conditions load flow studies are normally carried out on the assumption of perfect symmetry of network components by means of single phase (line) diagrams. The same assumption is often made for the harmonic frequencies, even though there is no specified guarantee from utilities of harmonic symmetry.

The harmonic currents produced by non-linear power plant are either specified in advance, or calculated more accurately for a base operating condition derived from a load flow solution of the complete network. These harmonic levels are then kept invariant throughout the solution. That is, the non-linearity is represented as a constant harmonic current injection, and a direct solution is possible.

In the absence of any other comparable distorting loads in the network, the effect of a given harmonic source is often assessed with the help of equivalent harmonic impedances. The single source concept is still widely used as the means to determine the harmonic voltage levels at points of common coupling and in filter design.

A common experience derived from harmonic field tests is the asymmetrical nature of the readings. Asymmetry, being the rule rather than the exception, justifies the need for multiphase harmonic models. The basic component of a multiphase algorithm is the multiconductor transmission line, which can be accurately represented at any frequency by means of an appropriate equivalent PI-model, including mutual effects as well as earth return, skin effect, etc. The transmission line models are then combined with the other network passive components to obtain three-phase equivalent harmonic impedances.

If the interaction between geographically separated harmonic sources can be ignored, the single source model can still be used to assess the distortion produced by each individual harmonic source. The principle of superposition is then invoked to derive the total harmonic distortion throughout the network. Any knowledge of magnitude and phase diversity between the various harmonic injections can then be used either in deterministic or probabilistic studies.

1.6 Iterative Methods

The increased power rating of modern HVdc and FACTS devices in relation to the system short circuit power means that the principle of superposition does not apply. The harmonic injection from each source will, in general, be a function of that from other sources and the system state. Accurate results can only be obtained by iteratively solving non-linear equations that describe the steady state as a whole. The system steady state is substantially, but not completely, described by the harmonic voltages throughout the network. In many cases, it can be assumed that there are no other frequencies present apart from the fundamental frequency and its harmonics. This type of analysis, the Harmonic Domain, can be viewed as a restriction of frequency domain modelling to integer harmonic frequencies but with all non-linear interactions modelled. Harmonic Domain modelling may also encompass a solution for three-phase load flow constraints, control variables, power electronic switching instants, transformer core saturation, etc.

There are two important aspects to the Harmonic Domain modelling of the power system:

(1) The derivation, form and accuracy of the non-linear equations used to describe the system steady state.
(2) The iterative procedure used to solve the non-linear equation set.

Many methods have been employed to obtain a set of accurate non-linear equations which describe the system steady state. After partitioning the system into linear regions and non-linear devices, the non-linear devices are described by isolated equations, given boundary conditions to the linear system. The system solution is then predominantly a solution for the boundary conditions for each non-linear device. Device modelling has been by means of time domain simulation to the steady state [12], analytic time domain expressions [11,13], waveshape sampling and FFT [14] and, more recently, by harmonic phasor analytic expressions [15].

In the past, Harmonic Domain modelling has been hampered by insufficient attention given to the solution method. Earlier methods used the Gauss–Seidel type fixed point interation, which frequently diverged. Improvements made since then have been to include linearising RLC components in the circuit to be solved in such a way as to have no effect on the solution itself [13,16]. A more recent approach has been to replace the non-linear devices at each iteration by a linear Norton equivalent, chosen to mimic the non-linearity as closely as possible, sometimes by means of a frequency coupled Norton admittance. The progression with these improvements to the fixed point iteration method is toward Newton-type solutions, as employed successfully in the load flow for many years. When the non-linear system to be solved is expressed in a form suitable for solution by Newton's method, the separate problems of device modelling and system solution are completely decoupled and the wide variety of improvements to the basic Newton method, developed by the numerical analysis community, can readily be applied.

1.7 References

1. Fourier, J B J (1822). *Théorie Analytique de la Chaleur* (book), Paris.
2. Arrillaga, J, Bradley, D and Bodger, P S, (1985). *Power System Harmonics*, J Wiley & Sons, London.
3. Chuah, L D and Lin P M, (1975) *Computer-aided Analysis of Electronic Circuits*, Englewood Cliffs, Prentice Hall, NJ.
4. Kuh, E S and Rohrer, R A, (1965). The state variable approach to network analysis, *Proc IEEE.*
5. Balabanian, N, Bickart, T A and Seshu, S, (1969). *Electrical Network Theory*, John Wiley & Sons, New York.
6. Arrillaga, J, Arnold, C P and Harker, B J, (1983). *Computer Modelling of Electrical Power Systems*, J Wiley & Sons, London.
7. Kulicke, B. (1979). Digital program NETOMAC zur Simulation Elecktromechanischer und Magnetischer Ausleighsvorgänge in Drehstromnetzen, *Electrizitatwirstschaft*, **78**, S.18–23.
8. Dommel, H W, Yan, A and Wei Shi, (1986). Harmonics from transformer saturation, *IEEE Trans*, **PWRD-1**(2) 209–215.
9. Aprille, T J, (1972). Two computer algorithms for obtaining the periodic response of non-linear circuits, Ph.D Thesis, University of Illinois at Urbana Champaign.

10. Usaola, J (1990). Regimen permanente de sistemas electricos de potencia con elementos no lineales mediante un procedimiento híbrido de analisis en los dominios del tiempo y de la frecuencia. Doctoral Thesis, Universidad Politecnica de Madrid.
11. Yacamini, R and de Oliveira, J C, (1980). Harmonics in multiple converter systems: a generalised approach, *IEE Proc B*, **127**(2), 96–106.
12. Arrillaga, J, Watson, N R, Eggleston, J F and Callaghan, C D, (1987). Comparison of steady state and dynamic models for the calculation of a.c./d.c. system harmonics, *Proc IEE*, **134C**(1), 31–37.
13. Carpinelli, G, *et al.*, (1994). Generalised converter models for iterative harmonic analysis in power systems, *Proc IEE General Transn. Distrib*, **141**(5), 445–451.
14. Callaghan, C and Arrillaga, J, (1989), A double iterative algorithm for the analysis of power and harmonic flows at ac–dc converter terminals, *Proc IEE*, **136**(6), 319–324.
15. Smith, B, *et al.*, (1995). A Newton solution for the harmonic phasor analysis of ac–dc converters, *IEEE PES Summer Meeting 95*, **SM 379–8**.
16. Callaghan, C and Arrillaga, J, (1990). Convergence criteria for iterative harmonic analysis and its application to static converters, *ICHPS IV*, Budapest, 38–43.

2

FOURIER ANALYSIS

2.1 Introduction

Fourier analysis is the process of converting time domain waveforms into their frequency components [1].

The Fourier series, which permits establishing a simple relationship between a time domain function and that function in the frequency domain, is derived in the first part of this chapter and its characteristics discussed with reference to simple waveforms.

More generally, the Fourier Transform and its inverse are used to map any function in the interval $-\infty$ to ∞ in either the time or frequency domain, into a continuous function in the inverse domain. The Fourier series, therefore, represents the special case of the Fourier Transform applied to a periodic signal.

In practice, data is often available in the form of a sampled time function, represented by a time series of amplitudes, separated by fixed time intervals of limited duration. When dealing with such data a modification of the Fourier Transform, the Discrete Fourier Transform, is used. The implementation of the Discrete Fourier Transform, by means of the Fast Fourier Transform algorithm, forms the basis of most modern spectral and harmonic analysis systems. The FFT is also a powerful numerical tool that enables the Harmonic Domain description of non-linear devices to be implemented in either the frequency or time domain, whichever is appropriate. The development of the Fourier and Discrete Fourier Transforms is also examined in this chapter along with the implementation of the Fast Fourier Transform.

The main sources of harmonic distortion are power electronic devices, which exercise controllability by means of multiple switching events within the fundamental frequency waveform. Although the standard Fourier method can still be used to analyse the complete waveforms, it is often advantageous to subdivide the power electronic switching into its constituent Fourier components; this is the transfer function technique, which is also described in this chapter.

2.2 Fourier Series and Coefficients [2,3]

The Fourier series of a periodic function $x(t)$ has the expression

$$x(t) = a_o + \sum_{n=1}^{\infty} \left(a_n \cos\left(\frac{2\pi nt}{T}\right) + b_n \sin\left(\frac{2\pi nt}{T}\right)\right). \tag{2.1}$$

This constitutes a frequency domain representation of the periodic function.

In this expression a_o is the average value of the function $x(t)$, whilst a_n and b_n, the coefficients of the series, are the rectangular components of the nth harmonic. The corresponding nth harmonic vector is

$$A_n \angle \phi_n = a_n + jb_n \tag{2.2}$$

with a magnitude:

$$A_n = \sqrt{a_n{}^2 + b_n{}^2}$$

and a phase angle

$$\phi_n = t_{an}^{-1}\left(\frac{b_n}{a_n}\right).$$

For a given function $x(t)$, the constant coefficient, a_o, can be derived by integrating both sides of equation (2.1) from $-T/2$ to $T/2$ (over a period T), i.e.

$$\int_{-T/2}^{T/2} x(t)\mathrm{d}t = \int_{-T/2}^{T/2} \left[a_o + \sum_{n=1}^{\infty}\left[a_o \cos\left(a_n \cos\left(\frac{2\pi nt}{T}\right) + b_n \sin\left(\frac{2\pi nt}{T}\right)\right]\right]\mathrm{d}t. \tag{2.3}$$

The Fourier series of the right-hand side can be integrated term by term, giving

$$\int_{-T/2}^{T/2} x(t)\mathrm{d}t = a_o\int_{-T/2}^{T/2} \mathrm{d}t + \sum_{n=1}^{\infty}\left[a_n\int_{-T/2}^{T/2} \cos\left(\frac{2\pi nt}{T}\right)\mathrm{d}t + b_n\int_{-T/2}^{T/2} \sin\left(\frac{2\pi nt}{T}\right)\mathrm{d}t\right]. \tag{2.4}$$

The first term on the right-hand side equals Ta_o, while the other integrals are zero. Hence, the constant coefficient of the Fourier series is given by

$$a_o = 1/T\int_{-T/2}^{T/2} x(t)\mathrm{d}t, \tag{2.5}$$

which is the area under the curve of x(t) from $-T/2$ to $T/2$, divided by the period of the waveform, T.

The a_n coefficients can be determined by multiplying Equation (2.1) by $\cos(2\pi mt/T)$, where m is any fixed positive integer, and integrating between $-T/2$ and $T/2$, as previously, i.e.

$$\int_{-T/2}^{T/2} x(t)\cos\left(\frac{2\pi mt}{T}\right)\mathrm{d}t = \int_{-T/2}^{T/2}\left[a_o + \sum_{n=1}^{\infty}\left[a_n \cos\left(\frac{2\pi nt}{T}\right) + b_n \sin\left(\frac{2\pi nt}{T}\right)\right]\right]$$
$$\cos\left(\frac{2\pi nt}{T}\right)\mathrm{dt} \tag{2.6}$$

$$= a_o \int_{-T/2}^{T/2} \cos\left(\frac{2\pi mt}{T}\right) \mathrm{d}t + \sum_{n=t}^{\infty} \left[a_n \int_{-T/2}^{T/2} \cos\left(\frac{2\pi nt}{T}\right) \times \right.$$

$$\left. \cos\left(\frac{2\pi mt}{T}\right) \mathrm{d}t + b_n \int_{-T/2}^{T/2} \sin\left(\frac{2\pi nt}{T}\right) \cos\left(\frac{2\pi mt}{T}\right) \mathrm{d}t \right]. \tag{2.7}$$

The first term on the right-hand side is zero, as are all the terms in b_n since $\sin(2\pi nt/T)$ and $\cos(2\pi mt/T)$ are orthogonal functions for all n and m.

Similarly, the terms in a_n are zero, being orthogonal, unless $m = n$. In this case, Equation (2.7) becomes

$$\begin{aligned} \int_{-T/2}^{T/2} x(t) \cos\left(\frac{2\pi mt}{T}\right) \mathrm{d}t &= a_n \int_{-T/2}^{T/2} \cos\left(\frac{2\pi nt}{T}\right) \mathrm{d}t \\ &= \frac{a_n}{2} \int_{-T/2}^{T/2} \cos\left(\frac{4\pi nt}{T}\right) \mathrm{d}t + \frac{a_n}{2} \int_{-T/2}^{T/2} \mathrm{d}t. \end{aligned} \tag{2.8}$$

The first term on the right-hand side is zero while the second term equals $a_n T/2$. Hence, the coefficients a_n can be obtained from

$$a_n = \frac{2}{T} \int_{-T/2}^{T/2} x(t) \cos\left(\frac{2\pi nt}{T}\right) \mathrm{d}t \qquad \text{for } n = 1 \rightarrow \infty. \tag{2.9}$$

To determine the coefficients b_n, Equation (2.1) is multiplied by $\sin(2\pi mt/T)$ and, by a similar argument to the above

$$b_n = \frac{2}{T} \int_{-T/2}^{T/2} x(t) \sin\left(\frac{2\pi nt}{T}\right) \mathrm{d}t \qquad \text{for } n = 1 \rightarrow \infty. \tag{2.10}$$

It should be noted that because of the periodicity of the integrands in Equations (2.5), (2.9) and (2.10), the interval of integration can be taken more generally as t and $t + T$.

If the function $x(t)$ is piecewise continuous (i.e. has a finite number of vertical jumps) in the interval of integration, the integrals exist and Fourier coefficients can be calculated for this function. Equations (2.5), (2.9) and (2.10) are often expressed in terms of the angular frequency as follows:

$$a_o = \frac{1}{2\pi} \int_{-\pi}^{\pi} x(\omega t) \mathrm{d}(\omega \mathrm{t}), \tag{2.11}$$

$$a_n = \frac{1}{\pi} \int_{-\pi}^{\pi} x(\omega t) \cos(n\omega t) \mathrm{d}(\omega \mathrm{t}), \tag{2.12}$$

$$b_n = \frac{1}{\pi} \int_{-\pi}^{\pi} x(\omega t) \sin(n\omega t \mathrm{d}(\omega \mathrm{t}), \tag{2.13}$$

so that

$$x(t) = a_o + \sum_{n=1}^{\infty} [a_n \cos(n\omega t) + b_n \sin(n\omega t)]. \tag{2.14}$$

2.3 Simplifications Resulting from Waveform Symmetry [2,3]

Equations (2.5), (2.9) and (2.10), the general formulae for the Fourier coefficients, can be represented as the sum of two separate integrals, i.e.

$$a_n = \frac{2}{T}\int_0^{T/2} x(t)\cos\left(\frac{2\pi nt}{T}\right)\mathrm{d}t + \frac{2}{T}\int_{-T/2}^{0} x(t)\cos\left(\frac{2\pi nt}{T}\right)\mathrm{d}t, \tag{2.15}$$

$$b_n = \frac{2}{T}\int_0^{T/2} x(t)\sin\left(\frac{2\pi nt}{T}\right)\mathrm{d}t + \frac{2}{T}\int_{-T/2}^{0} x(t)\sin\left(\frac{2\pi nt}{T}\right)\mathrm{d}t. \tag{2.16}$$

Replacing t by $-t$ in the second integral of Equation (2.15), with limits $(-T/2, 0)$

$$\begin{aligned} a_n &= \frac{2}{T}\int_0^{T/2} x(t)\cos\left(\frac{2\pi nt}{T}\right)\mathrm{d}t + \frac{2}{T}\int_{+T/2}^{0} x(-t)\cos\left(\frac{-2\pi nt}{T}\right)\mathrm{d}(-t) \\ &= \frac{2}{T}\int_0^{T/2} \left[x(t) + x(-t)\right]\cos\left(\frac{2\pi nt}{T}\right)\mathrm{d}t. \end{aligned} \tag{2.17}$$

Similarly,

$$b_n = \frac{2}{T}\int_0^{T/2} \left[x(t) - x(-t)\right]\sin\left(\frac{2\pi nt}{T}\right)\mathrm{d}t. \tag{2.18}$$

Odd symmetry:

The waveform has odd symmetry if

$$x(t) = -x(-t)$$

Then the a_n terms become zero for all n, while

$$b_n = \frac{4}{T}\int_0^{T/2} x(t)\sin\left(\frac{2\pi nt}{T}\right)\mathrm{d}t. \tag{2.19}$$

The Fourier series for an odd function will, therefore, contain only sine terms.

Even symmetry:

The waveform has even symmetry if

$$x(t) = x(-t).$$

In this case

$$b_n = 0 \qquad \text{for all } n$$

and

$$a_n = \frac{4}{T}\int_0^{T/2} x(t)\cos\left(\frac{2\pi nt}{T}\right)\mathrm{d}t. \tag{2.20}$$

The Fourier series for an even function will, therefore, contain only cosine terms.

Certain waveforms may be odd or even depending on the time reference position selected. For instance, the square wave of Figure 2.1, drawn as an odd function, can be transformed into an even function simply by shifting the origin (vertical axis) by $T/2$.

Halfwave symmetry:

A function $x(t)$ has halfwave symmetry if

$$x(t) = -x(t + T/2) \tag{2.21}$$

i.e. the shape of the waveform over a period $t + T/2$ to $t + T$ is the negative of the shape of the waveform over the period t to $t + T/2$. Consequently, the square wave function of Figure 2.1 has halfwave symmetry with $t = -T/2$.

Using Equation (2.9) and replacing (t) by $(t + T/2)$ in the interval $(-T/2, o)$

$$\begin{aligned} a_n &= \frac{2}{T}\int_0^{T/2} x(t)\cos\left(\frac{2\pi nt}{T}\right)\mathrm{d}t + \frac{2}{T}\int_{-T/2+T/2}^{0+T/2} x(t+T/2)\cos\left(\frac{2\pi n(t+T/2)}{T}\right)\mathrm{d}t \\ &= \frac{2}{T}\int_0^{T/2} x(t)\left[\cos\left(\frac{2\pi nt}{T}\right) - \cos\left(\frac{2\pi nt}{T} + n\pi\right)\right]\mathrm{d}t \end{aligned} \tag{2.22}$$

since by definition $x(t) = -x(t + T/2)$.

If n is an odd integer then

$$\cos\left(\frac{2\pi nt}{T} + n\pi\right) = -\cos\left(\frac{2\pi nt}{T}\right)$$

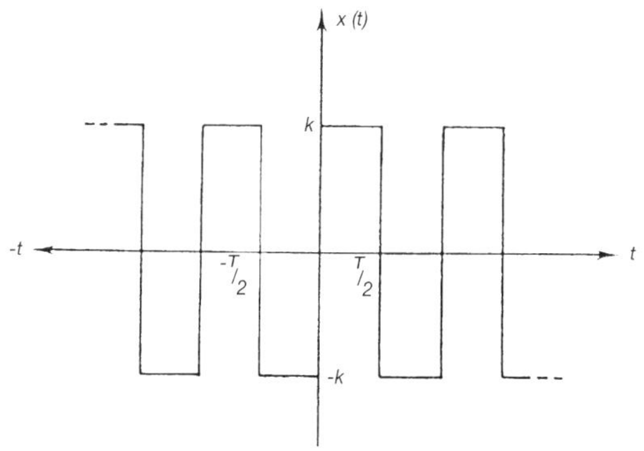

Figure 2.1 Square wave function

and

$$a_n = \frac{4}{T}\int_0^{T/2} x(t)\cos\left(\frac{2\pi nt}{T}\right)\mathrm{d}t. \tag{2.23}$$

However, if n is an even integer then,

$$\cos\left(\frac{2\pi nt}{T} + n\pi\right) = \cos\left(\frac{2\pi nt}{T}\right)$$

and

$$a_n = 0.$$

Similarly,

$$\begin{aligned} b_n &= \frac{4}{T}\int_0^{T/2} x(t)\sin\left(\frac{2\pi nt}{T}\right)\mathrm{d}t \qquad &&\text{for } n \text{ odd,} \\ &= 0 &&\text{for } n \text{ even.} \end{aligned} \tag{2.24}$$

Thus, waveforms which have halfwave symmetry, contain only odd order harmonics.

The square wave of Figure 2.1 is an odd function with halfwave symmetry. Consequently, only the b_n coefficients and odd harmonics will exist. The expression for the coefficients taking into account these conditions is

$$b_n = \frac{8}{T}\int_0^{T/4} x(t)\sin\left(\frac{2\pi nt}{T}\right)\mathrm{d}t, \tag{2.25}$$

which can be represented by a line spectrum of amplitudes inversely proportional to the harmonic order, as shown in Figure 2.2.

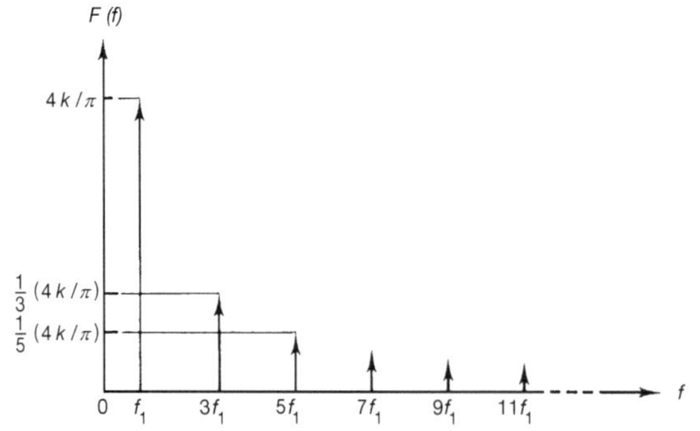

Figure 2.2 Line spectrum representation of a square wave

2.4 Complex Form of the Fourier Series

The representation of the frequency components as rotating vectors in the complex plane gives a geometrical interpretation of the relationship between waveforms in the time and frequency domains.

A uniformly rotating vector $A/2\mathrm{e}^{\mathrm{j}\phi}(X(f_n))$ has a constant magnitude $A/2$, and a phase angle ϕ , which is time varying according to

$$\phi = 2\pi ft + \theta, \tag{2.26}$$

where θ is the initial phase angle when $t = 0$.

A second vector $A/2e^{\mathrm{j}\phi}(X(-f_n))$ with magnitude $A/2$ and phase angle $-\phi$, will rotate in the opposite direction to $A/2e^{+j\phi}(X(f_n))$. This negative rate of change of phase angle can be considered as a negative frequency.

The sum of the two vectors will always lie along the real axis, the magnitude oscillating between A and $-$A according to

$$\frac{A}{2}\mathrm{e}^{\mathrm{j}\phi} + \frac{A}{2}\mathrm{e}^{-\mathrm{j}\phi} = A\cos\phi. \tag{2.27}$$

Thus, each harmonic component of a real valued signal can be represented by two half amplitude contra-rotating vectors as shown in Figure 2.3, such that

$$X(f_n) = X^*(-f_n) \tag{2.28}$$

where $X^*(-f_n)$ is the complex conjugate of $X(-f_n)$.

The sine and cosine terms of Equations (2.12) and (2.13) may, therefore, be solved into positive and negative frequency terms using the trigonometric identities

$$\cos(n\omega t) = \frac{\mathrm{e}^{\mathrm{j}n\omega t} + \mathrm{e}^{-\mathrm{j}n\omega t}}{2}, \tag{2.29}$$

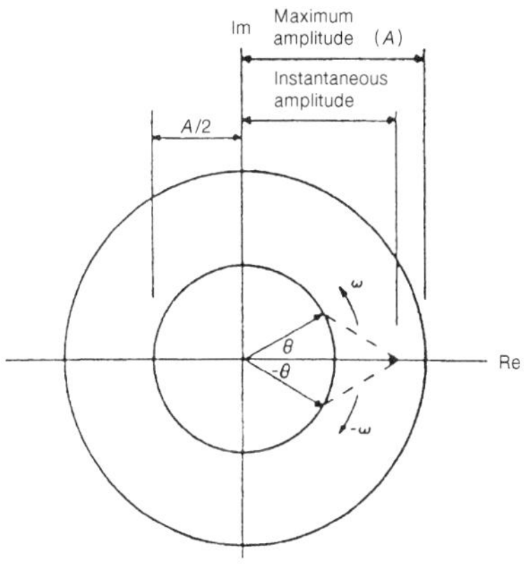

Figure 2.3 Contra-rotating vector pair producing a varying amplitude (pulsating) vector

$$\sin(n\omega t) = \frac{e^{jn\omega t} - e^{-jn\omega t}}{2j}. \tag{2.30}$$

Substituting into Equation (2.14) and simplifying yields

$$x(t) = \sum c_n e^{jn\omega t}, \tag{2.31}$$

where

$$\begin{aligned} c_n &= 1/2(a_n - jb_n), \qquad n > 0 \\ c_{-n} &= c_n \\ c_o &= a_o \end{aligned}$$

The c_n terms can also be obtained by complex integration

$$c_n = \frac{1}{\pi}\int_{-\pi}^{\pi} x(\omega t)e^{-jn\omega t}\, d(\omega t), \tag{2.32}$$

$$c_o = \frac{1}{2\pi}\int_{-\pi}^{\pi} x(\omega t)\, d(\omega t). \tag{2.33}$$

If the time domain signal $x(t)$ contains a component rotating at a single frequency nf, then multiplication by the unit vector $e^{-j2\pi ft}$, which rotates at a frequency $-nf$, annuls the rotation of the component, such that the integration over a complete period has a finite value. All components at other frequencies will continue to rotate after multiplication by $e^{-j2\pi nft}$, and will thus integrate to zero.

The Fourier Series is most generally used to approximate a periodic function by truncation of the series. In this case, the truncated Fourier series is the best trigonometric series expression of the function, in the sense that it minimizes the square error between the function and the truncated series. The number of terms required depends upon the magnitude of repeated derivatives of the function to be approximated. Repeatedly differentiating Equation (2.32) by parts, it can readily be shown that

$$c_n = \frac{1}{2\pi}\frac{1}{n^{m+1}}\int_{-\pi}^{\pi} f^{(m+1)}(\omega t)\, d(\omega t). \tag{2.34}$$

Consequently, the Fourier Series for repeatedly differentiated functions will converge faster than that for functions with low order discontinuous derivatives.

The complex Fourier series expansion is compatible with the Fast Fourier Transform, the method of choice for converting time domain data samples into a Nyquist rate limited frequency spectrum. The trigonometric Fourier expression can also be written as a series of phase-shifted sine terms by substituting

$$a_n \cos n\omega t + b_n \sin n\omega t = d_n \sin(n\omega t + \Psi_n) \tag{2.35}$$

into Equation (2.14), where

$$\begin{aligned} d_n &= \sqrt{a^2{}_n + b^2{}_n}, \\ \Psi_n &= \tan^{-1}\frac{b_n}{a_n}. \end{aligned} \tag{2.36}$$

Finally, the phase shifted sine terms can be represented as peak value phasors by setting

$$\Psi_n = d_n e^{j\Psi_n}, \tag{2.37}$$

so that

$$\begin{aligned} d_n \sin(n\omega t + \Psi_n) &= \mathrm{I}\{\Psi_n e^{jn\omega t}\} \\ &= |\Psi_n| \sin(n\omega t + \angle \Psi_n). \end{aligned} \tag{2.38}$$

The harmonic phasor Fourier series is, therefore,

$$f(t) = \sum_{n=0}^{\infty} \mathrm{I}\{\Psi_n e^{jn\omega t}\}. \tag{2.39}$$

which does not contain negative frequency components. Note that the dc term becomes

$$\begin{aligned} \Psi_0 &= \frac{a_o}{2} e^{j\pi/2} \\ &= j\frac{a_o}{2}. \end{aligned} \tag{2.40}$$

In practice, the upper limit of the summation is set to n_h, the highest harmonic order of interest.

2.5 Convolution of Harmonic Phasors

The point by point multiplication of two time domain waveforms is expressed in the harmonic domain by a discrete convolution of their Fourier series. When two harmonic phasors of different frequencies are convolved, the results are harmonic phasors at sum and difference harmonics. This is best explained by multiplying the corresponding sinusoids using the trigonometric identity for the product of sine waves, and then converting back to phasor form. Given two phasors, A_k and B_m, of harmonic orders k and m, the trigonometric identity for their time domain multiplication is:

$$\begin{aligned} &|A_k| \sin(k\omega t + \angle A_k)|B_m| \sin(m\omega t + \angle B_m) = \\ &\tfrac{1}{2}|A_k||B_m|\left[\sin\left((k-m)\omega t + \angle A_k - \angle B_m + \frac{\pi}{2}\right)\right. \\ &\left. - \sin\left((k+m)\omega t + \angle A_k + \angle B_m + \frac{\pi}{2}\right)\right]. \end{aligned} \tag{2.41}$$

Converting to phasor form:

$$\begin{aligned}
A_k \otimes B_m &= \tfrac{1}{2}|A_k||B_m|\left[e^{j(\angle A_k - \angle B_{m+}^{\pi}/2)}|_{(k-m)} - e^{j(\angle A_k - \angle B_{m+}^{\pi}/2)}|_{(k+m)}\right] \\
&= \tfrac{1}{2}\left[(|A_k|e^{j\angle A_k}|B_m|e^{-j\angle B_{m}e^{\pi}_{j}/2)}|_{(k-m)} - (|A_k|e^{j\angle A_k}|B_m|e^{j\angle B_m}e^{\pi j/2})_{k+m}\right] \\
&= \tfrac{1}{2}j\left[(A_k B^*{}_m)_{k-m} - (A_k B_m)_{k+m}\right].
\end{aligned} \tag{2.42}$$

If k is less than m, a negative harmonic can be avoided by conjugating the difference term. This leads to the overall equation:

$$A_k \otimes B_m = \begin{cases} \frac{1}{2}j(A_k B^*{}_m)_{(k-m)} - \frac{1}{2}j(A_k B_m)_{(k+m)} & \text{if } k \geqslant m \\ \frac{1}{2}j(A_k B^*{}_m)^*{}_{(m-k)} - \frac{1}{2}j(A_k B_m)_{(k+m)} & \text{otherwise.} \end{cases} \tag{2.43}$$

The multiplication of two non-sinusoidal periodic waveforms leads to a discrete convolution of their harmonic phasor Fourier series:

$$\begin{aligned}
f_a(t) f_b(t) &= \sum_{k=0}^{n_h} |A_k| \sin(k\omega t + \angle A_k) \sum_{m=0}^{n_h} |B_k| \sin(m\omega t + \angle B_m) \\
&= \sum_{k=0}^{n_h} \sum_{m=0}^{n_h} |A_k| \sin(k\omega t + \angle A_k) |B_k| \sin(m\omega t + \angle B_m).
\end{aligned} \tag{2.44}$$

Rewriting this in terms of phasors yields

$$F_A \otimes F_B = \sum_{k=0}^{n_h} \sum_{m=0}^{n_h} A_k \otimes B_m \,. \tag{2.45}$$

Equation (2.45) generates harmonic phasors of order up to $2n_h$, due to the sum terms. Substituting the equation for the convolution of two phasors, Equation (2.43), into (2.45) and solving for the lth order component yields:

$$(A \otimes B)_l = \tfrac{1}{2}j\left[\sum_{k=l}^{n_h} A_k B^*{}_{k-l} + \sum_{k=l}^{n_h} (A_k B^*{}_{k+l})^* - \sum_{k=0}^{n_h} A_k B^*{}_{l-k}\right], l > 0. \tag{2.46}$$

$$(A \otimes B)_l = \tfrac{1}{2}j\left[-A_0 B_0 + \sum_{k=0}^{n_h} A_k B^*{}_k\right], l = 0. \tag{2.47}$$

The convolution equations are non-analytic in the complex plane but are differentiable by decomposing into two real valued components (typically rectangular).

If negative frequencies are retained, the convolution is just the multiplication of 2 series

$$\begin{aligned}
f_{a(t)} f_{b(t)} &= \sum_{n=-n_h}^{n_h} c_{an} e^{jn\omega t} \sum_{l=-n_h}^{n_h} c_{bl} e^{jl\omega t} \\
&= \sum_{n=-n_h}^{n_h} \sum_{l=-n_h}^{n_h} c_{an} c_{bl}\, e^{j(l+n)\omega t}.
\end{aligned} \tag{2.48}$$

In practice, the discrete convolution can be evaluated faster using FFT methods.

2.6 The Fourier Transform [3,4]

Fourier analysis, when applied to a continuous, periodic signal in the time domain, yields a series of discrete frequency components in the frequency domain.

By allowing the integration period to extend to infinity, the spacing between the harmonic frequencies, ω, tends to zero and the Fourier coefficients, c_n, of equation (2.32) become a continuous function, such that

$$X(f) = \int_{-\infty}^{\infty} x(t)\, e^{-j2\pi ft} dt. \tag{2.49}$$

The expression for the time domain function $x(t)$ which is also continuous and of infinite duration, in terms of $X(f)$ is then:

$$x(t) = \int_{-\infty}^{\infty} X(f)\, e^{-j2\pi ft} df, \tag{2.50}$$

$X(f)$ is known as the spectral density function of $x(t)$.

Equations (2.49) and (2.50) form the Fourier Transform Pair. Equation (2.49) is referred to as the 'Forward Transform' and equation (2.50) as the 'Reverse' or 'Inverse Transform'. In general $X(f)$ is complex and can be written as

$$X(f) = ReX(f) + jImX(f) \tag{2.51}$$

The real part of $X(f)$ is obtained from

$$\begin{aligned} ReX(f) &= \tfrac{1}{2}[X(f) + X(-f)] \\ &= \int_{-\infty}^{\infty} x(t) \cos 2\pi ft dt. \end{aligned} \tag{2.52}$$

Similarly, for the imaginary part of $X(f)$

$$\begin{aligned} ImX(f) &= \tfrac{1}{2}j\,[X(f) - X(-f)] \\ &= -\int_{-\infty}^{\infty} x(t) \sin 2\pi\, ft dt. \end{aligned} \tag{2.53}$$

The amplitude spectrum of the frequency signal is obtained from

$$|X(f)| = [(ReX(f))^2 + (ImX(f))^2]^{\frac{1}{2}}. \tag{2.54}$$

The phase spectrum is

$$\phi(f) = \tan^{-1}\left[\frac{ImX(f)}{ReX(f)}\right]. \tag{2.55}$$

Using Equations (2.51) to (2.55), the inverse Fourier transform can be expressed in terms of the magnitude and phase spectra components.

$$x(t) = \int_{-\infty}^{\infty} |X(f)| \cos[2\pi ft - \phi(f)] df. \tag{2.56}$$

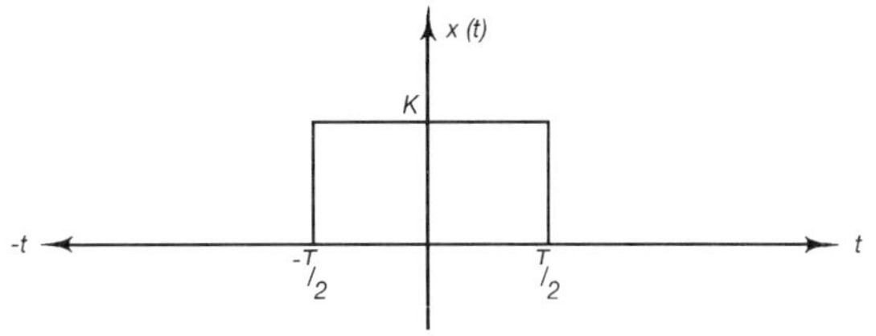

Figure 2.4 Rectangular function

As an example, let us consider a rectangular function such as Figure 2.4, defined by

$$\begin{aligned} x(t) &= K \text{ for } |t| \leqslant T/2 \\ &= 0 \text{ for } |t| > T/2, \end{aligned}$$

i.e. the function is continuous over all t but is zero outside the limits $(-T/2,\ T/2)$.

Its Fourier transform is

$$\begin{aligned} X(f) &= \int_{-\infty}^{\infty} x(t)\, \mathrm{e}^{-\mathrm{j}2\pi ft} \mathrm{d}t \\ &= \int_{-T/2}^{T/2} K\mathrm{e}^{-\mathrm{j}2\pi ft} \mathrm{d}t \\ &= \frac{-K}{\pi f} \cdot \frac{1}{2\mathrm{j}} [\mathrm{e}^{-\mathrm{j}\pi fT} - \mathrm{e}^{\mathrm{j}\pi fT}]. \end{aligned} \tag{2.57}$$

and using the identity

$$\sin\phi = \frac{1}{2\mathrm{j}}(\mathrm{e}^{\mathrm{j}\phi} - \mathrm{e}^{-\mathrm{j}\phi})$$

yields the following expression for the Fourier transform:

$$\begin{aligned} X(f) &= \frac{K}{\pi f} \sin(\pi f T) \\ &= KT\left[\frac{\sin(\pi f T)}{\pi f T}\right]. \end{aligned} \tag{2.58}$$

The term in brackets, known as the sinc function, is shown in Figure 2.5.

While the function is continuous, it has zero value at the points $f = n/T$ for $n = \pm 1, \pm 2, \ldots$ and the side lobes decrease in magnitude as $1/T$. This should be compared to the Fourier series of a periodic square wave which has discrete frequencies at odd harmonics. The interval $1/T$ is the effective bandwidth of the signal.

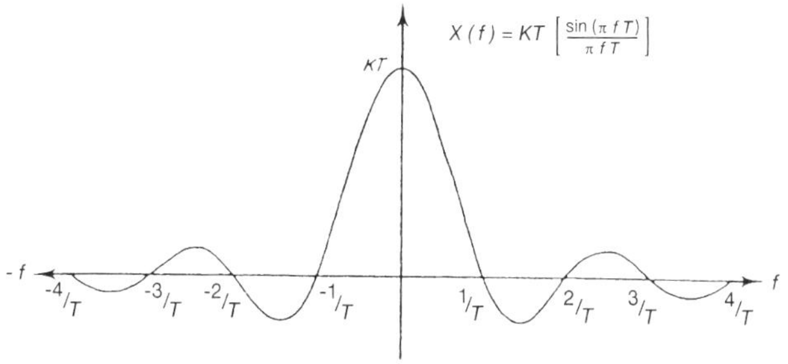

Figure 2.5 The sinc function, $\sin(\pi f T)/(\pi f T)$

2.7 Sampled Time Function [4,5]

With an increase in the digital processing of data, functions are often recorded by samples in the time domain. Thus, the signal can be represented as in Figure 2.6, where $f_s = 1/t_1$ is the frequency of the sampling. In this case, the Fourier transform of the signal is expressed as the summation of the discrete signal where each sample is multiplied by $e^{-j2\pi f n t_1}$; i.e.:

$$X(f) = \sum_{n=-\infty}^{\infty} x(nt_1)e^{-j2\pi f n t_1}. \tag{2.59}$$

The frequency domain spectrum, shown in Figure 2.7, is periodic and continuous.

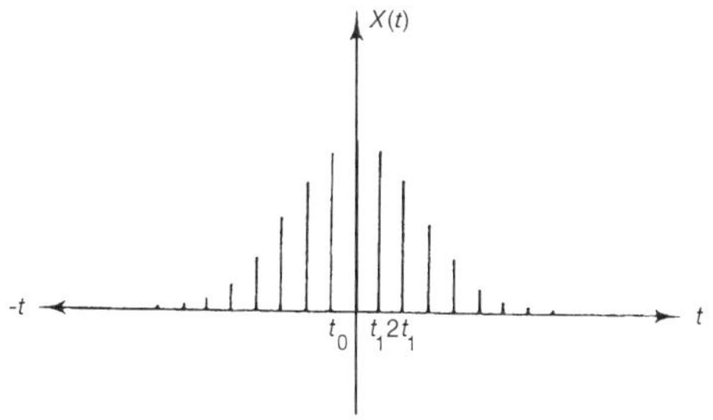

Figure 2.6 Sampled time domain function

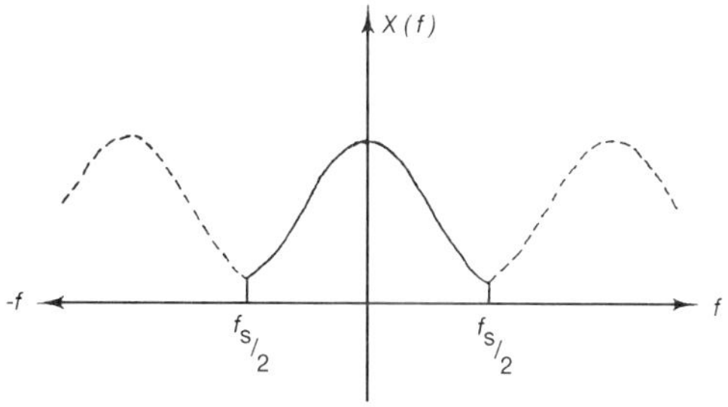

Figure 2.7 Frequency spectrum for discrete time domain function

The inverse Fourier transform is thus

$$x(t) = 1/f_s \int_{-f_s/2}^{f_s/2} X(f)\,\mathrm{e}^{\,\mathrm{j}2\pi f n t_1}\,\mathrm{d}f. \tag{2.60}$$

2.8 Discrete Fourier Transform [4,5]

In the case where the frequency domain spectrum is a sampled function, as well as the time domain function, we obtain a Fourier transform pair made up of discrete components

$$X(f_k) = 1/N \sum_{n=0}^{N-1} x(t_n)\mathrm{e}^{-\mathrm{j}2\pi kn/N} \tag{2.61}$$

and

$$x(t_n) = \sum_{k=0}^{N-1} X(f_k)\,\mathrm{e}^{\,\mathrm{j}2\pi kn/N}. \tag{2.62}$$

Both the time domain function and the frequency domain spectrum are assumed periodic as in Figure 2.8, with a total of N samples per period. It is in this discrete form that the Fourier Transform is most suited to numerical evaluation by digital computation.

Consider equation (2.61) rewritten as

$$X(f_k) = 1/N \sum_{n=0}^{N-1} x(t_n) W^{kn}. \tag{2.63}$$

where $W = \mathrm{e}^{-\mathrm{j}2\pi/N}$.

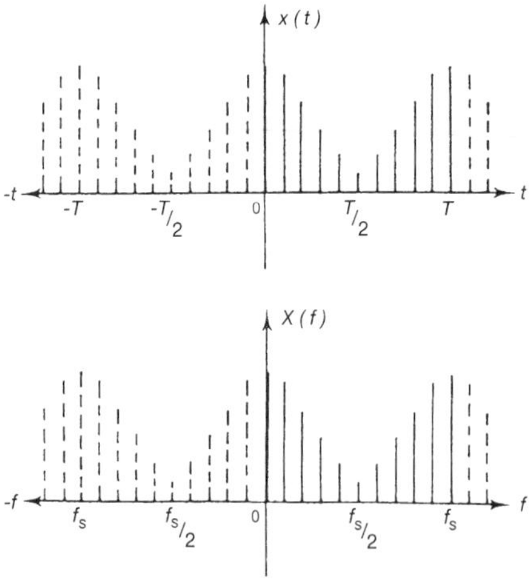

Figure 2.8 Discrete time and frequency domain function

Over all the frequency components, Equation (2.63) becomes a matrix equation.

$$\begin{bmatrix} X(f_0) \\ X(f_1) \\ \cdot \\ \cdot \\ \cdot \\ X(f_k) \\ \cdot \\ \cdot \\ X(f_{N-1}) \end{bmatrix} = 1/N \begin{bmatrix} 1 & 1 & \cdot\cdot\cdot & 1 & \cdot\cdot & 1 \\ 1 & W & \cdot\cdot\cdot & W^k & \cdot\cdot & W^{N-1} \\ \cdot & \cdot & \cdot & & & \cdot \\ \cdot & \cdot & \cdot & & & \cdot \\ \cdot & \cdot & \cdot & & & \cdot \\ 1 & W^k & \cdot\cdot\cdot & W^{k2} & \cdot\cdot & W^{k(N-1)} \\ \cdot & \cdot & & \cdot & \cdot & \cdot \\ \cdot & \cdot & & \cdot & \cdot & \cdot \\ 1 & W^{N-1} & \cdot\cdot\cdot & W^{(N-1)k} & \cdot\cdot & W^{(N-1)^2} \end{bmatrix} \cdot \begin{bmatrix} x(t_0) \\ x(t_1) \\ \cdot \\ \cdot \\ \cdot \\ x(t_k) \\ \cdot \\ \cdot \\ x(t_{N-1}) \end{bmatrix} \tag{2.64}$$

or in a condensed form

$$[X(f_k)] = 1/N[W^{kn}]\,[x(t_n)]. \tag{2.65}$$

In these equations, $[X(fk)]$ is a vector representing the N components of the function in the frequency domain, while $[x(t_n)]$ is a vector representing the N samples of the function in the time domain.

Calculation of the N frequency components from the N time samples, therefore, requires a total of N^2 complex multiplications to implement in the above form.

Each element in the matrix $[W^{kn}]$ represents a unit vector with a clockwise rotation of $2n/N(n = 0, 1, 2, \ldots, (N-1))$ introduced between successive components. Depending on the value of N, a number of these elements are the same.

For example, if $N = 8$ then

$$\begin{aligned} W &= e^{-j2\pi/8} \\ &= \cos\frac{\pi}{4} - j\sin\frac{\pi}{4}. \end{aligned}$$

As a consequence

$$\begin{aligned} W^0 &= -W^4 = 1 \\ W^1 &= -W^5 = (1\sqrt{2} - j1/\sqrt{2}) \\ W^2 &= -W^6 = -j \\ W^3 &= -W^7 = -(1/\sqrt{2} + j1/\sqrt{2}). \end{aligned}$$

These can also be thought of as unit vectors rotated through $\pm 0°$, $\pm 45°$, $\pm 90°$ and $\pm 135°$, respectively.

Further, W^8 is a complete rotation and hence equal to 1. The value of the elements of W^{kn} for $kn > 8$ can thus be obtained by subtracting full rotations, to leave only a fraction of a rotation, the values for which are shown above. For example, if $k = 5$ and $n = 6$, then $kn = 30$ and $W^{30} = W^{3\times 8+6} = W^6 = j$.

Thus, there are only 4 unique absolute values of W^{kn} and the matrix $[W^{kn}]$, for the case $N = 8$, becomes

$$\begin{bmatrix} 1 & 1 & 1 & 1 & 1 & 1 & 1 & 1 \\ 1 & W & -j & W^3 & -1 & -W & j & -W^3 \\ 1 & -j & -1 & j & 1 & -j & -1 & j \\ 1 & W^3 & j & W & -1 & -W^3 & -j & -W \\ 1 & -1 & 1 & -1 & 1 & -1 & 1 & -1 \\ 1 & -W & -j & -W^3 & -1 & W & j & W^3 \\ 1 & j & -1 & -j & 1 & j & -1 & -j \\ 1 & -W^3 & j & -W & -1 & W^3 & -j & W \end{bmatrix}$$

It can be observed that the dc component of the frequency spectrum, $X(f_o)$, obtained by the algebraic addition of all the time domain samples, divided by the number of samples, is the average value of all the samples

Subsequent rows show that each time sample is weighted by a rotation dependent on the row number. Thus, for $X(f_1)$ each successive time sample is rotated by $1/N$ of a revolution; for $X(f_2)$ each sample is rotated by $2/N$ revolutions, and so on.

The Nyquist frequency and aliasing [4]

With regard to equation (2.64) for the Discrete Fourier Transform and the matrix $[W^{kn}]$ it can be observed that for the rows $N/2$ to N, the rotations applied to each time sample are the negative of those in rows $N/2$ to 1. Frequency components above $k = N/2$ can be considered as negative frequencies, since the unit vector is being rotated through increments greater than π between successive components. In the example of $N = 8$, the elements of row 3 are successively rotated through $-\pi/2$. The elements of the row 7 are similarly rotated through $-3\pi/2$; or in negative frequency form through $\pi/2$. More generally, a rotation through

$$2\pi(N/2 + p)/N \text{ radians}$$
$$\text{for } p = 1,2,3, \ldots, (N/2 - 1) \qquad [\text{with } N \text{ even}]$$

corresponds to a negative rotation of

$$-2\pi(N/2 - p)/N \text{ radians.}$$

Hence, $-X(k)$ corresponds to $X(N - k)$ for $k = 1$ to $N/2$ as shown by Figure 2.9.

This is an interpretation of the sampling theorem which states that the sampling frequency must be at least twice the highest frequency contained in the original signal for a correct transfer of information to the sampled system.

The frequency component at half the sampling frequency is referred to as the Nyquist frequency.

The representation of frequencies above the Nyquist frequency as negative frequencies means that should the sampling rate be less than twice the highest frequency present in the sampled waveform then these higher frequency components can mimic components below the Nyquist frequency, introducing error into the analysis.

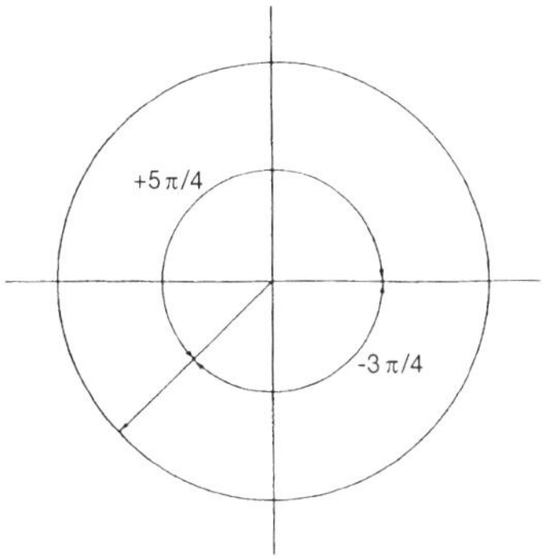

Figure 2.9 Correspondence of positive and negative angles

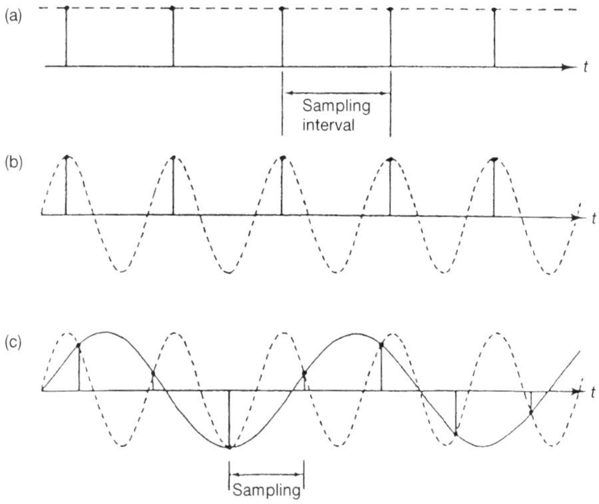

Figure 2.10 The effect of aliasing: (a) $x(t)=k$; (b) $x(t)=k\cos 2\pi nft$. For (a) and (b) both signals are interpreted as being dc. In (c) the sampling can represent two different signals with frequencies above and below the Nyquist or sampling rate

It is possible for high frequency components to complete many revolutions between samplings; however, since they are only sampled at discrete points in time, this information is lost.

This misinterpretation of frequencies above the Nyquist frequency, as being lower frequencies, is called 'aliasing' and is illustrated in Figure 2.10.

To prevent aliasing it is necessary to pass the time domain signal through a band limited low pass filter, the ideal characteristic of which is shown in Figure 2.11, with a cut-off frequency, f_c, equal to the Nyquist frequency.

Thus, if sampling is undertaken on the filtered signal and the Discrete Fourier Transform applied, the frequency spectrum has no aliasing effect and is an accurate representation of the frequencies in the original signal that are below the Nyquist frequency. However, information on those frequencies above the Nyquist frequency is lost due to the filtering process.

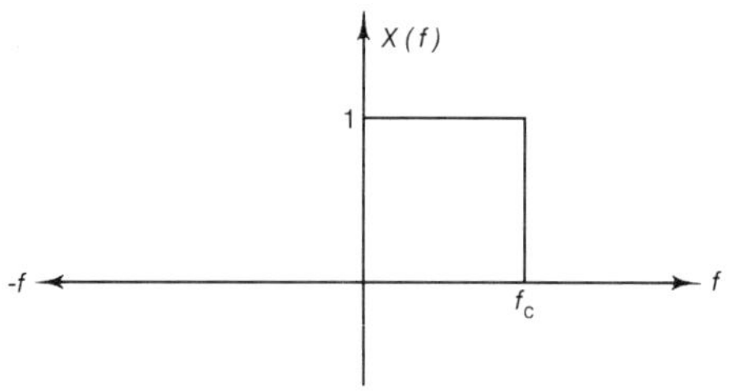

Figure 2.11 Frequency domain characteristics of an ideal low pass filter with cut-off frequency f_c

2.9 Fast Fourier Transform [4–7]

For large values of N, the computational time and cost of executing the N^2 complex multiplications of the Discrete Fourier Transform can become prohibitive.

Instead, a calculation procedure known as the Fast Fourier Transform, which takes advantage of the similarity of many of the elements in the matrix $[W^{kn}]$, produces the same frequency components using only $N/2 \log_2 N$ multiplications to execute the solution of equation (2.65). Thus, for the case $N = 1024 = 2^{10}$, there is a saving in computation time by a factor of over 200. This is achieved by factorising the $[W^{kn}]$ matrix of equation (2.65) into $\log_2 N$ individual or factor matrices such that there are only 2 non-zero elements in each row of these matrices, one of which is always unity. Thus, when multiplying by any factor matrix only N operations are required. The reduction in the number of multiplications required, to $(N/2)\log_2 N$, is obtained by recognising that:

$$W^{N/2} = -W^0$$
$$W^{(N+2)/2} = -W^1 \text{ etc.}$$

To obtain the factor matrices, it is first necessary to re-order the rows of the full matrix. If rows are denoted by a binary representation, then the re-ordering is by bit reversal.

For the example where $N = 8$; row 5, represented as 100 in binary (row 1 is 000), now becomes row 2, or 001 in binary. Thus, rows 2 and 5 are interchanged. Similarly, rows 4 and 7, represented as 011 and 110, respectively are also interchanged. Rows 1, 3, 6 and 8 have binary representations which are symmetrical with respect to bit reversal and hence remain unchanged.

The corresponding matrix is now

$$\begin{bmatrix} 1 & 1 & 1 & 1 & 1 & 1 & 1 & 1 \\ 1 & -1 & 1 & -1 & 1 & -1 & 1 & -1 \\ 1 & -\mathrm{j} & -1 & \mathrm{j} & 1 & -\mathrm{j} & -1 & \mathrm{j} \\ 1 & \mathrm{j} & -1 & -\mathrm{j} & 1 & \mathrm{j} & -1 & -\mathrm{j} \\ 1 & W & -\mathrm{j} & W^3 & -1 & -W & \mathrm{j} & -W^3 \\ 1 & -W & -\mathrm{j} & -W^3 & -1 & W & \mathrm{j} & W^3 \\ 1 & W^3 & \mathrm{j} & W & -1 & -W^3 & -\mathrm{j} & -W \\ 1 & -W^3 & \mathrm{j} & -W & -1 & W^3 & -\mathrm{j} & W \end{bmatrix}$$

This new matrix can be separated into $\log_2 8 (= 3)$ factor matrices.

$$
\begin{bmatrix}
1 & 1 & & & & & & \\
1 & -1 & & & & & & \\
 & & 1 & -\mathrm{j} & & & & \\
 & & 1 & \mathrm{j} & & & & \\
 & & & & 1 & W & & \\
 & & & & 1 & -W & & \\
 & & & & & & 1 & W^3 \\
 & & & & & & 1 & -W^3
\end{bmatrix}
\begin{bmatrix}
1 & & 1 & & & & & \\
 & 1 & & 1 & & & & \\
1 & & -1 & & & & & \\
 & 1 & & -1 & & & & \\
 & & & & 1 & & -\mathrm{j} & \\
 & & & & & 1 & & -\mathrm{j} \\
 & & & & 1 & & \mathrm{j} & \\
 & & & & & 1 & & \mathrm{j}
\end{bmatrix}
$$

$$
\begin{bmatrix}
1 & & & & 1 & & & \\
 & 1 & & & & 1 & & \\
 & & 1 & & & & 1 & \\
 & & & 1 & & & & 1 \\
1 & & & & -1 & & & \\
 & 1 & & & & -1 & & \\
 & & 1 & & & & -1 & \\
 & & & 1 & & & & -1
\end{bmatrix}
$$

As previously stated, each factor matrix has only two non-zero elements per row, the first of which is unity.

The re-ordering of the $[W^{kn}]$ matrix results in a frequency spectrum which is also re-ordered. To obtain the natural order of frequencies, it is necessary to reverse the previous bit-reversal.

In practice, a mathematical algorithm implicitly giving factor matrix operations is used for the solution of an FFT [8].

Using $N = 2^m$, it is possible to represent n and k by m bit binary numbers such that:

$$n = n_{m-1}2^{m-1} + n_{m-2}2^{m-2} + \ldots + 4n_2 + 2n_1 + n_0, \tag{2.66}$$

$$k = k_{m-1}2^{m-1} + k_{m-2}2^{m-2} + \ldots + 4k_2 + 2k_1 + k_0, \tag{2.67}$$

where $\quad n_i = 0,1$ and $k_i = 0,1$.

For $N = 8$:

$$n = 4n_2 + 2n_1 + n_0$$

and

$$k = 4k_2 + 2k_1 + k_0$$

where n_2, n_1, n_0 and k_2, k_1, k_0 are binary bits (n_2, k_2 most significant and n_0, k_0 least significant).

Equation (2.63) can now be re-written as:

$$X(k_2,k_1,k_0) = \sum_{n_2=0}^{1} \sum_{n_1=0}^{1} \sum_{n_0=0}^{1} 1/N \times (n_2,n_1,n_0)W. \tag{2.68}$$

Defining n and k in this way enables the computation of Equation (2.63) to be performed in three independent stages computing in turn:

$$A_1(k_0, n_1, n_0) = \sum_{n_2=0}^{1} 1/Nx(n_2,n_1,n_0)W^{4k_0n_2}, \tag{2.69}$$

$$A_2(k_0,k_1,n_0) = \sum_{n_1=0}^{1} A_1(k_0,n_1,n_0)W^{2(k_0+2k_1)n_1}, \tag{2.70}$$

$$A_3(k_0,k_1,k_2) = \sum_{n_0=0}^{1} A_2(k_0,k_1,n_0)W^{(k_0+2k_1+4k_2)}n_0. \tag{2.71}$$

From Equation (2.71) it is seen that the A_3 coefficients contain the required $X(k)$ coefficients but in reverse binary order.

Order of A_3 in binary form is $k_0k_1k_2$.
Order of $X(k)$ in binary form is $k_2k_1k_0$.

Hence

$$\begin{array}{lclclcl} & & \text{Binary} & & \text{Reversed} & & \\ A_3(3) & = & A_3(011) & = & X(110) & = & X(6) \\ A_3(4) & = & A_3(100) & = & X(100) & = & X(1) \\ A_3(5) & = & A3(101) & = & X(101) & = & X(5). \end{array}$$

2.10 Transfer Function Fourier Analysis [9,10]

An effective way of deriving the harmonic components of waveforms resulting from multiple periodic switching is by frequency domain based transfer functions.

The main application for the transfer function technique is the process of static power conversion where the conduction of the switching devices can be described by +1 for a connection from a phase to the positive dc rail, −1 for a connection to the negative dc rail and zero for no connection. For a three-phase static converter (Figure 2.12), three such functions are written, one for each phase. The spectrum for such a function can be easily written, and additional spectra in the transfer functions due to firing angle variation or commutation period variation can be incorporated. From these transfer functions, the converter dc voltage can be written in terms of the ac side voltage as

$$V_{dc} = \sum_{\Psi} Y_{\Psi dc} \cdot V_{\Psi} \tag{2.72}$$

and the ac current in terms of the dc side current as

$$I_{\Psi} = Y_{\Psi}ac \cdot I_{dc}, \tag{2.73}$$

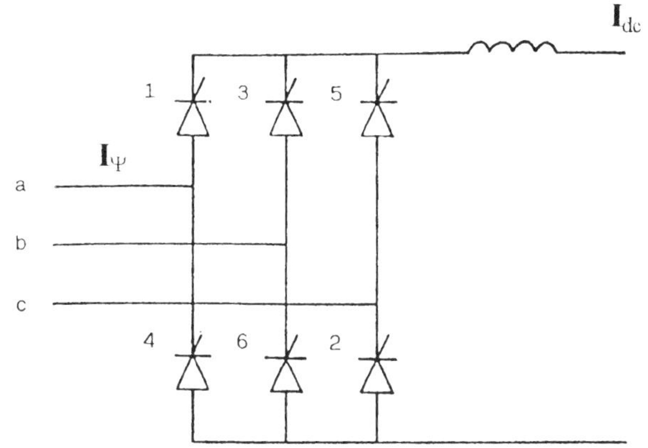

Figure 2.12 Three phase static converter

where Ψ is 0, 120 and 240 degrees, referring to phases a, b and c, and $Y_{\Psi dc}$ and $Y_{\Psi ac}$ are the transfer function to dc voltage and ac current, respectively.

By way of illustration, Figure 2.13 shows the six pulse ideal converter transfer function with a steady converter firing angle, related to each phase of the described voltage waveform, which written as a Fourier series is

$$Y_\Psi = \frac{2\sqrt{3}}{\pi}\sum_m (\pm)\frac{1}{m}\cos[m(\omega_{ot} - \alpha_o - \Psi)], \tag{2.74}$$

where $$(\pm) = \sin\left(\frac{m\pi}{2}\right) \quad \text{for } m = 1,5,7,11, \text{ etc.}$$

In general, the switched functions V_Ψ and I_{dc} in Equations (2.72) and (2.73) will contain any number of harmonics, i.e.

$$V_\Psi = V_1 \cos(\omega_{ot} - \Psi) + \sum_k V_k \cos(k\omega_{ot} + \delta_k \pm \Psi), \tag{2.75}$$

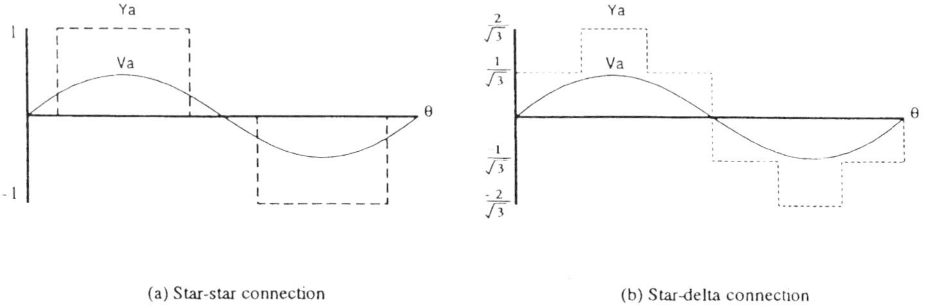

Figure 2.13 Transfer functions for ideal 6 pulse converters, phase a

$$I_{dc} = I_d + \sum_k I_k \cos(k\omega_{ot} + (k\omega_{ot}\delta_k). \tag{2.76}$$

The spectra of the dc voltage and ac current waveforms will then result from the multiplication of Expressions (2.74) by either (2.75) or (2.76).

An alternative to the multiplication of the component functions in the time domain is their convolution in the frequency domain. This alternative is used to calculate converter harmonic cross-modulation in Chapter 8.

The transfer function approach is essential to the derivation of the cyclo-converter frequency components, since in this case the frequency spectra of the output voltage and input current waveforms are related to both the main input and output frequencies. These waveforms contain frequencies which are not integer multiples of the main output frequency.

Each output phase of the basic cycloconverter is derived from a three-phase system via a 'positive' and a 'negative' static converter, as shown in Figure 2.14 [11].

By expressing the switching function as a phase-modulated harmonic series, a general harmonic series can be derived for the output voltage (or input current) waveform in terms of the independent variables.

By way of illustration, the quiescent voltage waveform of the positive converter shown in Figure 2.15, is given by

$$\begin{aligned}(v_p)_q = {} & V_N \sin\theta_i . F_1\left(\theta_i - \frac{\pi}{2}\right) + V_N \sin\left(\theta_i - \frac{2\pi}{3}\right) \cdot F_2\left(\theta_i - \frac{\pi}{2}\right) \\ & + V_N \sin\left(\theta_i + \frac{2\pi}{3}\right) \cdot F_3\left(\theta_i - \frac{\pi}{2}\right).\end{aligned} \tag{2.77}$$

The modulated firing control provides a 'to and fro' phase modulation $f(\theta_o)$ of the individual firings with respect to the quiescent firing.

In general, the value of $f(\theta_o)$ will oscillate symmetrically to and fro about zero, at a repetition frequency equal to the selected output frequency. The limits of control on either side of the quiescent point are then $\pm\pi/2$. Thus, the general expressions for the switching function of the positive and negative converters are

$$F\left(\theta_i - \frac{\pi}{2} + f(\theta_o)\right) \qquad \text{and} \qquad F\left(\theta_i + \frac{\pi}{2} - f(\theta_o)\right),$$

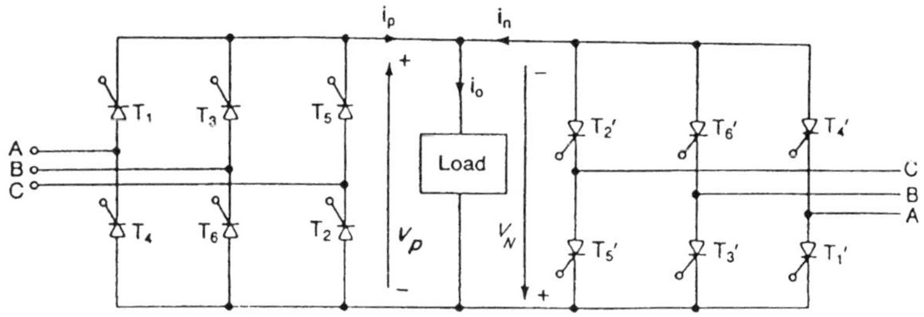

Figure 2.14 Basic cycloconverter

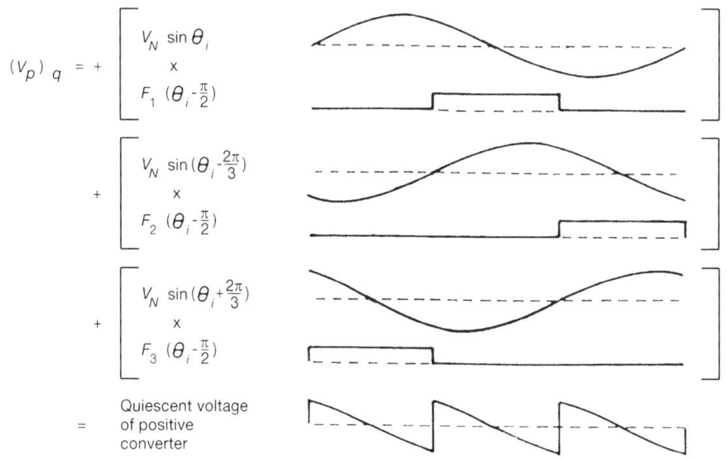

Figure 2.15 Derivation of voltage waveforms of the positive converter for quiescent ($\alpha = 90°$) operation

since the phase modulation of the firing angles of the positive and negative converters is equal but of opposite sign.

Moreover, it can be shown [11] that the optimum output waveform, i.e. the minimum r.m.s. distortion, is achieved when the firing angle modulating function is derived by the 'cosine wave crossing' control. Under this type of control the phase of firing of each thyristor is shifted with respect to the quiescent position by

$$f(\theta_o) = \sin^{-1} r \sin\theta_o, \tag{2.78}$$

where r is the ratio of amplitude of wanted sinusoidal component of output voltage to the maximum possible wanted component of output voltage, obtained with 'full' firing angle modulation.

For the derivation of the input current waveform it is more convenient to use two switching functions, i.e. the thyristor and the converter (the conducting half of the dual converter) switching functions. To simplify the description it is also necessary to make the following approximations: (i) the output current is purely sinusoidal; (ii) the source impedance (including transformer leakage) is neglected.

Considering first a single-phase output, illustrated in Figure 2.16, the current in each phase of the supply is given by

$$\begin{aligned} i_A =& I_o \sin(\theta_o + \phi_o) \cdot F_1\left(\theta_i - \frac{\pi}{2} + f(\theta_o)\right) \cdot F_p(\theta_o) \\ &+ I_o \sin(\theta_o + \phi_o) \cdot F_1\left(\theta_i + \frac{\pi}{2} - f(\theta_o)\right) \cdot F_N(\theta_o). \end{aligned} \tag{2.79}$$

From conventional Fourier analysis F_1, F_p and F_N can be expressed in terms of the following series:

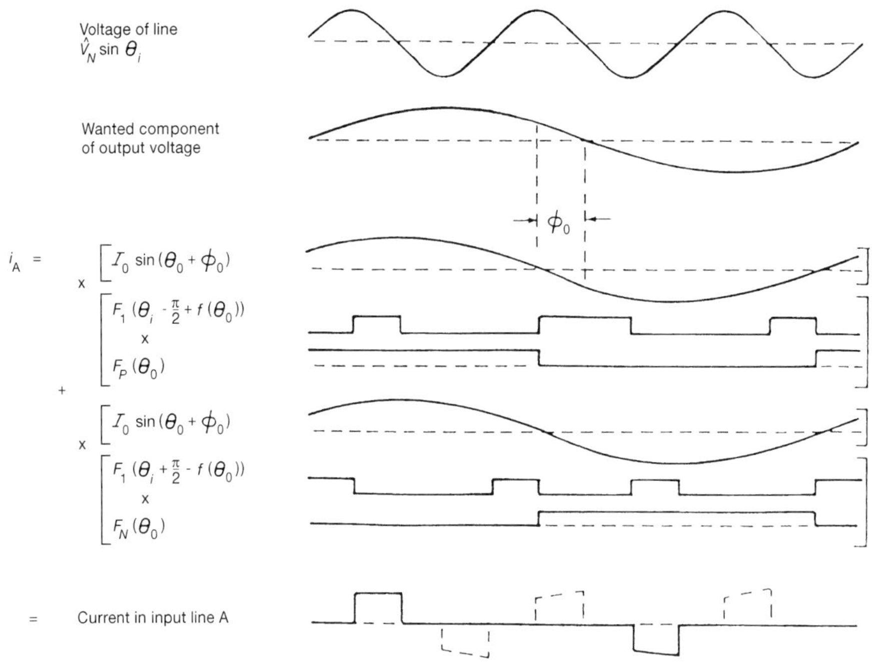

Figure 2.16 Derivation of the input line current of a cycloconverter. The input line current is shown in the bottom part of the figure as a continuous line for a single-phase load and as a broken line for a three-phase load

$$F_1\left(\theta_i \mp \frac{\pi}{2} \pm f(\theta_o)\right) = \frac{1}{3} + \frac{\sqrt{3}}{\pi}\left[\sin\left(\theta_i \pm \frac{\pi}{2} \mp f(\theta_o)\right) - \frac{1}{2}\cos 2\left(\theta_i \pm \frac{\pi}{2} \mp f(\theta_o)\right) - \frac{1}{4}\cos 4\left(\theta_i \pm \frac{\pi}{2} \mp f(\theta_o)\right)\right], \tag{2.80}$$

$$F_p(\theta_o) = \frac{1}{2} + \frac{2}{\pi}\left[\sin(\theta_o + \phi_o) + \frac{1}{3}\sin 3\,(\theta_o + \phi_o) + \frac{1}{5}\sin 5(\theta_o + \phi_o) + \dots\right], \tag{2.81}$$

$$F_N(\theta_o) = \frac{1}{2} - \frac{2}{\pi}\left[\sin(\theta_o + \phi_o) + \frac{1}{3}\sin 3(\theta_o + \phi_o) + \frac{1}{5}\sin 5(\theta_o + \phi_o) + \dots\right]. \tag{2.82}$$

Substituting in i_A and reducing

$$
\begin{aligned}
i_A = & I_o \sin(\theta_o + \phi_o)\Bigg\{\frac{1}{3} + \frac{\sqrt{3}}{\pi}\Bigg[\sin\theta_i \sin f(\theta_o) + \frac{1}{2}\cos 2\theta_i \cos 2f(\theta_o) \\
& - \frac{1}{4}\cos 4\theta_i \cos 4f(\theta_o) - \frac{1}{5}\sin 5\theta_i \sin 5f(\theta_o) + \ldots\Bigg] \\
& + \frac{4\sqrt{3}}{\pi^2}\Bigg[-\cos\theta_i \cos f(\theta_o) - \frac{1}{2}\sin 2\theta_i \sin 2f(\theta_o) + \frac{1}{4}\sin 4\theta_i \sin 4f(\theta_o) \\
& + \frac{1}{5}\cos 5\theta_i \cos 5f(\theta_o) \ldots\Bigg]\Bigg[\sin(\theta_o + \phi_o) + \frac{1}{3}\sin 3(\theta_o + \phi_o) \\
& \frac{1}{5}\sin 5(\theta_o + \phi_o) + \ldots\Bigg]\Bigg\}.
\end{aligned} \tag{2.83}
$$

In the above expression $f(\theta_o) = \sin^{-1} r \sin\theta_o$ (see Equation (2.78)) as explained above when the modulating function uses the cosine wave crossing control method.

In general, however, the output will also be three-phase and, assuming perfectly balanced input and output waveforms, each phase of the input will include the contribution of the three output currents, i.e. $i_A = i_{A1} i_{A2} + i_{A3}$ and the corresponding waveform is illustrated by a broken line in Figure 2.16.

2.11 Summary

The main Fourier concepts and techniques relevant to power system harmonic analysis have been described. These included the basic Fourier series, the Fourier Transform and its computer implementation in the form of the Fast Fourier Transform.

A Fourier-domain-based transfer function concept has also been introduced for the analysis of power electronic waveforms resulting from complex controls and multiple periodic switchings. The effectiveness of this technique will become apparent in Chapters 5 and 8.

2.12 References

1. Fourier, J B J, (1822). *Théorie Analytique de la Chaleur* (book).
2. Kreyszig, E, (1967). *Advanced Engineering Mathematics*, John Wiley and Sons Inc, 2nd Edition.
3. Kuo, F F, (1966). *Network Analysis and Synthesis*, John Wiley and Sons, Inc.
4. Brigham, E O, (1974). *The Fast Fourier Transform*, Prentice-Hall, Inc.
5. Cooley, J W and Tukey, J W, (1965). 'An algorithm for machine calculation of complex Fourier series', *Math Computation*, **19**, 297–301.
6. Cochran, W T, *et al*, (1967). What is the fast Fourier Transform. *Proc IEEE*, **10**, 1664–1677.
7. Bergland, G D, (1969). A guided tour of the fast Fourier Transform. *IEEE Spectrum*, July, 41–42.
8. Bergland, G D, (1968). A fast Fourier Transform algorithm for real-values series. *Numerical Analysis*, **11**(10), 703–710.

9. Stemmler, H, (1972). HVdc back to back interties on weak a.c. systems, second harmonic problems and solutions, *CIGRE Symposium*, **09-87**, no 300-08, 1–5.
10. Wood, A R, (1993). An analysis of non-ideal HVdc converter behaviour in the frequency domain, and a new control proposal, Ph.D. Thesis, University of Canterbury, New Zealand.
11. Pelly, B R, (1971). *Thyristor Phase Controlled Converters and Cycloconverters*, Wiley Interscience, New York.

3

TRANSMISSION SYSTEMS

3.1 Introduction

As the main vehicle of harmonic propagation, the transmission system must be accurately represented to predict the levels of waveform distortion throughout the power system. The following steps are used in the derivation of a multi-phase transmission system model:

- Definition of the components of the transmission system and their separation into homogeneous elements; typical elements in this context are an untransposed section of the transmission line, a cable, a series impedance and a shunt admittance.

- Selection of the location of observation points. If standing waves are to be displayed then observation points must be inserted at intervals of less than one tenth of a wavelength at the highest frequency of interest. Element data is then partitioned so that the observation points occur at the junctions between the component elements.

- Provision of element type data and those parameters necessary for the determination of the elements' electrical characteristics, such as the conductor type, their arrangement, earth resistivities, etc.

- Derivation of reduced equivalent impedance (admittance) matrices for the frequencies of interest.

Details of the method of calculation and the features used to improve computational efficiency are discussed in the following sections.

3.2 Network Subdivision

Although an element, or branch, is the basic component of a network, elements may be coupled and non-homogeneous, e.g. mutually coupled transmission lines with different tower geometries over the line length. To facilitate the inclusion of this type of element, a subsystem is defined as follows:

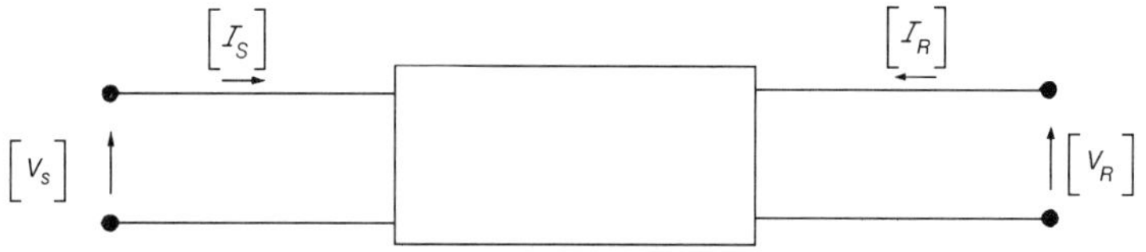

(a)

$$\begin{bmatrix} V_S \\ I_S \end{bmatrix} = \begin{bmatrix} [A][B] \\ [C][D] \end{bmatrix} \begin{bmatrix} V_R \\ I_R \end{bmatrix}$$

(b)

Figure 3.1 Two-port network transmission parameters: (a) multi-two-port network; (b) matrix transmission parameters

- A subsystem is the unit into which any part of the system may be divided such that no subsystem has any mutual coupling between its constituent branches and those of the rest of the system.
- The smallest unit of a subsystem is a single network element.
- The subsystem unit is retained for input data organisation. Data for any subsystem is input as a complete unit, the subsystem admittance matrix is formulated and then combined in the total system admittance matrix.
- Subsystem admittance matrices may be derived by finding, for each section, the *ABCD* or transmission parameters.

This procedure involves an extension of the usual two-port network theory to multi-two-port networks. Current and voltages are now matrix quantities as defined in Figure 3.1. The dimensions of the parameter matrices correspond to those of the section being considered, i.e. three, six, nine or twelve for one, two, three or four mutually coupled three-phase elements, respectively. All sections must contain the same number of mutually coupled three-phase elements, ensuring that all the parameter matrices are of the same order and that the matrix multiplications are executable. Uncoupled elements need to be considered as coupled ones with zero coupling to maintain correct dimensions for all matrices.

For the case of a non-homogeneous line with n different sections:

$$\begin{aligned} \begin{bmatrix} V_S \\ I_S \end{bmatrix} &= \begin{bmatrix} [A_1][B_1] \\ [C_1][D_1] \end{bmatrix} \times \begin{bmatrix} [A_2][B_2] \\ [C_2][D_2] \end{bmatrix} \times \ldots \times \begin{bmatrix} [A_n][B_n] \\ [C_n][D_n] \end{bmatrix} \begin{bmatrix} V_R \\ -I_R \end{bmatrix} \\ \begin{bmatrix} V_S \\ I_S \end{bmatrix} &= \begin{bmatrix} [A][B] \\ [C][D] \end{bmatrix} \begin{bmatrix} V_R \\ -I_R \end{bmatrix}. \end{aligned} \tag{3.1}$$

It must be noted that in general $[A] \neq [D]$ for a non-homogeneous line.

Once the resultant *ABCD* parameters have been found the equivalent nodal admittance matrix for the subsystem can be calculated from

$$[Y] = \begin{bmatrix} [D][B]^{-1} & [C]-[D][B]^{-1}[A] \\ [B]^{-1} & -[B]^{-1}\ \ [A] \end{bmatrix}. \tag{3.2}$$

If only input-output voltage information is required, the cascading approach described above is sufficient. However, if extra information along the line is required, appropriate fictitious nodes are created at specified points and/or at regular intervals, and the following nodal matrix equation is formed, inverted (factorized) and solved. The resultant vector provides the harmonic voltage profile along the line. This analysis applies to both homogeneous and non-homogeneous lines.

$$\begin{bmatrix} I_S \\ I_1 \\ I_2 \\ \vdots \\ I_n \\ I_R \end{bmatrix}_h = \begin{bmatrix} [Y_{SS}] & -[Y_{S1}] & & & & \\ -[Y_{1S}] & [Y_{SS}]+[Y_{11}] & -[Y_{12}] & & & \\ & -[Y_{21}] & [Y_{11}]+[Y_{22}] & \ddots & & \\ & & \ddots & \ddots & \ddots & \\ & & & \ddots & [Y_{nn}]+[Y_{RR}] & -[Y_{nR}] \\ & & & & -[Y_{Rn}] & [Y_{RR}] \end{bmatrix}_h \begin{bmatrix} V_S \\ V_1 \\ V_2 \\ \vdots \\ V_n \\ V_R \end{bmatrix}_h \tag{3.3}$$

3.3 Frame of Reference used in Three-phase System Modelling

Sequence components have long been used to enable convenient examination of the balanced power system under both balanced and unbalanced loading conditions.

The symmetrical component transformation is a general mathematical technique developed by Fortescue whereby any 'system of n vectors or quantities may be resolved when n is prime into n different symmetrical n phase systems' [1]. Any set of three-phase voltages or currents may therefore be transformed into three symmetrical systems of three vectors each. This in itself would not commend the method and the assumptions, which lead to the simplifying nature of symmetrical components, must be examined carefully.

Consider, as an example, the series admittance of a three-phase transmission line, shown in Figure 3.2, i.e. three mutually coupled coils. The admittance matrix relates the illustrated currents and voltages by

$$[I_{abc}] = [Y_{abc}][V_{abc}], \tag{3.4}$$

where

$$\begin{aligned} [I_{abc}] &= [I_a I_b I_c]^{\mathrm{T}} \\ [V_{abc}] &= [V_a V_b V_c]^{\mathrm{T}} \end{aligned} \tag{3.5}$$

and

$$[Y_{abc}] = \begin{bmatrix} y_{aa} & y_{ab} & y_{ac} \\ y_{ba} & y_{bb} & y_{bc} \\ y_{ca} & y_{cb} & y_{cc} \end{bmatrix} \tag{3.6}$$

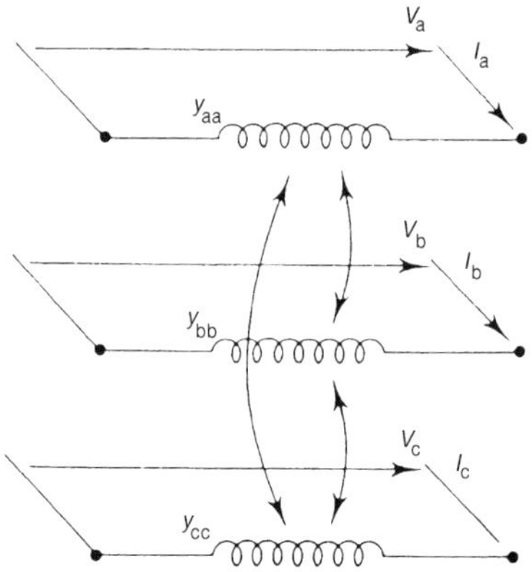

Figure 3.2 Admittance representation of a three-phase series element

By the use of the symmetrical components transformation the three coils of Figure 3.2 can be replaced by three uncoupled coils. This enables each coil to be treated separately with a great simplification of the mathematics involved in the analysis.

The transformed quantities (indicated by subscripts O12 for the zero, positive and negative sequences respectively) are related to the phase quantities by

$$[V_{012}] = [T_S]^{-1}[V_{abc}], \tag{3.7}$$

$$[I_{012}] = [T_S]^{-1}[I_{abc}] \tag{3.8}$$

$$= [T_S]^{-1}[Y_{abc}[T_S][V_{012}], \tag{3.9}$$

where $[T_S]$ is the transformation matrix.

The transformed voltages and currents are thus related by the transformed admittance matrix,

$$[Y_{012}] = [T_s]^{-1}[Y_{abc}][T_s]. \tag{3.10}$$

Assuming that the element is balanced, we have

$$\begin{aligned} y_{aa} &= y_{bb} = y_{cc}, \\ y_{ab} &= y_{bc} = y_{ca}, \\ y_{ba} &= y_{cb} = y_{ac}, \end{aligned} \tag{3.11}$$

and a set of invariant matrices $[T]$ exist. Transformation (3.10) will then yield a diagonal matrix $[y_{012}]$.

In this case, the mutually coupled three-phase system has been replaced by three uncoupled symmetrical systems. In addition, if the generation and loading may be assumed balanced, then only one system, the positive sequence system, has any current flow and the other two sequences may be ignored. This is essentially the situation with the single-phase harmonic penetration analysis.

In general, however, such an assumption is not valid. Unsymmetrical interphase coupling exists in transmission lines and to a lesser extent in transformers, and this results in coupling between the sequence networks.

If the original phase admittance matrix $[Y_{abc}]$ is in its natural unbalanced state then the transformed admittance matrix $[Y_{012}]$ is full. Therefore, current flow of one sequence will give rise to voltages of all sequences, i.e. the equivalent circuits for the sequence networks are mutually coupled. In this case, the problem of analysis is no simpler in sequence components than in the original phase components.

From the above considerations it is clear that the asymmetry inherent in transmission systems cannot be studied with any simplification by using the symmetrical component frame of reference.

With the use of phase coordinates the following advantages become apparent:

(1) Any system element maintains its identity.

(2) Features such as asymmetric impedances, mutual couplings between phases and between different system elements, and line transpositions are all readily considered.

(3) Transformer phase shifts present no problem.

Thus phase components are normally retained throughout the formation and solution of the admittance matrices in the following sections, while sequence components are used as an aid to interpretation of results.

Moreover, it will be shown in later chapters that iterative solutions involving static converters can be more efficient in sequence components due to the absence of zero sequence currents at the converter terminals.

3.4 Evaluation of Transmission Line Parameters

The lumped series impedance matrix $[Z]$ of a transmission line consists of three components, while the shunt admittance matrix $[Y]$ contains one.

$$[Z] = [Z_e] + [Z_g] + [Z_c] \tag{3.12}$$

$$[Y] = [Y_g] \tag{3.13}$$

where

$[Z_c]$ is the internal impedance of the conductors ($\Omega.\text{km}^{-1}$),

$[Z_g]$ is the impedance due to the physical geometry of the conductor's arrangement ($\Omega.\text{km}^{-1}$),

$[Z_e]$ is the earth return path impedance ($\Omega.\text{km}^{-1}$), and

$[Y_g]$ is the admittance due to the physical geometry of the conductor ($\Omega^{-1}\text{km}^{-1}$).

In multiconductor transmission all *primitive matrices* (the admittance matrices of the unconnected branches of the original network components) are symmetric and, therefore, the functions that define the elements need only be evaluated for elements on or above the leading diagonal.

3.4.1 Earth Impedance Matrix $[Z_e]$

The impedance due to the earth path varies with frequency in a non-linear fashion. The solution of this problem, under idealised conditions, has been given in the form of either an infinite integral or an infinite series [2].

As the need arises to calculate ground impedances for a wide spectrum of frequencies, the tendency is to select simple formulations aiming at a reduction in computing time, while maintaining a reasonable level of accuracy.

Consequently, what was originally a heuristic approach [3], is becoming the more favoured alternative, particularly at high frequencies. Based on Carson's work, the ground impedance can be concisely expressed as

$$z_e = 1000J(r,\theta)(\Omega.\text{km}^{-1}) \tag{3.14}$$

where

$$z_e \in [Z_e]$$

$$J(r,\theta) = \frac{\omega\mu_a}{\pi}\{P(r,\theta) + jQ(r,\theta)\}$$

$$r_{ij} = \sqrt{\frac{\omega\mu_a}{\rho}}\quad D_{ij}$$

$$D_{ij} = \sqrt{(h_i + h_j)^2 + d^2{}_{ij}} \quad ; \quad \text{for } i \neq j$$

$$D_{ij} = 2h_i \quad ; \quad \text{for } i = j$$

$$\theta_{ij} = \arctan\frac{d_{ij}}{h_i + h_j} \quad ; \quad \text{for } i \neq j$$

$$\theta_{ij} = 0 \quad ; \quad \text{for } i = j$$

$$\omega = 2\pi f(\text{rad.s}^{-1})$$

h_i = height of conductor i (m)

d_{ij} = horizontal distance between conductors i and j (m)

μ_a = permeability of free space $= 4\,\pi \times 10^{-7}$ H m^{-1}

ρ = earth resistivity (Ω.m).

Carson's solution to Equation (3.14) is defined by eight different infinite series which converge quickly for problems related to transmission line parameter calculation, but the number of required computations increases with frequency and separation of the conductors.

More recent literature has described closed form formulations for the numerical evaluation of line-ground loops, based on the concept of a mirroring surface beneath the earth at a certain depth. The most popular complex penetration model which has had more appeal is that of C. Dubanton [5], due to its simplicity and high degree of accuracy for the whole frequency span for which Carson's equations are valid.

Dubanton's formulae for the evaluation of the self and mutual impedances of conductors i and j are

$$Z_{ii} = \frac{j\omega\mu_o}{2\pi} \times \ln\frac{2(h_i + p)}{r_i} \tag{3.15}$$

$$Z_{ij} = \frac{j\omega\mu_o}{2\pi} \times \ln\frac{\sqrt{2(h_i + p)}}{\sqrt{(h_i - h_j)^2 + d_{ij}^2}} \tag{3.16}$$

where $p = 1/\sqrt{j\omega\mu_0\sigma}$ is the complex depth below the earth at which the mirroring surface is located.

An alternative and very simple formulation has been recently proposed by Acha [4], which for the purpose of harmonic penetration yields accurate solutions when compared to those obtained using Carson's equations. The following alternative formulation is used for the real and imaginary components of equation (3.14):

$$P = s_e - t_e r \tag{3.17}$$

$$Q = u_e - v_e \ln r \tag{3.18}$$

where the s_e, t_e, u_e and v_e coefficients are derived from accurate curve fitting of Carson's equations. For the calculation of line parameters for practical tower geometries, ground conductivities and frequencies of interest, $r = 2$ appears a reasonable maximum value to be considered, e.g. $r < 1.9$ for $\rho = 100\,\Omega$ m, $f = 3000$ Hz, and $d = 120$ m. Larger values of r are required only for calculating inductive coupling to distant cables. Coefficients calculated at steps of 0.5 in r produce very accurate results, except for the first section which is subdivided into two, i.e. $r < 0.20$ and $0.20 \leqslant r < 0.50$. Moreover, the exercise is only valid for a particular value of angle θ, but fittings at 15 degree intervals, with linear interpolation in-between have been found to be sufficiently accurate. The coefficients are given in Tables 3.1 to 3.4.

Once the values of r and θ have been computed, the nearest values in the tables are selected and inserted in Equations (3.17) and (3.18).

An example of the curve fitting approach and its comparison with Dubanton's solution is illustrated in Figure 3.3.

The error criteria used here is the difference between Carson's result and the approximate values of the real and imaginary part, relative to the magnitude of the Carson impedance, i.e.

$$E_P = \frac{(R_C - R_F)}{|Z_C|} \times 100\%$$

$$E_Q = \frac{(X_C - X_F)}{|Z_C|} \times 100\%$$

where

E_P and E_Q = coefficients of error for the P and Q terms

R_C and X_C = resistance and reactance calculated using Carson's equation

R_F and X_F = resistance and reactance calculated using curve fitting

$|Z_C|$ = magnitude of the Carson impedance

$= \sqrt{R_C^2 + X_C^2}$

3.4.2 Geometrical Impedance Matrix $[Z_g]$ and Admittance Matrix $[Y_g]$

If the conductors and the earth are assumed to be equipotential surfaces, the geometrical impedance can be formulated in terms of potential coefficients theory.

Table 3.1 Earth Impedance Coefficient s_e

r	θ 0°	15°	30°	45°	60°	75°	90°
0.2	0.3910	0.3911	0.3915	0.3922	0.3929	0.3937	0.3944
0.5	0.3796	0.3804	0.3829	0.3869	0.3922	0.3983	0.4044
1.0	0.3591	0.3606	0.3654	0.3734	0.3847	0.3993	0.4167
1.5	0.3293	0.3313	0.3373	0.3480	0.3643	0.3876	0.4195
2.0	0.3019	0.3037	0.3095	0.3201	0.3371	0.3631	0.4025
2.5	0.2774	0.2790	0.2838	0.2927	0.3074	0.3310	0.3695
3.0	0.2559	0.2570	0.2605	0.2670	0.2779	0.2959	0.3268
3.5	0.2371	0.2377	0.2398	0.2437	0.2502	0.2611	0.2803
4.0	0.2205	0.2207	0.2216	0.2230	0.2252	0.2287	0.2348
4.5	0.2058	0.2057	0.2054	0.2046	0.2029	0.1997	0.1932
5.0	0.1929	0.1924	0.1912	0.1886	0.1838	0.1741	0.1573
5.5	0.1809	0.1812	0.1781	0.1727	0.1646	0.1551	0.1274
6.0	0.1709	0.1703	0.1669	0.1611	0.1508	0.1358	0.1034
6.5	0.1617	0.1608	0.1570	0.1504	0.1386	0.1202	0.08458
7.0	0.1534	0.1522	0.1481	0.1407	0.1278	0.1072	0.07006
7.5	0.1460	0.1444	0.1398	0.1322	0.1184	0.09621	0.05893
8.0	0.1390	0.1374	0.1326	0.1244	0.1101	0.08729	0.05041
8.5	0.1327	0.1310	0.1261	0.1174	0.1027	0.07973	0.04383
9.0	0.1269	0.1252	0.1201	0.1111	0.09621	0.07329	0.03866
9.5	0.1216	0.1199	0.1147	0.1054	0.09044	0.06774	0.03451
10.0	0.1167	0.1149	0.1097	0.1003	0.08529	0.06293	0.03109
10.5	0.1121	0.1104	0.1051	0.09560	0.08067	0.05872	0.02821
11.0	0.1079	0.1062	0.1009	0.09133	0.07650	0.05501	0.02573
11.5	0.1024	0.1023	0.09693	0.08742	0.07274	0.05172	0.02357
12.0	0.1004	0.09867	0.09330	0.08382	0.06931	0.04878	0.02166

The self-potential coefficient Ψ_{ii} for the ith conductor and the mutual potential coefficient Ψ_{ij} between the ith and jth conductors are defined as follows,

$$\Psi_{ii} = \ln(2h_i/r_i) \tag{3.19}$$

$$\Psi_{ij} = \ln(D_{ij}/d_{ij}) \tag{3.20}$$

where r_i is the radius of the ith conductor (m) while the other variables are as defined earlier.

Potential coefficients depend entirely on the physical arrangement of the conductors and need only be evaluated once.

For practical purposes the air is assumed to have zero conductance and

$$[Z_g] = \mathrm{j}\omega K'[\Psi]\ \Omega/\mathrm{km} \tag{3.21}$$

where $[\Psi]$ is a matrix of potential coefficients
and $K' = 2 \times 10^{-4}$.

The lumped shunt admittance parameters $[Y]$ are completely defined by the inverse relation of the potential coefficients matrix, i.e.

Table 3.2 Earth Impedance Coefficient t_e

	θ						
r	0°	15°	30°	45°	60°	75°	90°
0.2	0.1892	0.1854	0.1739	0.1545	0.1268	0.09050	0.04560
0.5	0.1426	0.1418	0.1391	0.1338	0.1248	0.1100	0.08700
1.0	0.1042	0.1047	0.1064	0.1087	0.1112	0.1127	0.1107
1.5	0.07400	0.07500	0.07800	0.08320	0.09090	0.1014	0.1143
2.0	0.05560	0.05650	0.05940	0.06460	0.07280	0.08510	0.10320
2.5	0.04330	0.04410	0.04650	0.05080	0.05790	0.06910	0.08680
3.0	0.03470	0.03530	0.03710	0.04050	0.04610	0.05500	0.06970
3.5	0.02840	0.02890	0.03020	0.03280	0.03690	0.04340	0.05420
4.0	0.02370	0.02400	0.02500	0.02680	0.02970	0.03410	0.04120
4.5	0.02000	0.02024	0.02096	0.02222	0.02413	0.02688	0.03081
5.0	0.01712	0.01726	0.01781	0.01866	0.01987	0.02120	0.02282
5.5	0.01473	0.01501	0.01519	0.01552	0.01609	0.01734	0.01684
6.0	0.01291	0.01304	0.01316	0.01340	0.01359	0.01384	0.01247
6.5	0.01137	0.01144	0.01150	0.01161	0.01155	0.01122	0.009331
7.0	0.01009	0.01012	0.01013	0.01013	0.009886	0.009229	0.007094
7.5	0.009039	0.009005	0.008957	0.008912	0.008548	0.007662	0.005504
8.0	0.008108	0.008074	0.008000	0.007869	0.007433	0.006472	0.004367
8.5	0.007314	0.007278	0.007182	0.006994	0.006513	0.005528	0.003544
9.0	0.006632	0.006593	0.006481	0.006254	0.005748	0.004769	0.002935
9.5	0.006040	0.006000	0.005875	0.005623	0.005107	0.004153	0.002473
10.0	0.005524	0.005482	0.005349	0.005082	0.004565	0.003646	0.002113
10.5	0.005072	0.005028	0.004889	0.004614	0.004103	0.003226	0.001825
11.0	0.004672	0.004628	0.004486	0.004207	0.003706	0.002872	0.001589
11.5	0.004318	0.004273	0.004129	0.003851	0.003364	0.002573	0.001393
12.0	0.004002	0.003957	0.003813	0.003538	0.003066	0.002317	0.001227

$$[Y_g] = 1000\mathrm{j}\omega 2\pi\varepsilon_\mathrm{a}[\Psi]^{-1} \tag{3.22}$$

where ε_a = permittivity of free space = $8.854 \times 10^{-12}(\mathrm{F\,m^{-1}})$.

As $[Z_g]$ and $[Y_g]$ are linear functions of frequency, they need only be evaluated once and scaled for other frequencies.

3.4.3 Conductor Impedance Matrix $[Z_c]$

This term accounts for the internal impedance of the conductors. Both resistance and inductance have a non-linear frequency dependence. Current tends to flow on the surface of the conductor, this *skin effect* increases with frequency and needs to be computed at each frequency. An accurate result for a homogeneous nonferrous conductor of annular cross-section involves the evaluation of long equations based on the solution of Bessel functions, as shown in Equation (3.23).

$$Z_c = \frac{\mathrm{j}\omega\mu_\mathrm{o}}{2\pi}\,\frac{1}{x_e}\,\frac{J_\mathrm{o}(x_e)N'_\mathrm{o}(x_i) - N_\mathrm{o}(x_e)J'_\mathrm{o}(x_i)}{J'_\mathrm{o}(x_e)N'_\mathrm{o}(x_i) - N'_\mathrm{o}(x_e)J'_\mathrm{o}(x_i)} \tag{3.23}$$

Table 3.3 Earth Impedance Coefficient u_e

	θ						
r	0°	15°	30°	45°	60°	75°	90°
0.2	0.3795	0.3773	0.3710	0.3606	0.3467	0.32990	0.31100
0.5	0.4652	0.4613	0.4496	0.4302	0.4032	0.36890	0.32810
1.0	0.5018	0.4978	0.4856	0.4650	0.4356	0.39710	0.34930
1.5	0.5033	0.4992	0.4870	0.4663	0.4367	0.39790	0.34950
2.0	0.4855	0.4812	0.4679	0.4454	0.4132	0.37069	0.31730
2.5	0.4618	0.4569	0.4421	0.4167	0.3796	0.32930	0.26400
3.0	0.4370	0.4316	0.4150	0.3864	0.3438	0.28410	0.20270
3.5	0.4130	0.4071	0.3890	0.3574	0.3097	0.24130	0.14390
4.0	0.3906	0.3843	0.3649	0.3309	0.2792	0.20400	0.09390
4.5	0.3699	0.3633	0.3430	0.3073	0.2528	0.17300	0.05524
5.0	0.3505	0.3445	0.3238	0.2855	0.2299	0.14830	0.02789
5.5	0.3375	0.3255	0.3026	0.2730	0.2127	0.12860	0.010340
6.0	0.3202	0.3105	0.2882	0.2547	0.1957	0.11390	0.0002289
6.5	0.3051	0.2965	0.2746	0.2392	0.1816	0.10250	−0.004619
7.0	0.2916	0.2837	0.2620	0.2258	0.1697	0.09355	−0.006136
7.5	0.2790	0.2720	0.2508	0.2139	0.1594	0.08693	−0.005799
8.0	0.2681	0.2611	0.2401	0.2037	0.1507	0.08109	−0.004632
8.5	0.2580	0.2511	0.2303	0.1946	0.1431	0.07633	−0.003265
9.0	0.2486	0.2419	0.2213	0.1863	0.1364	0.07233	−0.002039
9.5	0.2400	0.2333	0.2131	0.1789	0.1304	0.06890	−0.001093
10.0	0.2320	0.2254	0.2056	0.1721	0.1250	0.06590	−0.0004458
10.5	0.2245	0.2180	0.1986	0.1658	0.1202	0.06323	−0.00005569
11.0	0.2175	0.2112	0.1921	0.1601	0.1157	0.06082	0.0001423
11.5	0.2110	0.2048	0.1860	0.1548	0.1117	0.05862	0.0002129
12.0	0.2049	0.1988	0.1804	0.1499	0.1079	0.05661	0.0002097

where

$x_e = \mathrm{j}\sqrt{\mathrm{j}\omega\mu_o\sigma_c}\, r_e$

$x_j = \mathrm{j}\sqrt{\mathrm{j}\omega\mu_o\sigma_c}\, r_i$

r_e = external radius of the conductor (m)

r_i = internal radius of the conductor (m)

J_o = Bessel function of the first kind and zero order

J'_o = derivative of the Bessel function of the second kind and zero order

N_o = Bessel function of the second kind and zero order

N'_o = derivative of the Bessel function of the second kind and zero order

σ_c = conductivity of the conductor material at the average conductor temperature.

The Bessel functions and their derivatives are solved, within a specified accuracy, by means of their associated infinite series. Convergence problems are frequently encountered at high frequencies and low ratios of conductor thickness to external radius i.e. $(r_e - r_i)/r_e$, necessitating the use of asymptotic expansions.

Table 3.4 Earth Impedance Coefficient v_e

r	θ 0°	15°	30°	45°	60°	75°	90°
0.2	0.4817	0.4822	0.48380	0.48640	0.49000	0.49420	0.499100
0.5	0.4248	0.4264	0.43140	0.43980	0.45180	0.46750	0.48690
1.0	0.3680	0.3699	0.37550	0.38550	0.40070	0.42240	0.45220
1.5	0.3073	0.3085	0.31260	0.32000	0.33220	0.35150	0.38190
2.0	0.2623	0.2628	0.26440	0.26750	0.27310	0.28310	0.30180
2.5	0.2278	0.2276	0.22690	0.22570	0.22420	0.22310	0.22440
3.0	0.2006	0.1998	0.19720	0.19240	0.18490	0.17360	0.15740
3.5	0.1788	0.1775	0.17340	0.16600	0.15390	0.13460	0.10370
4.0	0.1608	0.1592	0.15420	0.14490	0.12950	0.10470	0.06370
4.5	0.1459	0.1441	0.13840	0.12780	0.11040	0.08229	0.03572
5.0	0.1330	0.1316	0.12560	0.11330	0.09516	0.06582	0.01749
5.5	0.1247	0.1199	0.11260	0.10520	0.08436	0.05360	0.006572
6.0	0.1146	0.1110	0.10410	0.09442	0.07436	0.04499	0.0006278
6.5	0.1062	0.1033	0.09649	0.08578	0.06650	0.03860	−0.002085
7.0	0.09895	0.09640	0.08975	0.07866	0.06016	0.03382	−0.002899
7.5	0.09248	0.09040	0.08399	0.07252	0.05485	0.03040	−0.002728
8.0	0.08705	0.08499	0.07867	0.06746	0.05044	0.02751	−0.002149
8.5	0.08219	0.08018	0.07397	0.06307	0.04689	0.02521	−0.001492
9.0	0.07784	0.07586	0.06979	0.05922	0.04375	0.02335	−0.0009185
9.5	0.07391	0.07197	0.06605	0.05583	0.04103	0.02179	0.0004877
10.0	0.07034	0.06845	0.06269	0.05280	0.03864	0.02045	0.0002001
10.5	0.06710	0.06525	0.05965	0.05010	0.03652	0.01929	0.00003059
11.0	0.06413	0.06233	0.05689	0.04766	0.03464	0.01827	0.0000537
11.5	0.06141	0.05966	0.05437	0.04545	0.03295	0.01735	0.0000832
12.0	0.05891	0.05720	0.05206	0.04343	0.03142	0.01653	0.00008188

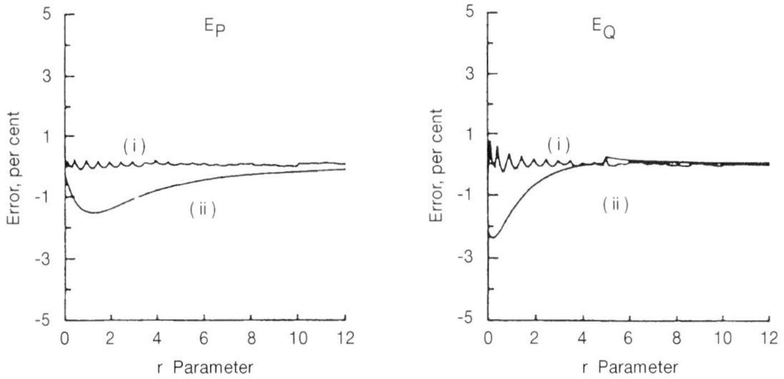

Figure 3.3 Relative errors in the calculation of the self-impedance of an earth return conductor. (i) Fitting technique, (ii) Dubanton's method

A new closed form solution has been proposed based on the concept of complex penetration, Semlyen [3]; unfortunately errors of up to 6.6% occur in the region of interest.

To overcome the difficulties of slow convergence of the Bessel function approach and the inaccuracy of the complex penetration method at relatively low frequency,

Table 3.5 Skin Effect Coefficients

$\sqrt{f/R_{dc}}$		Thickness/radius 1.0	0.8	0.6	0.4
25	s_c	0.9970	0.9976	0.9986	0.9994
	t_c	0.0004456	0.0003618	0.0002128	0.00009099
	u_c	−0.03649	−0.03392	−0.02756	−0.019150
	v_c	0.008132	0.007553	0.006132	0.004257
50	s_c	0.8230	0.8505	0.9069	0.9587
	t_c	0.006667	0.005599	0.003458	0.001528
	u_c	−0.3606	−0.3501	−0.3043	−0.2214
	v_c	0.02128	0.02027	0.01713	0.01223
75	s_c	0.3200	0.3421	0.4832	0.7254
	t_c	0.01678	0.01572	0.01175	0.006039
	u_c	−0.3981	−0.4635	−0.5621	−0.5279
	v_c	0.02248	0.02296	0.02253	0.01841
100	s_c	0.2127	0.1373	0.03979	0.2506
	t_c	0.01837	0.01863	0.01778	0.01233
	u_c	−0.09600	−0.1226	−0.3293	−0.6306
	v_c	0.01841	0.01844	0.01957	0.01989
150	s_c	0.2946	0.2830	0.1500	−0.1201
	t_c	0.01754	0.01719	0.01683	0.01609
	u_c	−0.05016	−0.02284	0.03587	−0.1790
	v_c	0.01791	0.01737	0.01591	0.01561
300	s_c	0.2783	0.2661	0.2375	0.1885
	t_c	0.01765	0.01730	0.01616	0.01408
	u_c	−0.03597	−0.03373	−0.03248	0.01406
	v_c	0.01783	0.01746	0.01635	0.01411

an alternative approach based upon curve fitting to the Bessel function formula has been proposed by Acha [4]. Equation (3.24) is used to approximate the internal impedance.

$$z_c = (s_c + t_c c)R_{dc} + \mathrm{j}(u_c + v_c c)R_{dc} \tag{3.24}$$

where

$$z_c \in [Z_c]$$
$$c = \sqrt{f/R_{dc}}$$
R_{dc} = direct current resistance of the conductor ($\Omega.\mathrm{km}^{-1}$)
s_c, t_c, u_c, v_c = curve fitting coefficients.

A maximum value of $c = 300$ and thickness to radius ratios of between 1.0 and 0.4 are considered in the derivation of the coefficients listed in Table 3.5, where linear interpolation can be used.

Based on a similar criteria as that used for the case of ground impedances, an assessment of the errors introduced by the curve fitting approach shows, in

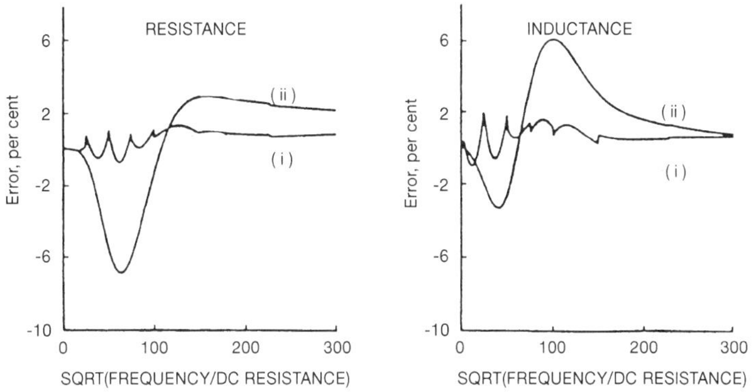

Figure 3.4 Relative errors in the calculation of conductor impedance. (i) Fitting technique (ii) Complex penetration method

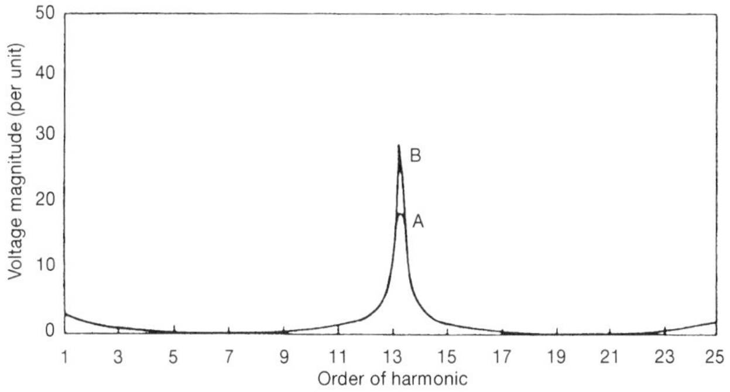

Figure 3.5 The effect of skin effect modelling: curve A, skin effect included; curve B, no skin effect

Figure 3.4, a maximum error of 2% for the real and reactive components of the internal impedance for a thickness ratio of 0.5.

The matrix $[Z_c]$ is diagonal, and normally computations for one phase conductor and one earth-wire are sufficient.

For long lines, skin effect resistance (R_{ac}/R_{dc}) ratios and their effect on the resonant voltage magnitudes are important. Because the series resistance of a transmission line is a small component of the series impedance when the transmission line is not at resonance, the harmonic voltages, shown in Figure 3.5, do not change to any significant extent when skin effect is included. At resonance the series resistance and shunt conductance become the dominant system components. Changes in the series resistance magnitude change the voltage peaks but do not affect the resonant frequency.

In Figure 3.5 the voltage calculated with skin effect is, at resonance, 50% higher than without skin effect; these results correspond with a R_{ac}/R_{dc} ratio of 2. In a single-phase model without ground return the ratio of voltages at resonance, with

Table 3.6 Corrections for Skin Effect in Overhead Lines

Company	Voltage (kV)	Harmonic Order	Resistance
NGC	400, 275 (Based on 0.4 sq.in. steel-core al. conductors)	$1 = h \leqslant 4.21$	$R_1\left(1+\dfrac{3.45h^2}{192+2.77h^2}\right)$
		$4.21 < h \leqslant 7.76$	$R_1(0.806 + 0.105\,h)$
		$h > 7.76$	$R_1(0.267 + 0.485\,\sqrt{h})$
	132		$R_1\left(1+\dfrac{0.646\,h^2}{192+0.518\,h^2}\right)$
EDF	400, 225	$1 = h \leqslant 4$	$R_1\left(1+\dfrac{3.45\,h^2}{192+2.77\,h^2}\right)$
		$4 < h < 8$	$R_1.(0.864 - 0.024\sqrt{h} + 0.105\,h)$
		$8 < h$	$R_1\ (0.267 + 0.485\sqrt{h})$
	150,90		$R_1\left(1+\dfrac{0.646\,h^2}{192+0.518\,h^2}\right)$

and without skin effect, is the same as the skin effect ratio. In a three-phase model the presence of shunt conductance and series resistance coupling between phases, and the different resonant frequencies of the phases, reduces the resonant peak voltages compared with single-phase modelling.

Skin effect is also taken into account in modelling the earth return as a conductor. The depth of penetration of the earth currents decreases with an increase in frequency or a decrease in earth resistivity. The series inductance decreases as a result of these changes.

As an alternative to the rigorous analysis described above, power companies often use approximations to the skin effect by means of correction factors. Typical corrections in current use by the NGC(UK) and EDF(France) are given in Table 3.6.

3.5 Single Phase Equivalent of a Transmission Line

3.5.1 Equivalent PI Model

The case of a single (phase) line is considered first to introduce the various concepts involved in the simplest possible way, before attempting a generalization to the more practical case of multiconductor transmission.

A transmission line consists of distributed inductance and capacitance, illustrated in Figure 3.6, which represent the magnetic and electrostatic conditions of the line, and resistance and conductance which represent the line losses.

Under perfectly balanced conditions, three-phase transmission lines can be represented by their single-phase positive sequence models and nominal PI circuits. For inclusion into an admittance matrix, it is necessary to use the admittances between busbars, and from busbars to earth as in Figure 3.7.

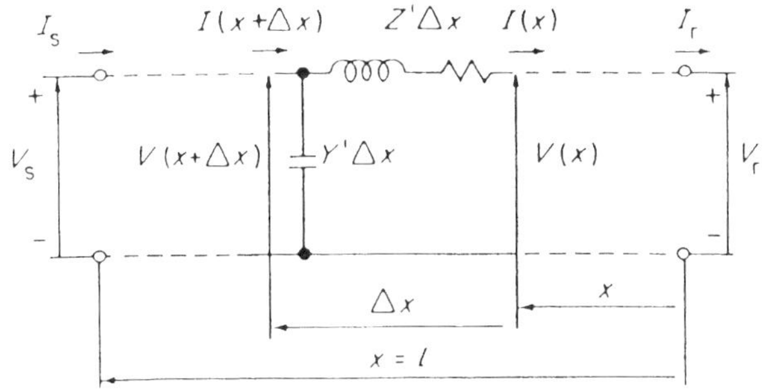

Figure 3.6 Distributed parameter transmission line. V, voltage; I, current; Z', series impedance per unit distance; Y' shunt admittance per unit distance; l, transmission line length

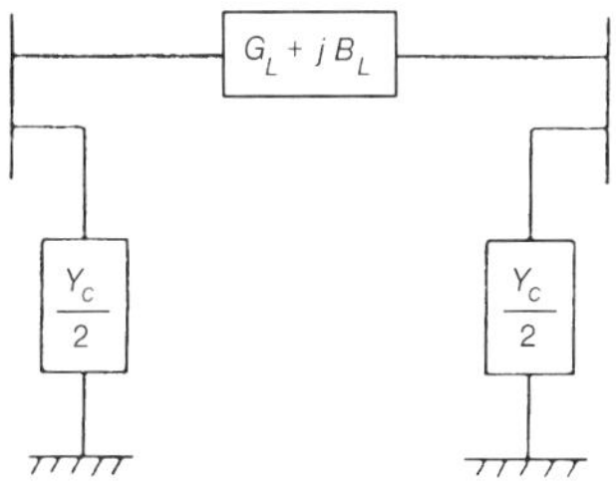

Figure 3.7 Admittance model of transmission line, where $G_L + jB_L = 1/(R_L + jX_L)$, $Y_C = 1/jX_C$, $X_L = \omega L$, and $X_C = 1/\omega C$. G, conductance; B, susceptance; X, reactance; ω, frequency (in radian per second)

For long lines a number of PI models are connected in series to improve the accuracy of voltages and currents, which are affected by standing wave effects. For example, a three-section PI model provides an accuracy to 1.2% for a quarter wavelength line (a quarter wavelength corresponds with 1500 and 1250 km at 50 and 60 Hz, respectively).

As the frequency increases, the number of nominal PI sections to maintain a particular accuracy increases proportionally, e.g. a 300 km line requires 30 nominal PI sections to maintain the 1.2% accuracy for the 50th harmonic. However, near resonance the accuracy departs significantly from an acceptable value.

The computational effort can be greatly reduced and the accuracy improved with the use of an equivalent PI model derived from the solution of the second order linear differential equations describing wave propagation along transmission lines [6]. With reference to Figure 3.6

$$\frac{d^2 V(x)}{dx^2} = Z'Y'V(x) \qquad \frac{d^2 I(x)}{dx^2} = Y'Z'I(x) \tag{3.25}$$

where $Z' = r + j2\pi fL$ is the series impedance per unit length and $Y' = g + j2\pi fC$ is the shunt admittance per unit length.

The solution of wave Equations (3.25) at a distance x from the sending end of the line is:

$$V(x) = \exp(-\gamma.x)V_i + \exp(\gamma.x)V_r \tag{3.26}$$

$$I(x) = (Z')^{-1}\gamma[\exp(-\gamma.x)V_i - \exp(\gamma.x)V_r] \tag{3.27}$$

where $\gamma = \sqrt{Z'Y'} = \alpha + \mathrm{j}\beta$ is the propagation constant and V_i and V_r the forward and reverse travelling voltages, respectively.

Depending on the problem in hand, e.g. if the evaluation of terminal quantities only is required, it may be more convenient to formulate a solution using two-port matrix equations.

Considering a homogeneous multiconductor line of length l,

$$V_s = V_i + V_r \tag{3.28}$$

$$I_s = (1/Z_o)(V_i - V_r) \qquad \text{for } x = 0 \tag{3.29}$$

where $Z_o = \sqrt{Z'/Y'}$ is the characteristic impedance

and

$$V_R = \exp(-\gamma l)V_i + \exp(\gamma l)V_r \tag{3.30}$$

$$I_R = (1/Z_o)[\exp(-\gamma l)V_i - \exp(\gamma l)V_r] \qquad \text{for } x = l. \tag{3.31}$$

Thus, from the equations at $x = 0$,

$$2V_i = V_s + Z_o I_s \tag{3.32}$$

$$2V_r = V_s - Z_o I_s \tag{3.33}$$

so that substitution in Equations (3.30) and (3.31) at $x = l$ yields

$$V_R = \cosh(\gamma l)V_S - \sinh(\gamma l)Z_o I_s \tag{3.34}$$

$$I_R = -Y_o \sinh(\gamma l)V_s + Y_o \cosh(\gamma l)Z_o I_s. \tag{3.35}$$

The following relationships can be written for the circuit of Figure 3.8

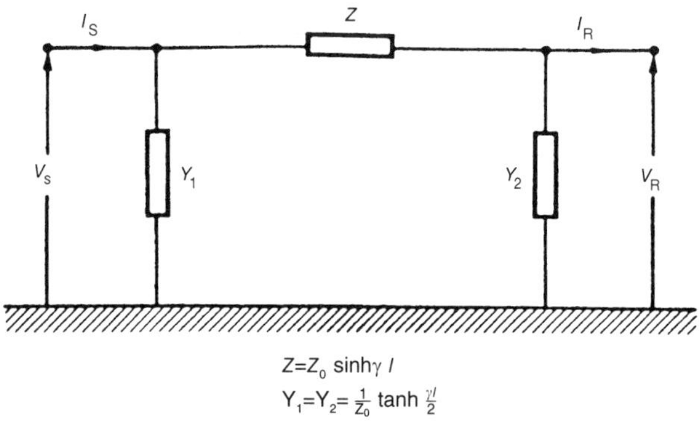

Figure 3.8 The equivalent PI model of a long transmission line

$$V_R = V_s - Z[I_s - Y_1 V_s] \tag{3.36}$$

$$I_R = I_s - Y_1 V_s - Y_2 V_R. \tag{3.37}$$

Equation (3.36) can also be written

$$V_R = V_s(1 + ZY_1) - ZI_s \tag{3.38}$$

and Equation (3.37), after substituting V_R, changes to

$$I_R = I_s(1 + ZY_2) - V_s(Y_1 + Y_2 + ZY_1 Y_2). \tag{3.39}$$

Equating the last two equations with Equations (3.34) and (3.35)

$$1 + ZY_1 = 1 + ZY_2 = \cosh(\gamma l) \tag{3.40}$$

$$Z = Z_o \sinh(\gamma l) \tag{3.41}$$

$$Y_1 + Y_2 + ZY_1 Y_2 = \frac{1}{Z_o}\sinh(\gamma l). \tag{3.42}$$

These equations are satisfied by choosing

$$Z = Z_o \sinh(\gamma l) \tag{3.43}$$

$$Y_1 = Y_2 = \frac{1}{Z_o}\frac{\cosh(\gamma l) - 1}{\sinh(\gamma l)} = \frac{1}{Z_o}\tanh\frac{(\gamma l)}{2}. \tag{3.44}$$

Therefore, Equations (3.43) and (3.44) constitute the series and shunt elements (shown in Figure 3.9) of the equivalent PI circuits of the line.

As an example, the series and shunt components of a 220 kV, 230 km line, are plotted in Figure 3.10 against frequency. These were calculated using geometric mean distances and three equal length transposition sections [7]. The shunt resistance and shunt reactance are formed by inverting the shunt admittance.

The series and shunt reactances are the predominant components of the impedances; both have a period of 1300 Hz. The line length of 230 km corresponds with one wavelength at this frequency. The series reactance increases from its inductive 50 Hz value up to a maximum at 325 Hz (the quarter wavelength frequency) and then decreases, passing through zero at 650 Hz (the half wavelength frequency). Between the half and full wavelength frequencies the series reactance is capacitive. By contrast, the shunt reactance is capacitive and large at fundamental frequency, reducing in magnitude to zero at the half wavelength frequency. Beyond this it becomes inductive.

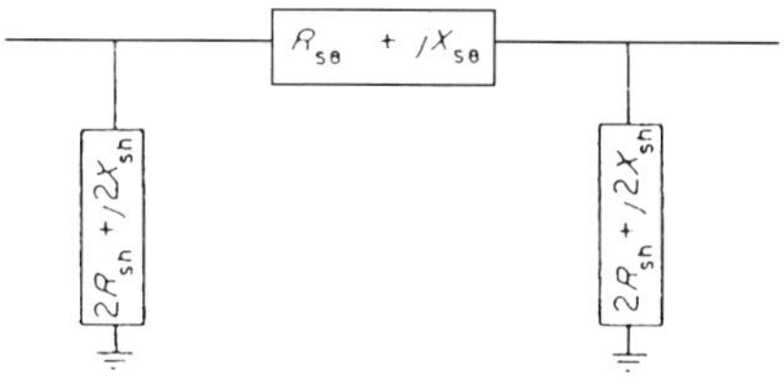

Figure 3.9 Equivalent PI impedances

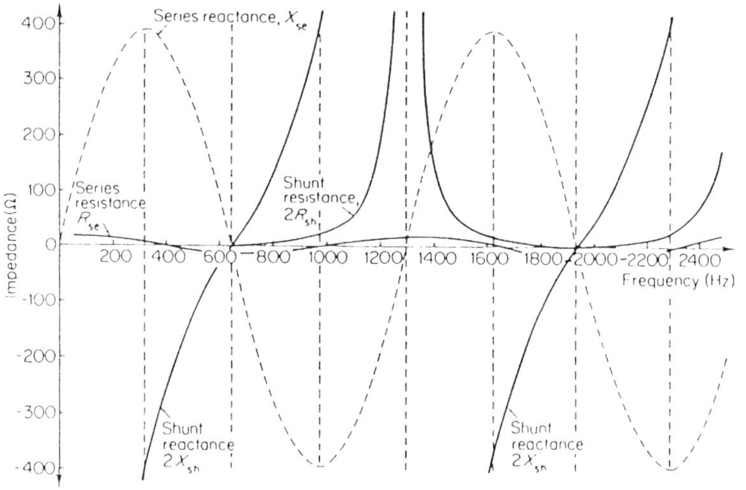

Figure 3.10 Impedance versus frequency for the equivalent PI model (skin effect included)

The series resistance is small at audiofrequencies. This is to be expected in a system designed to transmit power at fundamental frequency with minimum losses. Also, the peak magnitudes increase slowly as frequency increases. Since the series resistance does not get appreciably larger over the audio-frequency range, the attenuation does not increase significantly. Thus currents with frequencies in this range will propagate large distances on the power system. The negative resistances are a mathematical artifice and are not physically measurable. However, they give the correct terminal conditions for a distributed parameter transmission line.

The shunt resistance, which is normally considered to be zero in a nominal PI model, has considerable effect at resonant frequencies and, as can be observed from Figure 3.10, becomes very large as the wavelength frequency is similar to the case of a series and parallel resonating tuned circuit. In Figure 3.10, the series and shunt reactances are equal in magnitude but of opposite sign at 325 Hz, i.e. there is a series resonance (or node) with a low purely resistive impedance. This effect is better illustrated in Fig. 3.11, where the impedance of the open-circuited line is plotted. In this case the low impedance magnitude (series resonance) only contains the series and shunt resistances and occurs at the odd quarter wavelength frequencies.

At 650 Hz, although both the series and shunt reactances are small, the transmission line has a high impedance equivalent to a parallel resonating tuned circuit. This condition is called an antinode and can also be observed in Figure 3.11. These parallel resonances occur at the half wavelength frequencies.

Low impedance at the odd quarter wavelength frequencies and large impedance at the half wavelength frequencies, indicate the low level of attenuation of audio-frequency signals. The addition of other system components such as loads and generators must provide the harmonic damping.

The upper asymptote or maximum impedance is $Z_0 \coth\alpha l$ and the lower asymptote or minimum impedance is $Z_0 \tan\alpha l$.

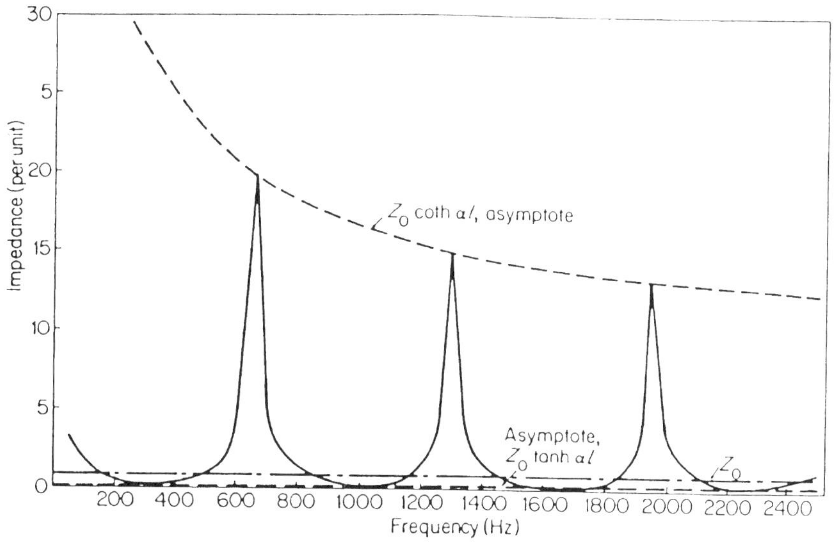

Figure 3.11 Impedance versus frequency for the open circuited Islington to Kikiwa transmission line (skin effect included)

The lower asymptote is small in value and slowly increases with frequency, while the upper asymptote decreases from an infinite value as frequency increases. For large frequencies these two asymptotes approach a value equal to the characteristic impedance.

Due to the standing wave effect of voltages and currents on transmission lines, the maximum value of these are likely to occur at points other than at the receiving end or sending end busbars. These local maxima could result in insulation damage, overheating or electromagnetic interference. It is thus important to calculate the maximum values of currents and voltages along a line and the points at which these occur.

Knowing the receiving end current and voltage, for each harmonic frequency, the current and voltage at any point on the line can be calculated for each frequency by using the following equations [8]:

$$I(x) = \frac{I_R}{2Z_o}[(Z_R + Z_0)e^{\gamma x} + (Z_0 - Z_R)e^{-\gamma x}], \tag{3.45}$$

$$V(x) = \frac{I_R}{2}[(Z_R + Z_0)e^{\gamma x} + (Z_R - Z_0)e^{-\gamma x}], \tag{3.46}$$

where x is the distance from the receiving end, I_R is the receiving end current, V_R is the receiving end voltage, and $Z_R = V_R/I_R$. The points on the transmission line at which these are maximum, are obtained by considerations of the currents and voltages as forward (incident) and backward (reflected) travelling waves with respect to the receiving end.

For example, consider the current Equation (3.45). The incident current at the receiving end is

$$I_R^+ = \frac{(Z_R + Z_0)}{2Z_0} I_R \tag{3.47}$$

and the reflected current at the receiving end is

$$I_R^- = \frac{(Z_0 - Z_R)}{2Z_0} I_R \tag{3.48}$$

The angles associated with these currents, at any point along the line, are given by

$$\theta^+ = \theta_R^+ + \beta x, \qquad \theta^- = \theta_R^- - \beta_x \tag{3.49}$$

where θ_R^+, θ_R^- are the angles of the current at the receiving end.

The current will be a maximum for θ^+ equal to θ^-. Thus

$$\theta_R^+ + \beta x = \theta_R^- - \beta x \tag{3.50}$$

or

$$x = \frac{\theta_R^- - \theta_R^+}{2\beta} \tag{3.51}$$

The current will also have local maximum at intervals of one half wavelength along the line.

While the total r.m.s. voltage and current (over the fundamental and all harmonics) are of greatest importance, the location of the maximum total r.m.s. voltage and current will most likely be dominated by that harmonic which is closest to a resonant frequency of the system.

3.6 Multiconductor Transmission Line

3.6.1 Nominal PI Model

The impedance of a three-phase transmission line with an overhead earth wire is illustrated in Figure 3.12. Each conductor has resistance, inductance, and capacitance, and is mutually coupled to the others.

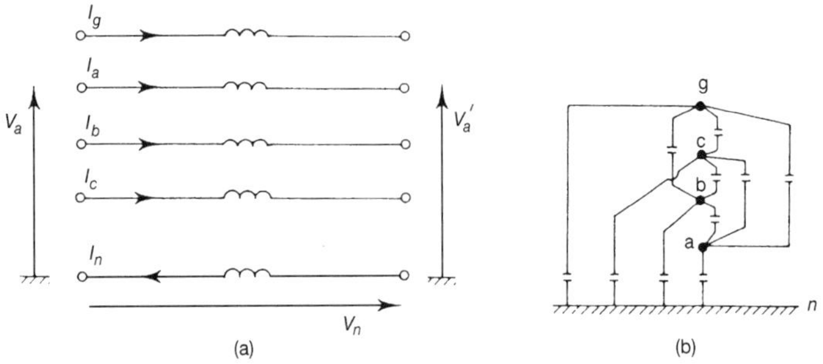

Figure 3.12 (a) Three-phase transmission series impedance equivalent and (b) three-phase transmission shunt impedance equivalent

With respect to Figure 3.12, the following equation can be written for the series impedance equivalent of phase a:

$$\begin{aligned} V_a - V'_a = {} & I_a(R_a + j\omega L_a) + I_b(j\omega L_{ab}) + I_c(j\omega L_{ac}) \\ & + j\omega L_{ag} I_g - j\omega L_{an} I_n + V_n, \end{aligned} \tag{3.52}$$

where

$$V_n = I_n(R_n + j\omega L_n) - I_a j\omega L_{na} - I_b j\omega L_{nb} - I_c j\omega L_{nc} - I_g j\omega L_{ng} \tag{3.53}$$

and substituting

$$I_n = I_a + I_b + I_c + I_g \tag{3.54}$$

gives

$$\begin{aligned} V_a - V'_a = {} & I_a(R_a + j\omega L_a) + I_b j\omega L_{ab} + I_c j\omega L_{ac} \\ & + j\omega L_{ag} I_g - j\omega L_{an}(I_a + I_b + I_c + I_g) + V_n. \end{aligned} \tag{3.55}$$

Regrouping and substituting for V_n, i.e.

$$\begin{aligned} \Delta V_a = {} & V_a - V'_a \\ = {} & I_a(R_a + j\omega L_a - j\omega L_{an} + R_n + j\omega L_n - j\omega L_{na}) \\ & + I_b(j\omega L_{ab} - j\omega L_{an} + R_n + j\omega L_n - j\omega L_{nb}) \\ & + I_c(j\omega L_{ac} - j\omega L_{an} + R_n + j\omega L_n - j\omega L_{nc}) \\ & + I_g(j\omega L_{ag} - j\omega L_{an} + R_n + j\omega L_n - j\omega L_{ng}) \end{aligned} \tag{3.56}$$

$$\begin{aligned} \Delta V_a = {} & I_a(R_a + j\omega L_a - 2j\omega L_{an} + R_n + j\omega L_n) \\ & + I_b(j\omega L_{ab} - j\omega L_{bn} - j\omega L_{an} + R_n + j\omega L_n) \\ & + I_c(j\omega L_{ac} - j\omega L_{cn} - j\omega L_{an} + R_n + j\omega L_n) \\ & + I_g(j\omega L_{ag} - j\omega L_{gn} - j\omega L_{an} + R_n + j\omega L_n) \end{aligned} \tag{3.57}$$

or

$$\Delta V_a = Z_{aa-n} I_a + Z_{ab-n} I_b + Z_{ac-n} I_c + Z_{ag-n} I_g \tag{3.58}$$

and writing similar equations for the other phases and earth wire, the following matrix equation results:

$$\begin{bmatrix} \Delta V_a \\ \Delta V_b \\ \Delta V_c \\ \Delta V_g \end{bmatrix} \begin{bmatrix} Z_{aa-n} Z_{ab-n} Z_{ac-n} & Z_{ag-n} \\ Z_{ba-n} Z_{bb-n} Z_{bc-n} & Z_{bg-n} \\ Z_{ca-n} Z_{cb-n} Z_{cc-n} & Z_{cg-n} \\ Z_{ga-n} Z_{gb-n} Z_{gc-n} & Z_{gg-n} \end{bmatrix} \begin{bmatrix} I_a \\ I_b \\ I_c \\ I_g \end{bmatrix} \tag{3.59}$$

Usually we are interested only in the performance of the phase conductors, and it is more convenient to use a three-conductor equivalent for the transmission line. This is achieved by writing matrix equation (3.59) in partitioned form as follows:

$$\begin{bmatrix} \Delta V_{abc} \\ \Delta V_g \end{bmatrix} = \begin{bmatrix} Z_A & Z_B \\ Z_C & Z_D \end{bmatrix} \begin{bmatrix} I_{abc} \\ I_g \end{bmatrix} \tag{3.60}$$

From (3.60)

$$[\Delta V_{abc}] = [Z_A][I_{abc}] + [Z_B][I_g] \tag{3.61}$$

$$[\Delta V_g] = [Z_C][I_{abc}] + [Z_D][I_g] \tag{3.62}$$

From Equations (3.60) and (3.62), and assuming that the earth wire is at zero potential,

$$[\Delta V_{abc}] = [Z_{abc}][I_{abc}], \tag{3.63}$$

where

$$[Z_{abc}] = [Z_A] - [Z_B][Z_D]^{-1}[Z_C] = \begin{bmatrix} Z'_{aa-n} & Z'_{ab-n} & Z'_{ac-n} \\ Z'_{ba-n} & Z'_{bb-n} & Z'_{bc-n} \\ Z'_{ca-n} & Z'_{cb-n} & Z'_{cc-n} \end{bmatrix} \tag{3.64}$$

With reference to Figure 3.12(b), the potentials of the line conductors are related to the conductor charges by the matrix equation [9]

$$\begin{bmatrix} V_a \\ V_b \\ V_c \\ V_g \end{bmatrix} = \begin{bmatrix} P_{aa} & P_{ab} & P_{ac} & P_{ag} \\ P_{ba} & P_{bb} & P_{bc} & P_{bg} \\ P_{ca} & P_{cb} & P_{cc} & P_{cg} \\ P_{ga} & P_{gb} & P_{gc} & P_{gg} \end{bmatrix} \begin{bmatrix} Q_a \\ Q_b \\ Q_c \\ Q_g \end{bmatrix} \tag{3.65}$$

considerations as for the series impedance matrix, lead to

$$[V_{abc}] = [P'_{abc}][Q_{abc}] \tag{3.66}$$

where $[P'_{abc}]$ is a 3×3 matrix which includes the effects of the earth wire. The capacitance matrix of the transmission line of Figure 3.12 is given by

$$[C'_{abc}] = [P'_{abc}]^{-1} = \begin{bmatrix} C_{aa} & -C_{ab} & -C_{ac} \\ -C_{ba} & C_{bb} & -C_{bc} \\ -C_{ca} & -C_{cb} & C_{cc} \end{bmatrix} \tag{3.67}$$

The series impedance and shunt admittance lumped-PI model representation of the three-phase line is shown in Figure 3.13(a) and its matrix equivalent is illustrated in Figure 3.13(b). These two matrices can also be represented by compound admittances [10] (Figure 3.13(c)).

Using the compound component concept, the nodal injected currents of Figure 3.13(c) are related to the nodal voltages by the equation

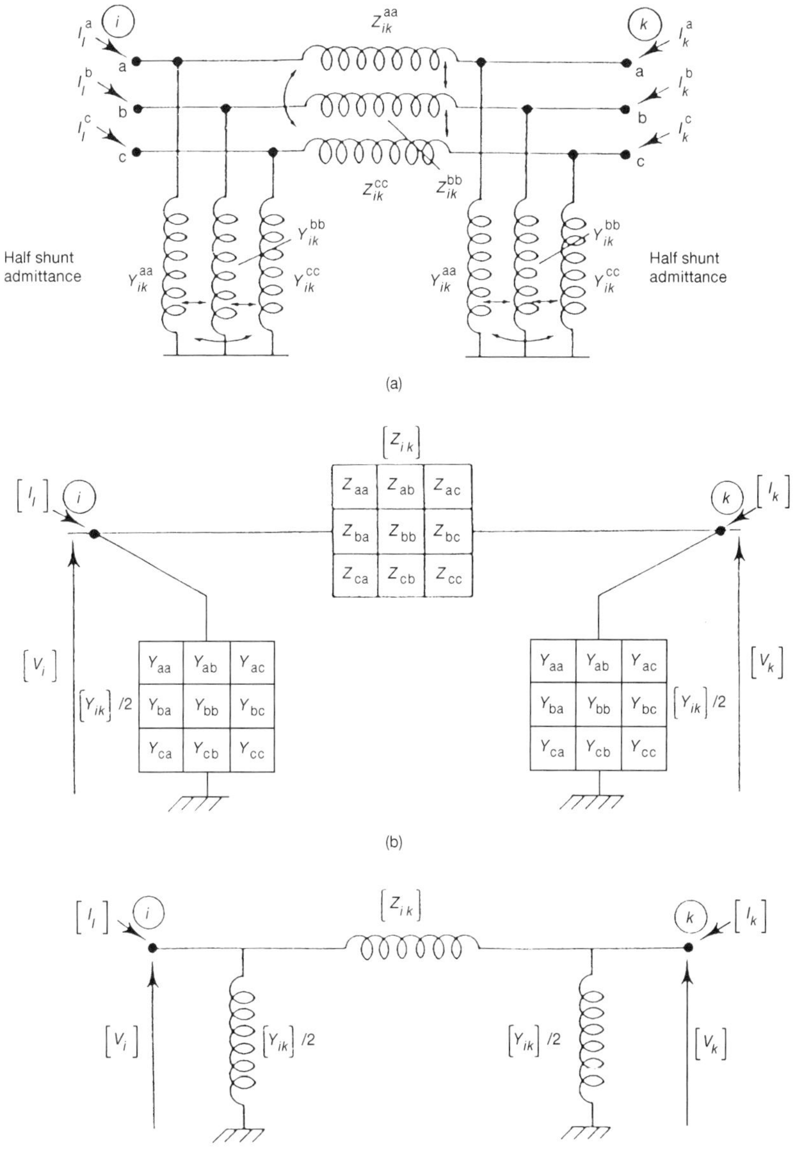

Figure 3.13 Lumped PI model of a short three-phase line series impedance: (a) full circuit representation; (b) matrix equivalent; (c) using three-phase compound admittances

$$\begin{bmatrix} [I_i] \\ [I_k] \end{bmatrix} = \begin{bmatrix} [Z]^{-1} + [Y]/2 & -[Z]^{-1} \\ -[Z]^{-1} & [Z]^{-1} + [Y]/2 \end{bmatrix} \begin{bmatrix} [V_i] \\ [V_k] \end{bmatrix} \tag{3.68}$$

$$6 \times 1 \qquad\qquad 6 \times 6 \qquad\qquad 6 \times 1$$

This forms the element admittance matrix representation for the short line between busbars i and k in terms of 3×3 matrix quantities.

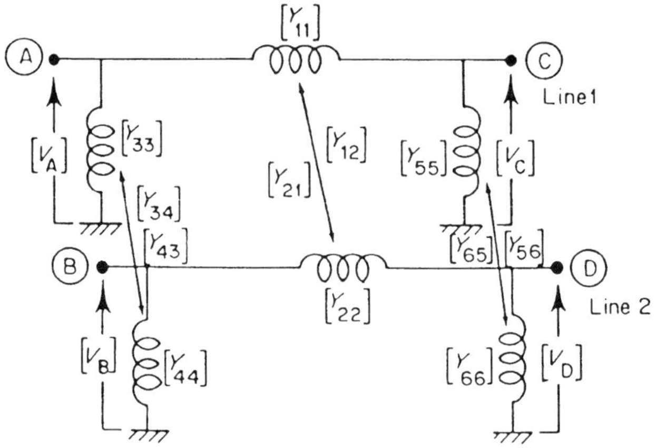

Figure 3.14 Two-coupled three-phase lines

3.6.2 Mutually Coupled Three-phase Lines

When two or more transmission lines occupy the same right of way for a considerable length, the electrostatic and electromagnetic coupling between those lines must be taken into account.

Consider the simplest case of two mutually coupled three-phase lines. The two coupled lines are considered to form one subsystem composed of four system busbars. The coupled lines are illustrated in Figure 3.14, where each element is a 3×3 compound admittance and all voltages and currents are 3×1 vectors.

The coupled series elements represent the electromagnetic coupling while the coupled shunt elements represent the capacitive or electrostatic coupling. These coupling parameters are lumped in a similar way to the standard line parameters.

With the admittances labelled as in Figure 3.14, and applying the rules of linear transformation for compound networks, the admittance matrix for the subsystem is defined as follows:

$$\underset{12\times 1}{\begin{bmatrix} I_A \\ I_B \\ I_C \\ I_D \end{bmatrix}} = \underset{12\times 12}{\begin{bmatrix} Y_{11}+Y_{33} & Y_{12}+Y_{34} & -Y_{11} & -Y_{12} \\ Y_{12}^{T}+Y_{34}^{T} & Y_{22}+Y_{44} & -Y_{12}^{T} & -Y_{22} \\ -Y_{11} & -Y_{12} & Y_{11}+Y_{55} & Y_{12}+Y_{56} \\ -Y_{22}^{T} & -Y_{22} & Y_{12}^{T}+Y_{56}^{T} & Y_{22}+Y_{66} \end{bmatrix}} \cdot \underset{12\times 1}{\begin{bmatrix} V_A \\ V_B \\ V_C \\ V_D \end{bmatrix}} \tag{3.69}$$

It is assumed here that the mutual coupling is bilateral. Therefore $Y_{21} = Y^{T}{}_{12}$, etc.

The subsystem may be redrawn as in Figure 3.15. The pairs of coupled 3×3 compound admittances are now represented as a 6×6 compound admittance. The matrix representation is also shown. Following this representation and the labelling

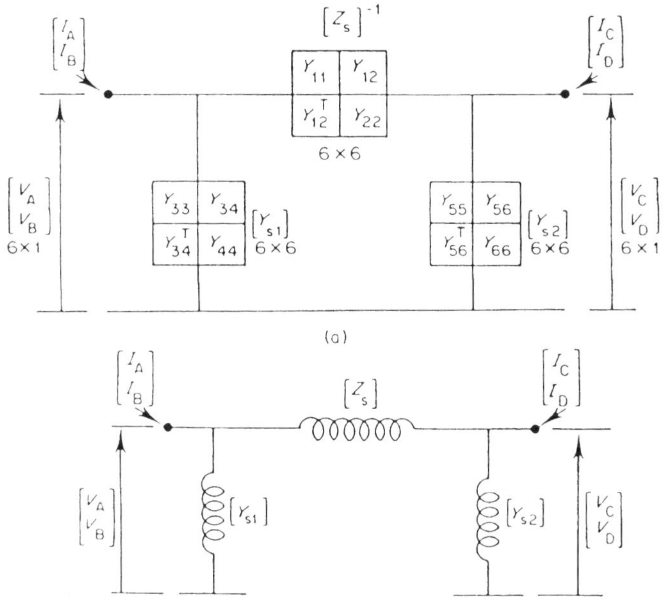

Figure 3.15 A 6 × 6 compound admittance representation of two coupled three-phase lines: (a) 6 × 6 matrix representation; (b) 6 × 6 compound admittance representation

of the admittance block in the figure, the admittance matrix may be written in terms of the 6 × 6 compound coils as

$$\underset{12\times 1}{\begin{bmatrix} \begin{bmatrix} I_A \\ I_B \end{bmatrix} \\ \begin{bmatrix} I_C \\ I_D \end{bmatrix} \end{bmatrix}} = \underset{12\times 12}{\begin{bmatrix} [Z_s]^{-1} + [Y_{s1}] & -[Z_s]^{-1} \\ -[Z_s]^{-1} & [Z_s]^{-1} + [Y_{s2}] \end{bmatrix}} \underset{12\times 1}{\begin{bmatrix} \begin{bmatrix} V_A \\ V_B \end{bmatrix} \\ \begin{bmatrix} V_C \\ V_D \end{bmatrix} \end{bmatrix}} \tag{3.70}$$

This is clearly identical to Equation (3.69) with the appropriate matrix partitioning.

The representation of Figure 3.15 is more concise and the formation of Equation (3.70) from this representation is straightforward, being exactly similar to that which results from the use of 3 × 3 compound admittances for the normal single three-phase line.

The data which must be available, to enable coupled lines to be treated in a similar manner to single lines, is the series impedance and shunt admittance matrices. These matrices are of order 3 × 3 for a single line, 6 × 6 for two coupled lines, 9 × 9 for three and 12 × 12 for four coupled lines.

Once the matrices $[Z_s]$ and $[Y_s]$ are available, the admittance matrix for the subsystem is formed by application of Equation (3.70).

When all the busbars of the coupled lines are distinct, the subsystem may be combined directly into the system admittance matrix. However, if the busbars are

Figure 3.16 Mutually coupled parallel transmission lines

not distinct then the admittance matrix as derived from Equation (3.70) must be modified. This is considered in the following section.

3.6.3 Consideration of Terminal Connections

The admittance matrix as derived above must be reduced if there are different elements in the subsystem connected to the same busbar. As an example, consider two parallel transmission lines as illustrated in Figure 3.16.

The admittance matrix derived previously related the currents and voltages at the four busbars $A1$, $A2$, $B1$ and $B2$. This relationship is given by

$$\begin{bmatrix} I_{A1} \\ I_{A2} \\ I_{B1} \\ I_{B2} \end{bmatrix} = [Y_{A1A2B1B2}] \begin{bmatrix} V_{A1} \\ V_{A2} \\ V_{B1} \\ V_{B2} \end{bmatrix} \tag{3.71}$$

The nodal injected current at busbar A, I_A, is given by

$$I_A = I_{A1} + I_{A2}; \tag{3.72}$$

similarly

$$I_B = I_{B1} + I_{B2}. \tag{3.73}$$

Also, from inspection of Figure 3.16

$$V_A = V_{A1} = V_{A2}, \qquad V_B = V_{B1} = V_{B2}. \tag{3.74}$$

The required matrix equation relates the nodal injected currents, I_A and I_B, to the voltages at these busbars. This is readily derived from Equation (3.71) and the conditions specified above. This is simply a matter of adding appropriate rows and columns, and yields

$$\begin{bmatrix} I_A \\ I_B \end{bmatrix} = [Y_{AB}] \begin{bmatrix} V_A \\ V_B \end{bmatrix} \tag{3.75}$$

where $[Y_{AB}]$ is the required nodal admittance matrix for the subsystem.

It should be noted that the matrix in Equation (3.71) must be retained as it is required in the calculation of the individual line currents.

3.6.4 Equivalent PI Model

In the case of multiconductor transmission lines, the nominal PI series impedance and shunt admittance matrices per unit distance, $[Z']$ and $[Y']$ respectively, are square and their size is fixed by the number of mutually coupled conductors.

The derivation of the equivalent PI model for harmonic penetration studies from the nominal PI matrices, is similar to that of the single phase lines, except that it involves the evaluation of hyperbolic functions of the propagation constant which is now a matrix:

$$[\gamma] = ([Z'][Y'])^{\frac{1}{2}}. \tag{3.76}$$

There is no direct way of calculating sinh or tanh of a matrix, thus a method using eigenvalues and eigenvectors, called *modal analysis* is employed [11].

The basis of such a method for a multiconductor line is as follows: Consider the second order linear differential equations describing wave propagation along a single transmission line (section 3.5.1). These can be expanded for the multiconductor line in the form of Equation (3.77), where the matrices are of order m, the number of phases involved:

$$\left[\frac{\mathrm{d}^2 V}{\mathrm{d}x^2}\right] = [Z'][Y'][V], \tag{3.77}$$

$$\left[\frac{\mathrm{d}^2 I}{\mathrm{d}x^2}\right] = [Y'][Z'][I]. \tag{3.78}$$

It should be noted that in this case the matrix products $[Z'][Y']$ and $[Y'][Z']$ are not equal, except in special cases [11]. By transforming phase voltages to modal voltages, and by choosing the proper transformation matrix $[T_U]$, Equation (3.75) can be changed to

$$[\mathrm{d}^2 V_{mode}/\mathrm{d}x^2] = [\Lambda][V_{mode}] \tag{3.79}$$

where $[\Lambda]$ is now a diagonal matrix; the elements of $[\Lambda]$ are the eigenvalues of the matrix product $[Z'][Y']$, and the transformation matrix $[T_v]$ is the matrix of eigenvectors of that matrix product. Equation (3.78) can be diagonalized as well, with the same diagonal matrix $[\Delta]$, i.e.

$$[\mathrm{d}^2 I_{mode}/\mathrm{d}x^2] = [\Delta][I_{mode}]. \tag{3.80}$$

With the diagonalized Equations (3.79) and (3.80), an m-phase line can now be studied as if it consisted of m single-phase lines, similar to the symmetrical component approach, except that the zero, positive and negative sequence networks now become the mode-1, mode-2 and mode-3 networks. In each mode, the single-phase long line series impedance and shunt admittance of Figure 3.6 are used. The propagation constant of each mode is simply

$$\gamma_{mode-i} = \sqrt{\gamma_i} \tag{3.81}$$

where γ_i is the ith eigenvalue or ith element in $[\Lambda]$. The modal series impedance and shunt admittance are not directly available, but must be computed from

$$[Z'_{mode}] = [T_U]^{-1}[Z'][T_i] \text{ and } [Y'_{mode}] = [T_i]^{-1}[Y'][T_U], \tag{3.82}$$

with both modal matrices being diagonal. $[Y'_{mode}]$ may no longer be purely imaginary even though only shunt capacitance is modelled. This will depend on how the transformation matrices were normalised. For steady-state analysis at one particular frequency, this causes no problems. Once Z_{series} and Y_{shunt} have been calculated for each mode, the representation in phase quantities is easily obtained by transforming back, with

$$[Z_{series}] = [T_U][Z_{series-mode}][T_i]^{-1} \text{ and } [Y_{shunt}] = [T_i][Y_{shunt-mode}][T_U]^{-1} \tag{3.83}$$

becoming the values of the equivalent PI model which will accurately represent the untransposed line.

In expanded form, the following are expressions for the series impedance and shunt admittance of the equivalent PI model [12]:

$$[Z]_{EPM} = l[Z'][M]\left[\frac{\sinh\gamma l}{\gamma l}\right][M]^{-1} \tag{3.84}$$

where l is the transmission line length, $[Z]_{EPM}$ is the equivalent PI series impedance matrix, $[M]$ is the matrix of normalised eigenvectors,

$$\left[\frac{\sinh\gamma l}{\gamma l}\right] = \begin{bmatrix} \frac{\sinh\gamma_1 l}{\gamma_1 l} & 0 & \cdots & 0 \\ 0 & \frac{\sinh\gamma_2 l}{\gamma_2 l} & \cdots & 0 \\ \vdots & \vdots & & \vdots \\ 0 & 0 & & \frac{\sinh\gamma_j l}{\gamma_j l} \end{bmatrix} \tag{3.85}$$

and γ_j is the jth eigenvalue for $j/3$ mutually coupled circuits. Similarly

$$[Y]_{EPM} = l[M]\left[\frac{\tanh(\gamma l/2)}{\gamma l/2}\right][M]^{-1}[Y'], \tag{3.86}$$

where $[Y]_{EPM}$ is the equivalent PI shunt admittance matrix.

An illustration of the relative accuracies achieved by the equivalent PI and nominal PI models is displayed in Figure 3.17.

This figure shows the per unit positive sequence voltages of a 230 km, 220 kV line, the parameter information for which is shown in Figure 3.18. Standing wave effects show the different accuracies provided by the two models, with the resonant frequencies of the nominal PI model approaching those of the equivalent PI model as the number of sections increases from three to six. Accuracy of the normal PI model decreases as the frequency increases and this can be observed at the wavelength frequency.

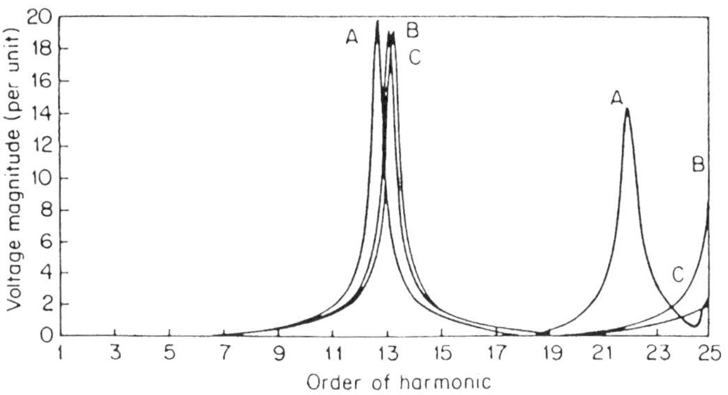

Figure 3.17 Comparison of the equivalent and nominal PI transmission line models at 5 Hz intervals: curve A, three sections; curve B, six sections; curve C, equivalent PI

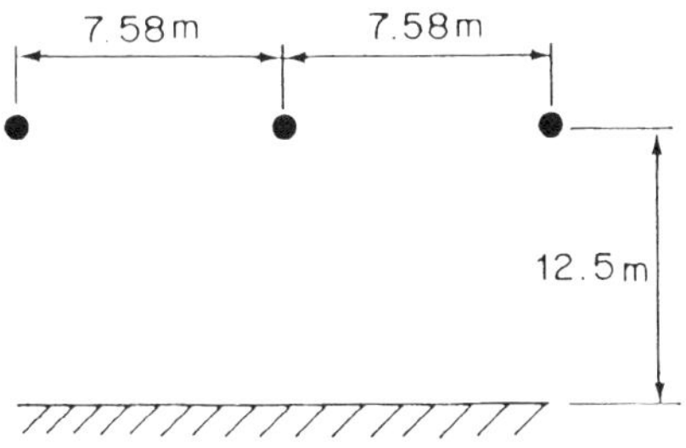

Figure 3.18 Conductor information for the Islington to Kikiwa line: conductor type, Zebra (54/3.18 + 7/3.18); length, 230 km; resistivity, 100 Ω m

Cascading of nominal PI circuits requires a large number of sections for long lines at higher frequencies, to achieve acceptable accuracy. The equivalent PI model avoids the problems of determining the number of sections needed and round-off error that accumulates in this situation.

Computer derivation of the correction factors for conversion from the nominal PI to the equivalent PI model, and their incorporation into the series impedance and shunt admittance matrices, is carried out as indicated in the structure diagram of Figure 3.19. The LR2 algorithm of Wilkinson and Reinsch [13] is used for accurate calculations in the derivation of the eigenvalues and eigenvectors.

3.7 Three-phase Transformer Models

The basic two-winding transformer is shown in Figure 3.20. Its primitive network, on the assumption that the flux paths are symmetrically distributed between all windings, is represented by the equation:

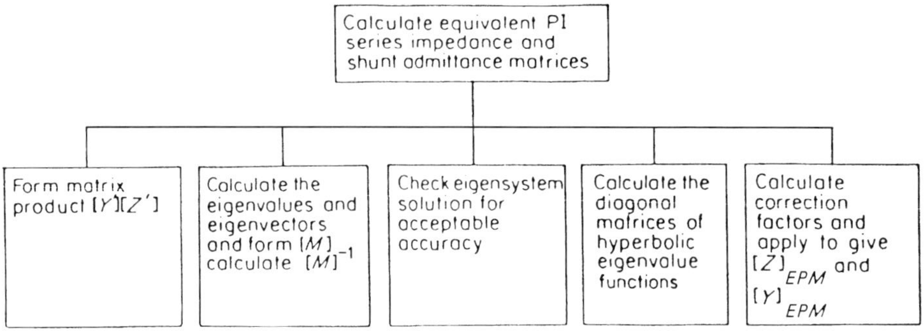

Figure 3.19 Structure diagram for calculation of the equivalent PI model

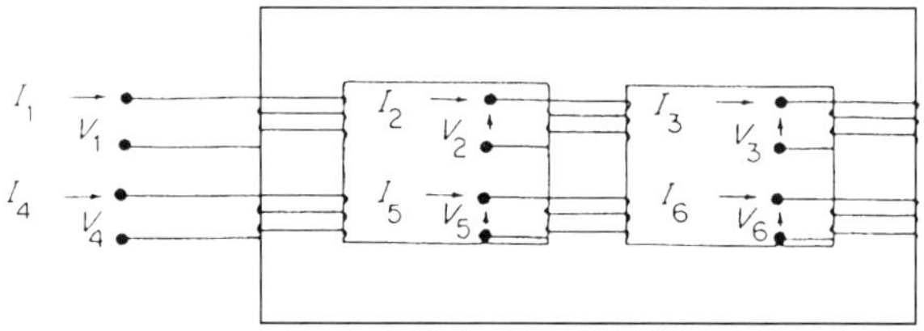

Figure 3.20 Diagrammatic representation of a two-winding transformer

$$\begin{bmatrix} I_1 \\ I_2 \\ I_3 \\ I_4 \\ I_5 \\ I_6 \end{bmatrix} = \begin{bmatrix} y_p & y'_m & y'_m & -y_m & y''_m & y''_m \\ y'_m & y_p & y'_m & y''_m & -y_m & y''_m \\ y'_m & y'_m & y_p & y''_m & y''_m & -y_m \\ -y_m & y''_m & y''_m & y_s & y'''_m & y'''_m \\ y''_m & -y_m & y''_m & y'''_m & y_s & y'''_m \\ y''_m & y''_m & -y_m & y'''_m & y'''_m & y_s \end{bmatrix} \begin{bmatrix} V_1 \\ V_2 \\ V_3 \\ V_4 \\ V_5 \\ V_6 \end{bmatrix} \tag{3.87}$$

where y'_m is the mutual admittance between primary coils, y''_m is the mutual admittance between primary and secondary coils on different cores, and y'''_m is the mutual admittance between secondary coils.

If a tertiary winding is also present, the primitive network consists of nine (instead of six) coupled coils and its mathematical model will be a 9×9 admittance matrix.

The interphase coupling can usually be ignored (e.g. the case of three single-phase separate units) and all the primed terms are effectively zero.

The connection matrix [C] between the primitive network and the actual transformer buses is derived from the transformer connection.

By way of example consider the Wye G-Delta connection of Figure 3.21. The following connection matrix applies:

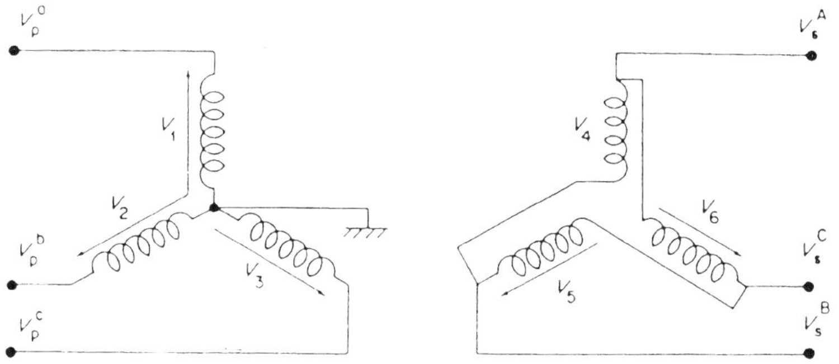

Figure 3.21 Network connection diagram for a Wye G-delta transformer

$$\begin{bmatrix} V_1 \\ V_2 \\ V_3 \\ V_4 \\ V_5 \\ V_6 \end{bmatrix} = \begin{bmatrix} 1 & 0 & 0 & 0 & 0 & 0 \\ 0 & 1 & 0 & 0 & 0 & 0 \\ 0 & 0 & 1 & 0 & 0 & 0 \\ 0 & 0 & 0 & 1 & -1 & 0 \\ 0 & 0 & 0 & 0 & 1 & -1 \\ 0 & 0 & 0 & -1 & 0 & 1 \end{bmatrix} \begin{bmatrix} V_p^a \\ V_p^b \\ V_p^c \\ V_s^A \\ V_s^B \\ V_s^C \end{bmatrix} \tag{3.88}$$

or

$$[V]_{Branch} = [C][V]_{node} \tag{3.89}$$

We can also write

$$[Y]_{NODE} = [C]^t[Y]_{PRIM}[C] \tag{3.90}$$

and using $[Y]_{PRIM}$ from equation (3.87)

$[Y]_{NODE} =$

$$\begin{bmatrix} y_p & y'_m & y'_m & -(y_m + y''_m) & (y_m + y''_m) & 0 \\ y'_m & y_p & y'_m & 0 & -(y_m + y''_m) & (y_m + y''_m) \\ y'_m & y'_m & y_p & (y_m + y''_m) & 0 & -(y_m + y''_m) \\ -(y_m + y''_m) & 0 & (y_m + y''_m) & 2(y_s - y'''_m) & -(y_s - y'''_m) & -(y_s - y'''_m) \\ (y_m + y''_m) & -(y_m + y''_m) & 0 & -(y_s - y'''_m) & 2(y_s - y'''_m) & -(y_s - y'''_m) \\ 0 & (y_m + y''_m) & -(y_m + y''_m) & -(y_s - y'''_m) & -(y_s - y'''_m) & 2(y_s - y'''_m) \end{bmatrix} \begin{matrix} a \\ b \\ c \\ A \\ B \\ C \end{matrix} \tag{3.91}$$

If the primitive admittances are expressed in per unit the upper right and lower left quadrants of matrix (3.91) must be divided by $\sqrt{3}$ and the lower right quadrant by 3. Then, in the absence of interphase coupling the nodal admittance matrix equation of the Wye G-delta connection becomes:

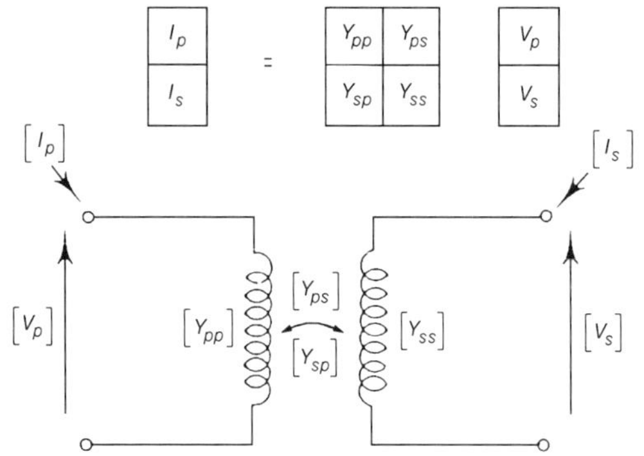

Figure 3.22 Two-winding three-phase transformer as two coupled compound coils

$$\begin{bmatrix} I_p^a \\ I_p^b \\ I_p^c \\ I_s^A \\ I_s^B \\ I_s^C \end{bmatrix} = \begin{bmatrix} y & & & -y/\sqrt{3} & y/\sqrt{3} & \\ & y & & & -y/\sqrt{3} & y/\sqrt{3} \\ & & y & y/\sqrt{3} & & -y/\sqrt{3} \\ -y/\sqrt{3} & & y/\sqrt{3} & 2/3y & -1/3y & -1/3y \\ y/\sqrt{3} & -y/\sqrt{3} & & -1/3y & 2/3y & -1/3y \\ & y/\sqrt{3} & -y/\sqrt{3} & -1/3y & -1/3y & 2/3y \end{bmatrix} \begin{bmatrix} V_p^a \\ V_p^b \\ V_p^c \\ V_s^A \\ V_s^B \\ V_s^C \end{bmatrix} \tag{3.92}$$

where y is the transformer leakage admittance in per unit, bearing in mind the modification suggested earlier to take into account the increase of resistance with frequency.

In general, any two-winding three-phase transformer may be represented by two coupled compound coils as shown in Figure 3.22 where $[Y_{sp}] = [Y_{ps}]^{\mathrm{T}}$.

If the parameters of the three phases are assumed balanced, all the common three-phase connections can be modelled by three basic submatrices. The submatrices $[Y_{pp}]$,$[Y_{ps}]$, etc, are given in Table 3.7 for the common connections in terms of the following matrices:

Table 3.7 Characteristic submatrices used in forming the transformer admittance matrices

Transformer Connection		Self-Admittance		Mutual Admittance
Bus P	Bus S	Y_{pp}	Y_{ss}	Y_{ps}, Y_{sp}
Wye G	Wye G	Y_{I}	Y_{I}	$-Y_{\mathrm{I}}$
Wye G	Wye	Y_{I}	$Y_{\mathrm{II}/3}$	$-Y_{\mathrm{II}/3}$
Wye G	Delta	Y_{I}	Y_{II}	Y_{III}
Wye	Wye	$Y_{\mathrm{II}/3}$	$Y_{\mathrm{II}/3}$	$-Y_{\mathrm{II}/3}$
Wye	Delta	$Y_{\mathrm{II}/3}$	Y_{II}	Y_{III}
Delta	Delta	Y_{II}	Y_{II}	$-Y_{\mathrm{II}}$

$$Y_1 = \begin{bmatrix} y_t & & \\ & y_t & \\ & & y_t \end{bmatrix} Y_{11} = \begin{bmatrix} 2y_t & -y_t & -y_t \\ -y_t & 2y_t & -y_t \\ -y_t & -y_t & 2y_t \end{bmatrix} Y_{111} = \begin{bmatrix} -y_t & y_t & \\ & -y_t & y_t \\ y_t & & -y_t \end{bmatrix}$$

For transformers with neutrals connected through an impedance, an extra coil is added to the primitive network for each unearthed neutral and the primitive admittance increases in dimension. However, by noting that the injected current in the neutral is zero, these extra terms can be eliminated from the connected network admittance matrix.

Once the admittance matrix has been formed for a particular connection it represents a simple subsystem composed of the two busbars interconnected by the transformer.

3.8 Line Compensating Plant

3.8.1 Shunt Elements

Shunt reactors and capacitors are used in a transmission system for reactive power control. The data for these elements are usually given in terms of their rated megavolt-amps and rated kilovolts. The equivalent phase admittance in per unit is calculated from these data.

The coupled admittances to ground at bus k are formed into a 3×3 admittance matrix as shown in Figure 3.23, and this reduces to the compound admittance representation indicated. The admittance matrix is incorporated directly into the system admittance matrix, contributing only to the self-admittance of the particular bus.

While provision for off-diagonal terms exists, the admittance matrix for shunt elements is usually diagonal, as there is normally no coupling between the components of each phase.

Consider, as an example, the three-phase capacitor bank shown in Figure 3.24. A 3×3 matrix representation similar to that for a line section is illustrated.

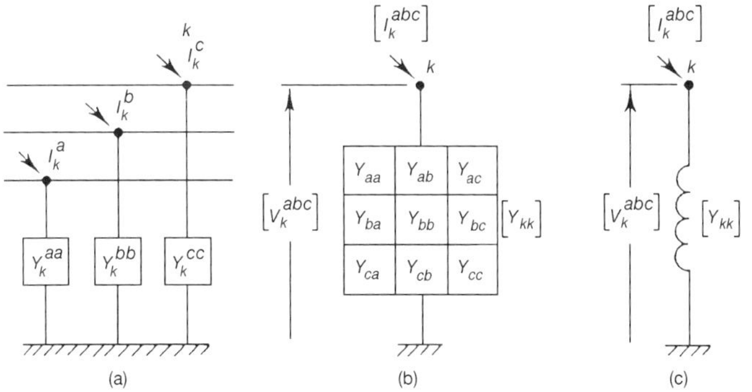

Figure 3.23 Representation of a shunt element: (a) coupled admittance, (b) admittance matrix, (c) compound admittance

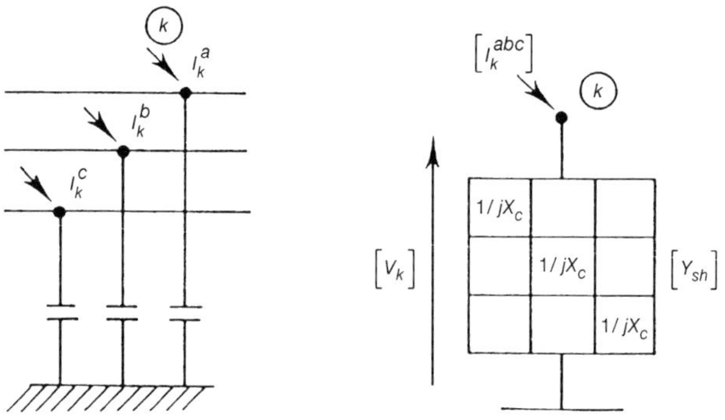

Figure 3.24 Representation of a shunt capacitor bank

The element admittance matrix will be diagonal and proportional to frequency. The megavolt-amp rating at fundamental frequency (Q) and the nominal voltage (V) are used to calculate the capacitive reactance, i.e. $x_c = V^2/nQ$. The presence of any series inductance in the capacitor banks is ignored.

In terms of ABCD parameters (described in section 3.2), the matrix equation of a shunt element is:

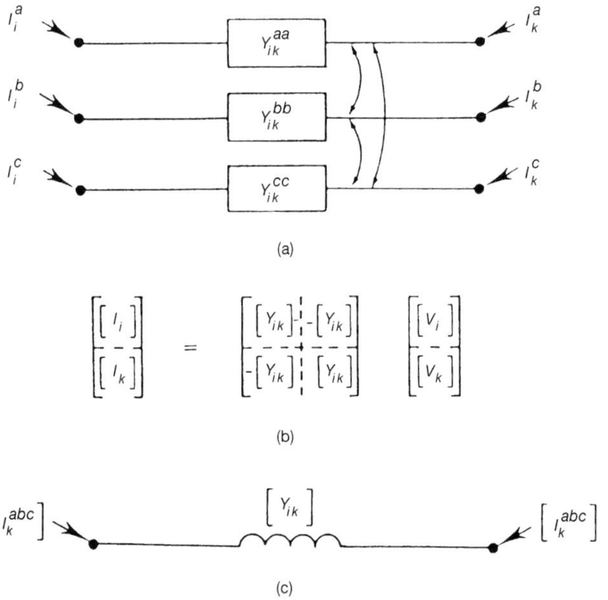

Figure 3.25 Representation of a series element: (a) coupled admittances; (b) admittance matrix; (c) compound admittance

$$\begin{bmatrix} V_s \\ I_s \end{bmatrix} = \begin{bmatrix} [U] & \\ [Y_{sh}] & [U] \end{bmatrix} \times \begin{bmatrix} V_r \\ -I_r \end{bmatrix} \tag{3.93}$$

where $[Y_{sh}] =$ Diag(shunt admittance of each phase).
$[U] =$ Identity matrix.

3.8.2 Series Elements

Series elements are connected directly between two buses and for modelling purposes they constitute a subsystem in the network subdivision.

A three-phase coupled series admittance between two busbars i and k is shown in Figure 3.25, as well as its reduced nodal admittance matrix (Figure 3.25(b)) and compound admittance (Figure 3.25(c)).

The series capacitor, used for transmission line reactance compensation, is an example of an uncoupled series element; in this case the admittance matrix is diagonal. For a lumped series element, the ABCD parameter matrix equation is:

$$\begin{bmatrix} V_s \\ I_s \end{bmatrix} = \begin{bmatrix} [U] & [Z_{se}] \\ & [U] \end{bmatrix} \times \begin{bmatrix} V_r \\ -I_r \end{bmatrix} \tag{3.94}$$

where $[Z_{se}] =$ Diag(series impedance of each phase)
$[U] =$ Identity matrix.

3.9 Underground and Submarine Cables [14,15]

A unified solution similar to that of overhead transmission is difficult for underground cables because of the great variety in their construction and layouts.

The cross section of a cable, although extremely complex can be simplified to that of Figure 3.26 and its series per unit length harmonic impedance is calculated by the following set of loop equations.

$$-\begin{bmatrix} dV_1/dx \\ dV_2/dx \\ dV_3/dx \end{bmatrix} = \begin{bmatrix} Z'_{11} & Z'_{12} & 0 \\ Z'_{21} & Z'_{22} & Z'_{23} \\ 0 & Z'_{32} & Z'_{33} \end{bmatrix} \begin{bmatrix} I_1 \\ I_2 \\ I_3 \end{bmatrix} \tag{3.95}$$

where

Z'_{11} is the sum of the following three component impedances,
$Z'_{core\text{-}outside}$: internal impedance of the core with the return path outside the core
$Z'_{core\text{-}insulation}$: impedance of the insulation surrounding the core
$Z'_{sheath\text{-}inside}$: internal impedance of the sheath with the return path inside the sheath.

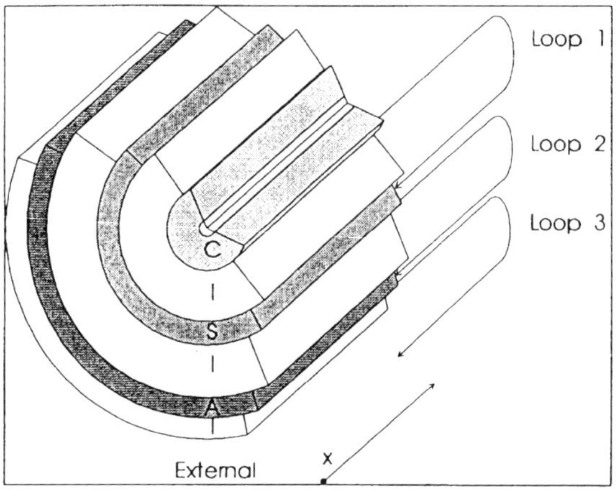

Figure 3.26 Cable cross-section

Similarly

$$Z'_{22} = Z'_{sheath\text{-}outside} + Z'_{sheath/armour\text{-}insulation} + Z'_{armour\text{-}inside} \tag{3.96}$$

and

$$Z'_{33} = Z'_{armour\text{-}outside} + Z'_{armour/earth\text{-}insulation} + Z'_{earth\text{-}inside}, \tag{3.97}$$

with analogous definitions as for Z'.

The coupling impedances $Z'_{12} = Z'_{21}$ and $Z'_{23} = Z'_{32}$ are negative because of opposing current directions (I_2 in negative direction in loop 1, and I_3 in negative direction in loop 2), i.e.

$$Z'_{12} = Z'_{21} = -Z'_{sheath\text{-}mutual} \tag{3.98}$$

$$Z'_{23} = Z'_{32} = -Z'_{armour\text{-}mutual} \tag{3.99}$$

where

$Z'_{sheath\text{-}mutual}$: mutual impedance (per unit length) of the tubular sheath between inside loop 1 and the outside loop 2.

$Z'_{armour\text{-}mutual}$: mutual impedance (per unit length) of the tubular armour between the inside loop 2 and the outside loop 3.

Finally, $Z'_{13} = Z'_{31} = 0$ because loop 1 and loop 3 have no common branch.

The impedances of the insulation are given by

$$Z'_{insulation} = j\omega \frac{\mu}{2\pi} \ln \frac{r_{outside}}{r_{inside}} \qquad \text{in } \Omega/\text{m} \tag{3.100}$$

where μ : permeability of insulation in H/m

$r_{outside}$: outside radius of insulation

r_{inside} : inside radius of insulation.

If the insulation is missing, e.g. between armour and earth, then Z' insulation $= 0$.

The internal impedances and the mutual impedance of a tubular conductor are a function of frequency, and can be derived from Bessel and Kelvin functions.

$$Z'_{tube\text{-}inside} = \frac{\sqrt{\mathrm{j}}.\omega.\mu}{2\pi.mq.D}[I_o(\sqrt{\mathrm{j}}\,mq)K_1(\sqrt{\mathrm{j}}\,mr) + K_o(\sqrt{\mathrm{j}}\,mq)I_1(\sqrt{\mathrm{j}}\,mr)] \quad (3.101a)$$

$$Z'_{tube\text{-}outside} = \frac{\sqrt{\mathrm{j}}.\omega.\mu}{2\pi\ mr\ \mathrm{D}}[I_o(\sqrt{\mathrm{j}}\,mr)K_1(\sqrt{\mathrm{j}}\,mq) + K_o(\sqrt{\mathrm{j}}\,mr)I_1(\sqrt{\mathrm{j}}\,(mq)] \quad (3.101b)$$

$$Z'_{tube\text{-}mutual} = \frac{\omega.\mu}{2\pi\ mq\ mr\ \mathrm{D}} \quad (3.101c)$$

with $$D = I_1(\sqrt{\mathrm{j}}\,mr)K_1(\sqrt{\mathrm{j}}\,mq) - I_1(\sqrt{\mathrm{j}}\,mq)K_1(\sqrt{\mathrm{j}}\,mr) \quad (3.101d)$$

where

$$mr = \sqrt{K\frac{1}{1-s^2}} \quad (3.102)$$

$$mq = \sqrt{K\frac{s^2}{1-s^2}} \quad (3.103)$$

with

$$K = \frac{8\pi.10^{-4}.f.\mu_r}{R'_{dc}} \quad (3.104)$$

$$s = \frac{q}{r} \quad (3.105)$$

q = inside radius

r = outside radius

R'_{dc} = dc resistance in Ω/km

The only remaining term is $Z'_{earth\text{-}inside}$ in Equation (3.97) which is the earth return impedance for underground cables, or the sea return impedance for submarine cables. The earth return impedance can be calculated approximately with equation (3.101a) by letting the outside radius go to infinity. This approach, also used by Bianchi and Luoni [15] to find the sea return impedance is quite acceptable considering the fact that sea resistivity and other input parameters are not known accurately.

Equation (3.95) is not in a form compatible with the solution used for overhead conductors, where the voltages with respect to local ground and the actual currents in the conductors are used as variables. Equation (3.95) can easily be brought into such a form by introducing the appropriate terminal conditions, namely with

$$V_1 = V_{core} - V_{sheath} \qquad I_1 = I_{core}$$
$$V_2 = V_{sheath} - V_{armour} \qquad I_2 = I_{core} + I_{sheath}$$

and

$$V_3 = V_{armour} \qquad I_3 = I_{core} + I_{sheath} + I_{armour}$$

Equation (3.95) can be rewritten as

$$-\begin{bmatrix} \mathrm{d}V_{core}/\mathrm{d}x \\ \mathrm{d}V_{sheath}/\mathrm{d}x \\ \mathrm{d}V_{armour}/\mathrm{d}x \end{bmatrix} = \begin{bmatrix} Z'_{cc} & Z'_{cs} & Z'_{ca} \\ Z'_{sc} & Z'_{ss} & Z'_{sa} \\ Z'_{ac} & Z'_{as} & Z'_{aa} \end{bmatrix} \begin{bmatrix} I_{core} \\ I_{sheath} \\ I_{armour} \end{bmatrix} \tag{3.106}$$

where

$$Z'_{cc} = Z'_{11} + 2Z'_{12} + Z'_{22} + 2Z'_{23} + Z'_{33}$$
$$Z'_{cs} = Z'_{sc} = Z'_{12} + Z'_{22} + 2Z'_{23} + Z'_{33}$$
$$Z'_{ca} = Z'_{ac} = Z'_{sa} = Z'_{as} = Z'_{23} + Z'_{33}$$
$$Z'_{ss} = Z'_{22} + 2Z'_{23} + Z'_{33}$$
$$Z'_{aa} = Z'_{33}$$

Because a good approximation for many cables having bonding between the sheath and the armour and the armour earthed to the sea is $V_{sheath} = V_{armour} =$ zero, the system can be reduced to

$$-\mathrm{d}V_{core}/\mathrm{d}x = ZI_{core} \tag{3.107}$$

where Z is a reduction of the impedance matrix of Equation (3.106).

Similarly, for each cable the per unit length harmonic admittance is calculated, i.e.

$$-\begin{bmatrix} \mathrm{d}I_1/\mathrm{d}x \\ \mathrm{d}I_2/\mathrm{d}x \\ \mathrm{d}I_3/\mathrm{d}x \end{bmatrix} = \begin{bmatrix} \mathrm{j}\omega C'_1 & 0 & 0 \\ 0 & \mathrm{j}\omega C'_2 & 0 \\ 0 & 0 & \mathrm{j}\omega C'_3 \end{bmatrix} \begin{bmatrix} V_1 \\ V_2 \\ V_3 \end{bmatrix} \tag{3.108}$$

where $C'_i = 2\pi\, \varepsilon_o\, \varepsilon_r / l_n(r/q)$. Therefore, when converted to core, sheath and armour quantities,

$$-\begin{bmatrix} \mathrm{d}I_{core}/\mathrm{d}x \\ \mathrm{d}I_{sheath}/\mathrm{d}x \\ \mathrm{d}I_{armour}/\mathrm{d}x \end{bmatrix} = \begin{bmatrix} Y'_1 & -Y'_1 & 0 \\ -Y'_1 & Y' + Y'_2 & -Y'_2 \\ 0 & -Y'_2 & Y'_2 + Y'_3 \end{bmatrix} \begin{bmatrix} V_{core} \\ V_{sheath} \\ V_{armour} \end{bmatrix} \tag{3.109}$$

where $Y'_1 = \mathrm{j}\omega l_i$. If, as before, $V_{sheath} = V_{armour} =$ zero, equation (3.109) reduces to

$$-\mathrm{d}I_{core}/\mathrm{d}x = Y'_1 V_{core} \tag{3.110}$$

Therefore, for frequencies of interest, the cable per unit length harmonic impedance, Z', and admittance, Y', are calculated with both the zero and positive

Table 3.8 Corrections for Skin Effect in Cables

Company	Voltage (kV)	Harmonic Order	Resistance
NGC	400, 275 (Based on 2.5 sq.in. conductor at 5 in. spacing between centres)	$h \geqslant 1.5$	$0.74\ R_1\ (0.267 + 1.073\sqrt{h})$
	132	$h \geqslant 2.35$	$R_1\ (0.187 + 0.532\sqrt{h})$
EDF	400, 225	$h \geqslant 2$	$0.74\ R1\ (0.267 + 1.073\sqrt{h})$
	150, 90	$h \geqslant 2$	$R_1\ (0.187 + 0.532\sqrt{h})$

sequence values being equal to the Z in Equation (3.107), and the Y' in Equation (3.110), respectively.

In the absence of rigorous computer models, such as described above, power companies often use approximations to the skin effect by means of correction factors. Typical corrections used by the NGC(UK) and EDF(France) are given in Table 3.8.

3.10 Examples of Application of the Models

3.10.1 Harmonic Flow in a Homogeneous Transmission Line [16,17]

A 230 km 220 kV line of flat configuration is used as the first test system; the parameters of this line are shown in Figure 3.27. A three-dimensional graphic representation is used to provide simultaneous information of the harmonic levels along the line. At each harmonic (up to the 25th harmonic), one per unit positive sequence current is injected at the Islington end of the line. The voltages caused by this current injection are, therefore, the same as the calculated impedance, i.e. V_+ gives Z_{++}, V_- gives Z_{+-} and V_0 gives Z_{+0} (the subscripts +, −, 0 refer to the positive, negative and zero sequences, respectively).

Figures 3.28, 3.29 and 3.30 illustrate the effect of two extreme cases of line termination (at Kikiwa), i.e. the line open-circuited and short-circuited, respectively. The difference in harmonic magnitudes along the line are due to standing wave effects and shifting of the resonant frequencies caused by line terminations.

Figure 3.28 indicates the existence of high voltage levels at both ends of the open-circuited line at the half wavelength frequencies. The 25th harmonic clearly illustrates the standing wave effect, with voltage maxima and minima alternating at quarter of the wavelength intervals.

At any particular frequency, a peak voltage at a point in the line will indicate the presence of a peak current of the same frequency at a point about a quarter wavelength away. This is clearly seen in Figure 3.29.

When the line is short-circuited at the extreme end, the harmonic current penetration is completely different, as shown in Figure 3.30(a). The high current levels at the receiving end of the line are due to the short-circuit condition. Figure 3.11 shows that the resonant maxima decrease as frequency increases. However, this

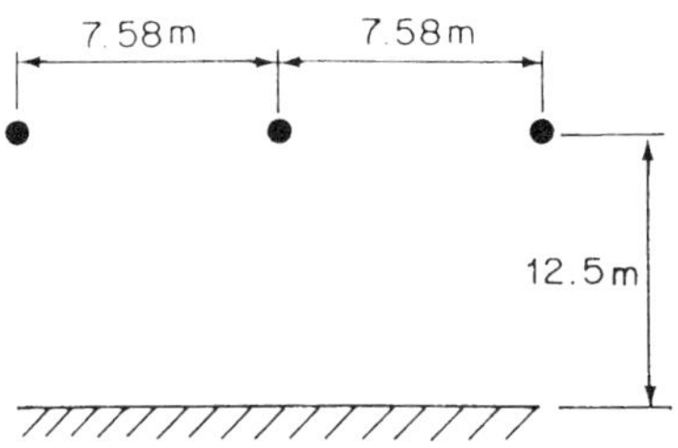

Figure 3.27 Conductor information for the Islington to Kikiwa line: conductor type, Zebra (54/3.18 + 7/3.18); length, 230 km; resistivity, 100 Ω m

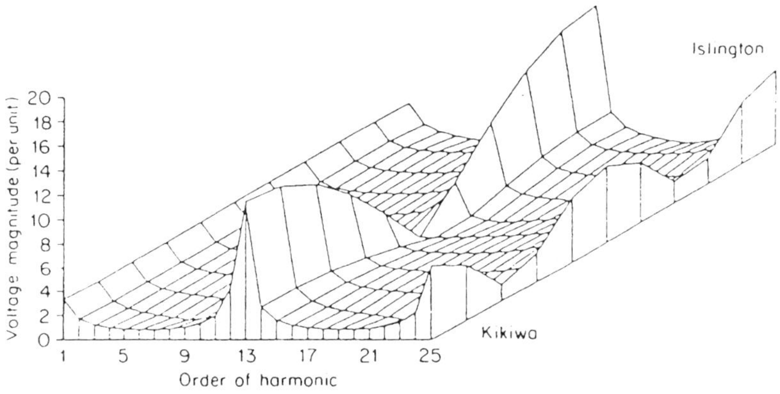

Figure 3.28 Positive sequence voltage versus frequency along the open-ended Islington to Kikiwa line

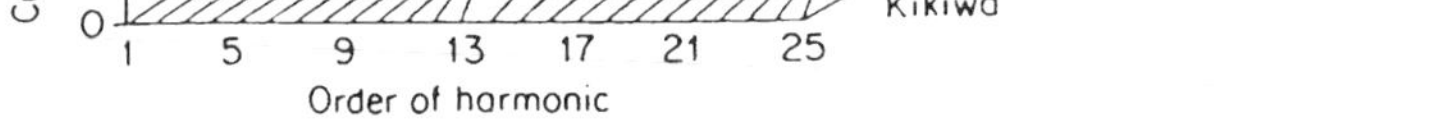

Figure 3.29 Positive sequence current along the open-ended line for a 1 per unit positive sequence current injection at Islington

does not appear to be the case in Figure 3.30(a). The reason is that the points plotted correspond only to harmonic frequencies and resonances do not fall exactly on these frequencies; i.e. the peak-magnitudes at non-harmonic frequencies can be greater than the values plotted in the figure.

Coupling Between Harmonic Sequences It is the zero sequence penetration, rather than the positive sequence, that provides relevant information for the assessment of possible harmonic interference in neighbouring telephone systems. The presence of

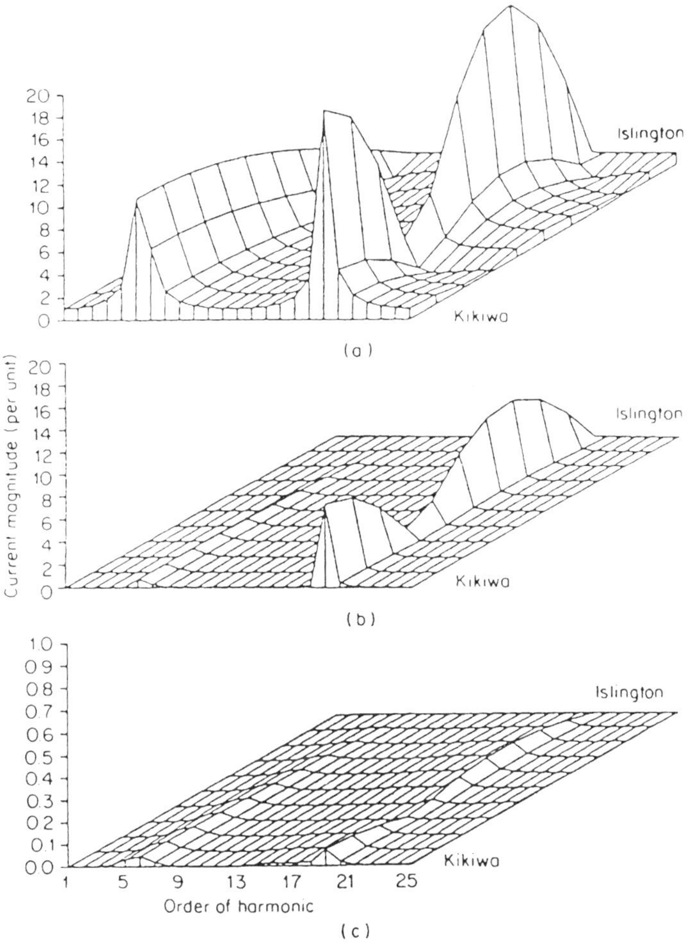

Figure 3.30 Sequence currents along the short circuited line for a 1 per unit positive sequence current injection at Islington: (a) positive sequence current; (b) negative sequence current; (c) zero sequence current

zero sequence in a transmission line connected to a converter bridge is entirely due to asymmetries in either the converter ac plant components or the transmission line itself.

In Figure 3.30 the locations of maximum zero sequence current (Figure 3.30(c)) coincide with those of the positive sequence (Figure 3.30(a)), and the highest level produced in the test line, about 10% of the injected positive sequence current, occurs at the 19th harmonic, at the Kikiwa end of the short-circuited line. However, the levels of zero sequence current are low (notice the scale change between positive and zero sequence plots).

Differences in Phase Voltages In conventional harmonic analysis using single-phase positive sequence models, a transmission line is assumed to have one resonant

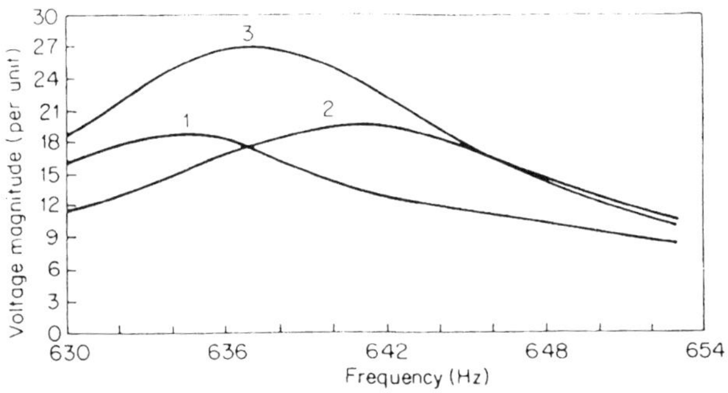

Figure 3.31 Three-phase resonant frequencies of the Islington to Kikiwa line with a 1 per unit positive sequence current injection (skin effect included)

frequency. However, the use of the three-phase algorithm to model the Islington-Kikiwa unbalanced transmission line shows that the resonant frequencies are different for each phase. In this case, the spread of frequencies can be seen from Figure 3.31 to be approximately 6 Hz.

The different magnitudes of the resonant frequencies (up to 30%) of the three phases, partly explains the problems encountered with correlating single-phase modelling and measurements on the physical network. The results clearly indicate that harmonics in the transmission system are unbalanced and three-phase in nature.

Effect of Mutual Coupling in Double Circuits The unbalanced behaviour of double circuit lines is well documented at fundamental frequency [18,19]. The three-phase harmonic penetration algorithm is used in this section to determine the importance of modelling mutual coupling at harmonic frequencies. The line used is shown in Figure 3.32.

Figure 3.33 displays the harmonic impedance (Z_{++}, Z_{00}) seen from the point of harmonic injection, both for a 1 per unit positive sequence current and 1 per unit

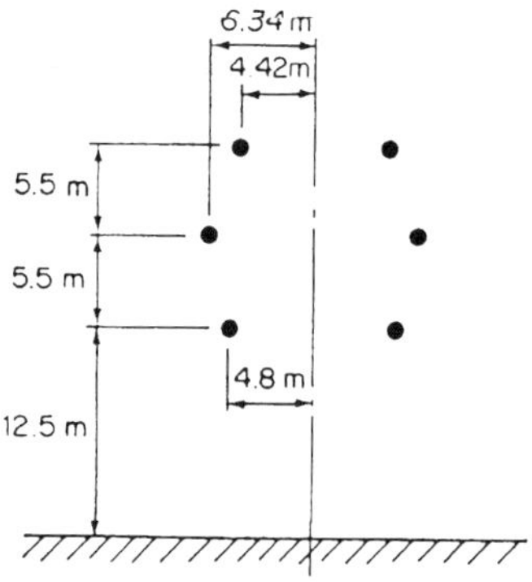

Figure 3.32 Line geometry of a double circuit line. Length, 167 km; earth resitivity, 100 Ω m; two conductors per bundle; bundle spacing, 0.45 m; conductor, 30/3.71 + 7/3.71

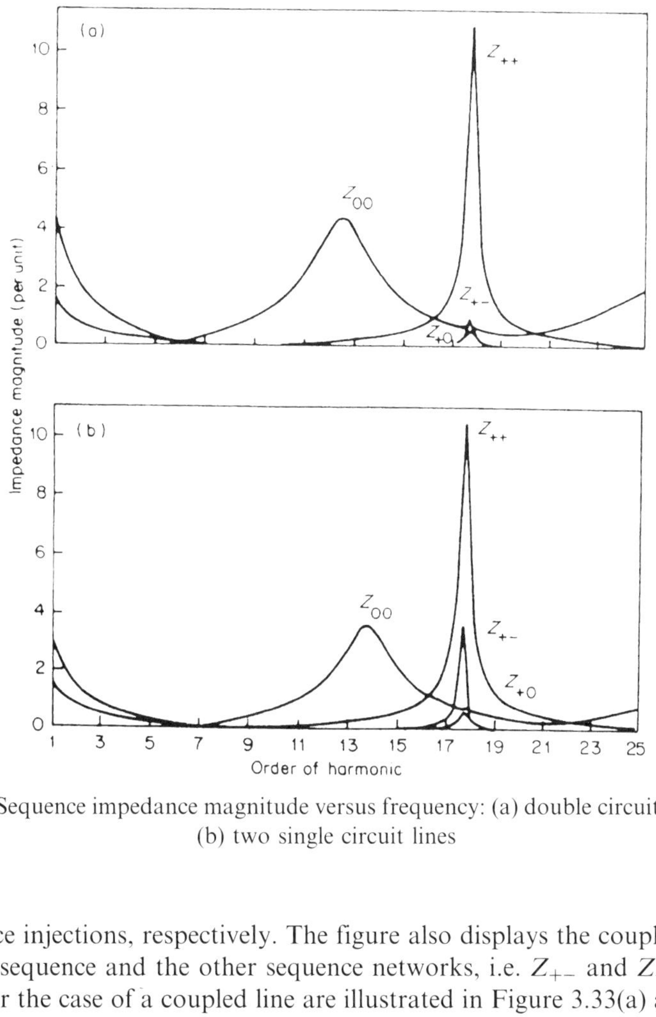

Figure 3.33 Sequence impedance magnitude versus frequency: (a) double circuit coupled line; (b) two single circuit lines

zero sequence injections, respectively. The figure also displays the coupling between the positive sequence and the other sequence networks, i.e. Z_{+-} and Z_{+0}.

Results for the case of a coupled line are illustrated in Figure 3.33(a) and those of two single circuit lines in Figure 3.33(b).

The magnitudes and resonant frequencies of the Z_{++} and Z_{+0} impedances are not affected by the modelling of mutual coupling. However, the level of Z_{+-} has changed substantially at resonance showing appreciable imbalance. Moreover, the magnitude and resonant frequency of Z_{00} is very different in the two cases.

Robinson [20] reported that telephone interference caused by zero sequence currents did not coincide with high levels of power system harmonics. This can partly be explained by the different resonant frequencies of the Z_{++} and Z_{00} observed.

3.10.2 Harmonic Analysis of Transmission Line with Transpositions [21]

The conductor geometries of high voltage transmission lines produce considerable impedance asymmetry, which in turn causes corresponding voltage imbalance at the far end of the line.

It is generally accepted that, for practical distances, the effect of line asymmetry can be eliminated by the use of phase transpositions dividing the line into three, or multiples of three equal lengths. Accordingly, transpositions are often used in long distance transmission as a means of balancing the fundamental frequency impedances of the line.

In fundamental frequency studies the effect of transpositions is generally accounted for by averaging the distributed parameters of the three transposed sections and using them in a single nominal or equivalent PI-circuit. Such a method, however, assumes that the line geometry is perfectly symmetrical at all points, whereas the transpositions occur at two discrete distances, at different points on the standing wave.

The series impedance and shunt equivalent matrices are combined into one admittance matrix that represents the transposed section, i.e.

$$Y' = \begin{bmatrix} Y'_{SS} & -Y'_{SR} \\ -Y'_{RS} & Y'_{RR} \end{bmatrix} \tag{3.111}$$

where

$$[Y_{SS}] = [Y_{RR}] = [Z]^{-1} + \tfrac{1}{2}[Y]$$
$$[Y_{SR}] = [Y_{SR}] = [Z]^{-1}$$

The admittance parameters for the individual sections are then transformed into A′, B′, C′, D′ parameters, such that they can be cascaded, i.e.

$$[A] = [A_1][A_2][A_3] \tag{3.112}$$

Finally, the resultant transmission parameters A, B, C, D are converted back into an admittance matrix which properly represents the effects of transpositions.

The nodal admittance matrix equation of the three-phase transmission line may be written as

$$\begin{bmatrix} I_S \\ I_R \end{bmatrix} = \begin{bmatrix} Y_{SS} & -Y_{SR} \\ -Y_{RS} & Y_{RR} \end{bmatrix} \begin{bmatrix} V_S \\ V_R \end{bmatrix} \tag{3.113}$$

where I_S, I_R, V_S and V_R are vectors of a size determined by the number of coupled conductors.

Applying a partial inversion algorithm to Equation (3.113), the following matrix of inverse hybrid parameters is obtained.

$$\begin{bmatrix} I_S \\ V_R \end{bmatrix} = \begin{bmatrix} Y_{SS} - Y_{SR}Y_{RR}^{-1}Y_{RS} & -Y_{SR}Y_{RR}^{-1} \\ Y_{RR}^{-1}Y_{RS} & Y_{RR}^{-1} \end{bmatrix} \begin{bmatrix} V_S \\ I_R \end{bmatrix} \tag{3.114}$$

or

$$V_R = Y_{RR}^{-1}Y_{RS}V_S + Y_{RR}^{-1}I_R \tag{3.115}$$

Two different cases are of interest and will be used in the following sections. The first relates to a harmonic voltage excited open-ended line, specified as $V_S = 1$ p.u.

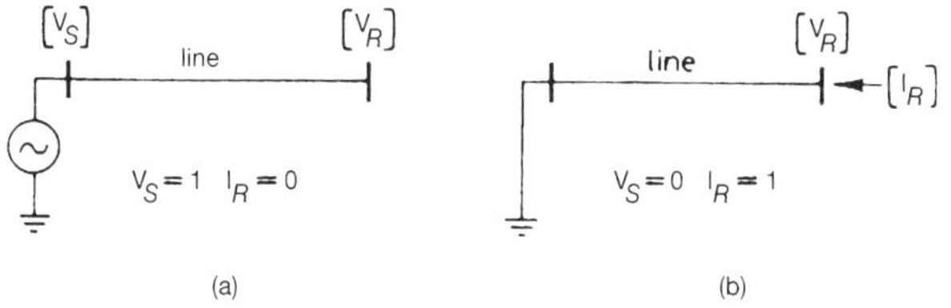

Figure 3.34 Diagram of terminal conditions (a) voltage source and open-ended line; (b) current source and short-circuited line. Reproduced from [21] by permission of IEE

and $I_R = 0$. This case produces the highest voltage harmonic levels and must, therefore, be considered for design purposes. The second important case is the harmonic current excited short-circuited ended line, specified as $V_S = 0$ and $I_R = 1$ p.u. which is more likely to be of practical interest. These two cases are illustrated by the simplified diagrams of Figures 3.34(a) and (b), respectively.

Effect of transpositions with Voltage Excitation Harmonic voltage sources are thought to be of little significance at the moment and are generally ignored when assessing harmonic distortion. However, under non-ideal system conditions, synchronous generators can produce harmonic EMFS [22]. Moreover, some power electronic circuits operate as harmonic voltage sources [23,24]. It is thus appropriate to consider the effectiveness of transpositions in the presence of harmonic as well as power frequency voltage sources.

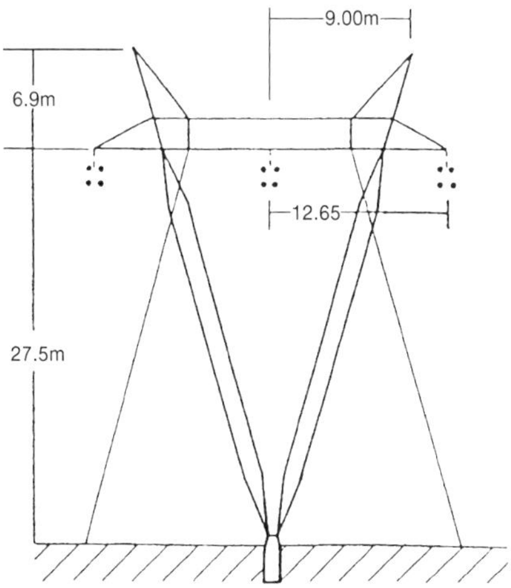

Figure 3.35 Test Line. Details of the test line: The test line, shown in Figure 3.35, is of flat configuration without earth wires and the main parameters are: Nominal voltage = 500 kV; Conductor type: Panther (30/3.00 + 7/3.00 ACSR); Resistivity = 100 Ω/m; Equal distances between transpositions and the natural impedance matrix. Reproduced from [21] by permission of IEE

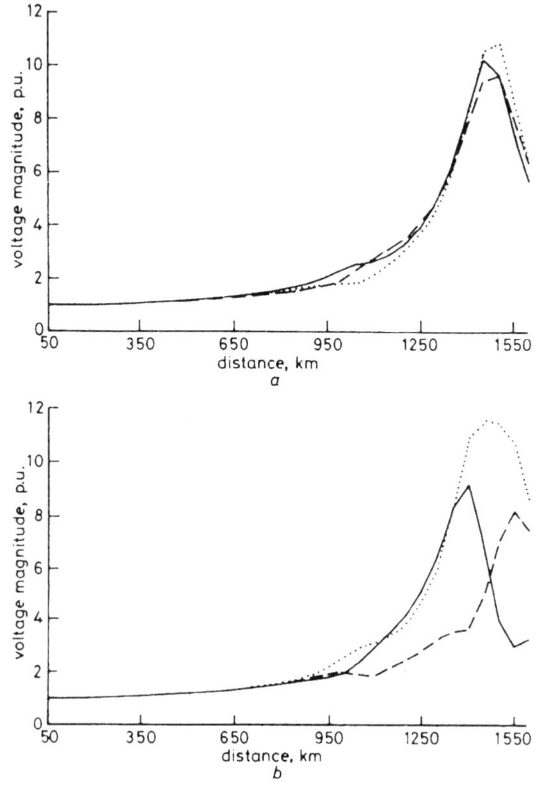

Figure 3.36 Fundamental frequency three-phase voltages at the end of the test line (open-circuited) versus line distance. Reproduced from [21] by permission of IEE

(a) without transpositions	(b) with transpostions
—— R	—— Y
------ Y	------ B
········ B	········ R

The test line is fed from 1 p.u. voltage sources at fundamental and harmonic frequencies. It is realised that the presence of 1 p.u. harmonic voltage sources is unrealistic, but such a figure provides a good reference for comparability between the effects at different frequencies. The expected harmonic voltage levels are likely to be about 1–3% of the fundamental and, therefore, the results plotted in later figures should be scaled down proportionally.

Open-ended Line The fundamental frequency behaviour of the open-ended line is illustrated in Figures (3.36(a) and (b) for the line untransposed and transposed, respectively). In each case, the receiving end voltages are plotted for line distances varying from 50 to 1500 km.

These figures indicate that in the absence of voltage compensation, the effectiveness of transpositions is limited to line distances under one-eighth of the

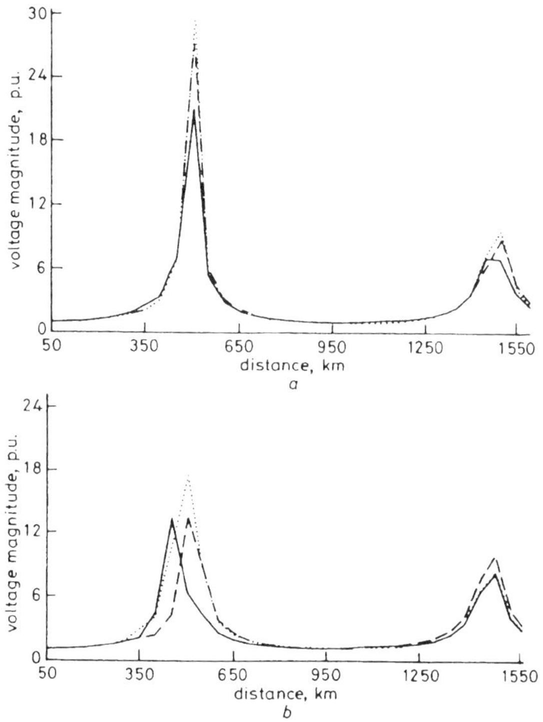

Figure 3.37 Three-phase third harmonic voltages at the end of the test line (open-circuited) versus line distance). See Key for Figure 3.36. Reproduced from [21] by permission of IEE

wavelength (i.e. 750 km at 50 Hz). For distances approaching the quarter wavelength, the transposed line produces considerably greater imbalance than the untransposed.

Although such transmission distances are impractical without compensation, the results provide some indication of the behaviour to be expected with shorter lines at harmonic frequencies. Such behaviour is exemplified in Figure 3.37 which corresponds to the case of a line excited by 1 p.u. 3rd harmonic voltage. However, the results plotted in Figure 3.37 obtained at 50 km intervals, are not sufficiently discriminating around the points of resonance. Thus the region of resonant distances has been expanded in Figure 3.38 to illustrate more clearly the greatly increased imbalance caused by the transpositions. The resonant peaks of the three phases occur at very different distances, e.g. Figure 3.37(b) shows 50 km diversity between the peaks. Therefore, for a given line distance the resonant frequencies will vary, thus increasing the risk of a resonant condition.

It is also interesting to note the dramatic voltage amplification which occurs for electrical distances equal to the first quarter wavelength. Figure 3.38 shows a peak of 35 per unit for the 3rd harmonic when the line is 500 km long and the 5th harmonic peak (not shown) reaches 45 per unit at about 300 km.

Figure 3.37(a) and (b) also show the effect of attenuation with distance, i.e. the considerable reduction of the peaks at resonant distances at the odd quarters of wavelength other than the first. Such attenuation is caused by the series and shunt resistive components of the equivalent PI-model.

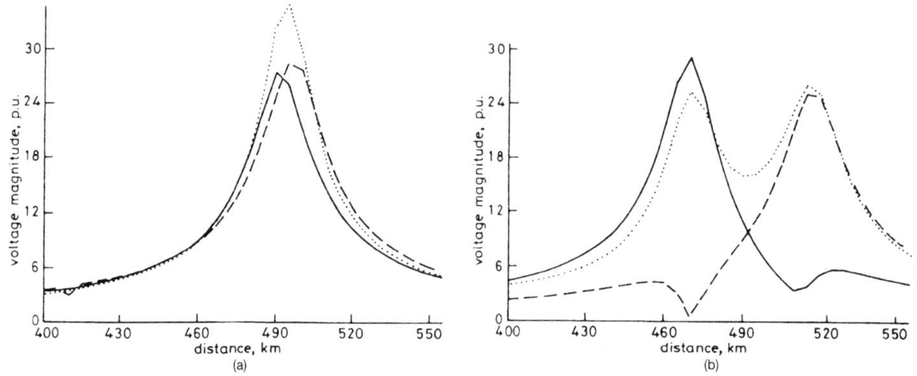

Figure 3.38 Results of Figure 3.37 expanded in the region of resonance. For Key see Figure 3.36. Reproduced from [21] by permission of IEE

The immediate effect of the transpositions is the compensation of geometrical line asymmetry. This can only result in electrical symmetry if the average currents in each of the transposed sections are similar. Thus the deterioration of voltage balance current standing wave along the line. The improved symmetry of the phase voltages at the three quarter wave distance, seen in Figure 3.37, is due to the averaging effect produced by the third harmonic standing wave, as illustrated by the idealized waveforms of Figure 3.39.

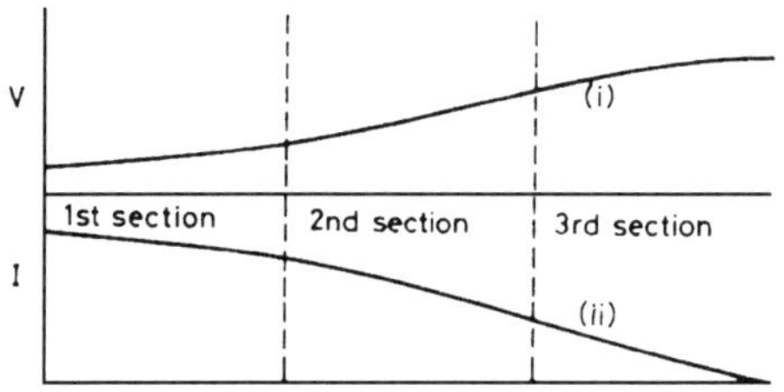

Figure 3.39(a) Standing waves along a line of quarter wavelength (i) voltage wave (ii) current wave. Reproduced from [21] by permission of IEE

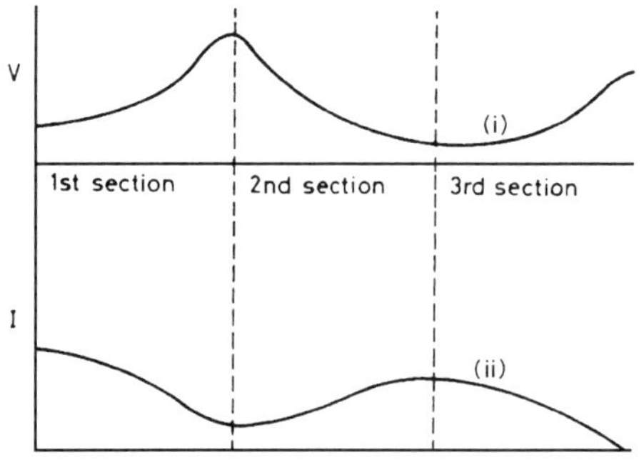

Figure 3.39(b) Third harmonic standing waves along a line of three-quarter wavelengths. (i) voltage wave (ii) current wave. Reproduced from [21] by permission of IEE

From the above discussion the effectiveness of transpositions should improve as the voltage and current profile throughout the line becomes more uniform, i.e. closer to the natural loading condition, which is discussed in the next section.

Line Loaded If an ideal (uncoupled and unattenuated) line is loaded with its characteristic impedance, the sending end voltage will be sustained throughout the line, provided that the phase angle difference between the sending and receiving end voltages is kept below 45° (or 750 km at 50 Hz). To assess the effectiveness of transpositions with loading, the test line was loaded with its characteristic impedance calculated at 50 Hz. It must be noted that in a coupled multiconductor line such impedance is a matrix, of which only the diagonal elements are being used for the loading. Furthermore, the three diagonal elements are different and are also frequency dependent. We cannot therefore expect to see the uniform 1 p.u. voltage predicted by conventional theory.

Results for the fundamental frequency, plotted in Figure 3.40, illustrate that the effectiveness of transpositions is limited to distances up to about 750 km. For longer lines, similarly to the open-ended line case, transpositions are not effective, although the per unit voltage imbalance of the loaded transposed line (Figure 3.40(b)) is greatly reduced as compared with that of the open line (Figure 3.36(b)).

Up to the first quarter wavelength the effect of natural (fundamental frequency) loading on the harmonic voltages is very similar to the fundamental frequency. For this particular loading conditions the effectiveness of transpositions is limited to distances of about 350 and 200 km for the 3rd and 5th harmonics, respectively. Beyond those distances the transposed lines produce higher levels of imbalance. Subsequent harmonic peaks are seen to reduce rapidly with loading. By way of example, the 5th harmonic voltages without and with transpositions are shown in Figures 3.41(a) and (b), respectively.

The harmonic behaviour of a loaded transmission line without and with transpositions is illustrated in Figures 3.42(a) and (b), respectively. This figure displays the variation of 5th harmonic voltage at the receiving end of a 250 km line with one per unit voltage injection at the sending end. The level of imbalance of the untransposed line (Figure 3.42(a)) shows a gradual increase up to about the natural load (1 p.u. admittance) and very little change thereafter. In contrast, Figure 3.42(b)

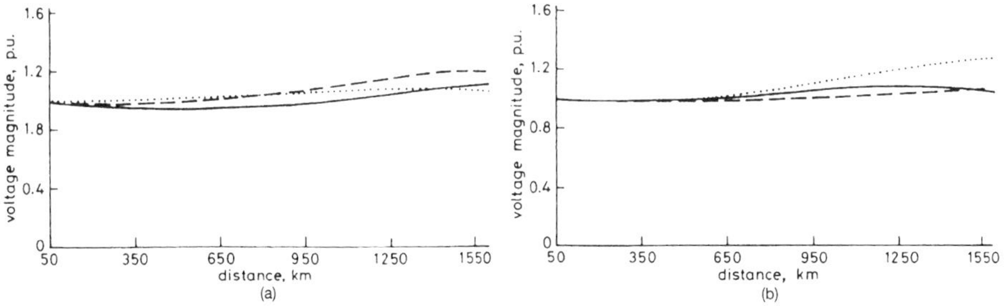

Figure 3.40 Fundamental frequency three-phase voltages at the end of the test line (loaded with the characteristic impedance). For Key see Figure 3.36. Reproduced from [21] by permission of IEE

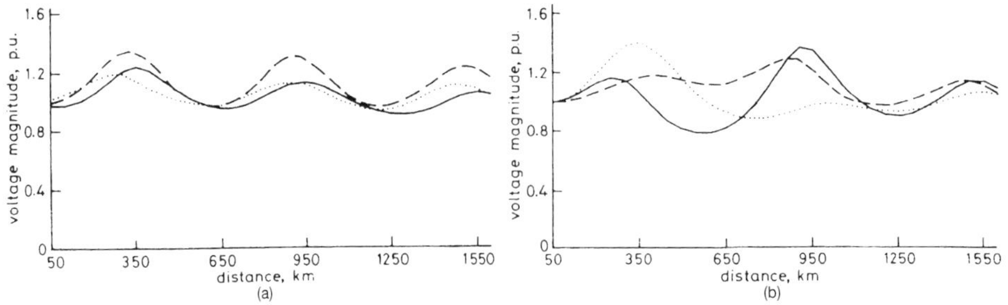

Figure 3.41 Three-phase fifth harmonic voltages at the end of the test line (loaded with the characteristic impedance). For Key see Figure 3.36. Reproduced from [21] by permission of IEE

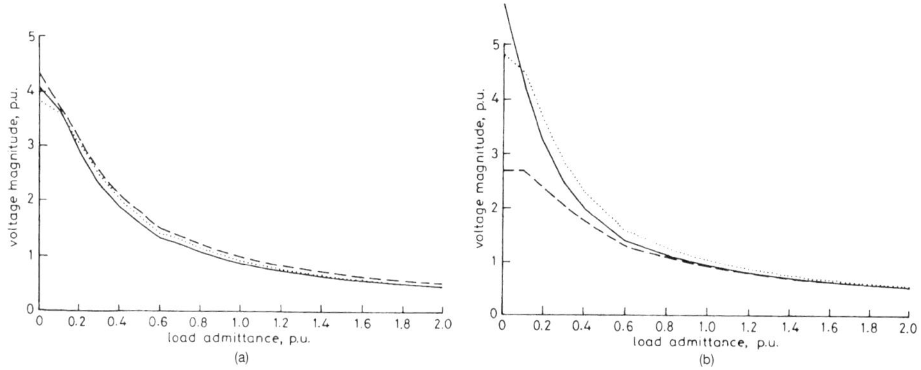

Figure 3.42 Three-phase fifth harmonic voltages at the end of a 250 km test line versus loading admittance (referred to the characteristic admittance). For Key see Figure 3.36. Reproduced from [21] by permission of IEE

illustrates a dramatic increase in the voltage imbalance as the load reduces from the natural level (1 p.u. admittance) to the open circuit condition. A qualitative justification for this behaviour has been made in Figure 3.39. As the line load increases above the natural level, Figure 3.42(b) shows that effectiveness of the transposition increases.

Considering the relatively insignificant levels of harmonic voltage excitation expected from a well-designed system, the resulting voltage distortion in a transposed or untransposed load line is not expected to cause problems, except when the line is lightly loaded. With harmonic current excitation the situation may be quite different, and its effect is examined next.

Effect of Transpositions with Current Excitation The main cause of power system harmonic distortion is the large static power converter, such as used in HVdc transmission and in the metal reduction industry. Because of their large dc smoothing inductance compared to the ac system impedance, static converters can be considered as current sources on the ac side and voltage sources on the dc side [25].

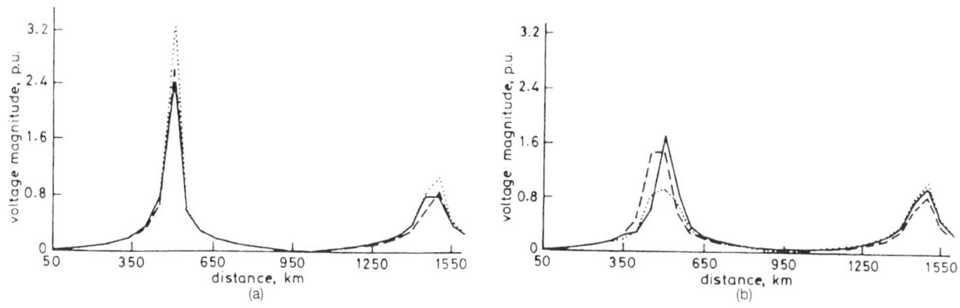

Figure 3.43 Three-phase third harmonic voltages caused by 1 p.u. third harmonic current at the point of harmonic current injection. For Key see Figure 3.36. Reproduced from [21] by permission of IEE

Thus, the harmonic modelling of a long transmission line feeding a static converter is basically that of Figure 3.34(b), i.e. a harmonic current source at the receiving end of the line with the sending end shorted to ground through a relatively low impedance.

The harmonic voltages at the point of current harmonic injection follow the same pattern as those of the open circuit line with harmonic voltage excitation. This is clearly illustrated in Figure 3.43 for a case of 3rd harmonic current injection. Similarly to the voltage excited open line, substantial voltage distortion results when the line length is close to a quarter wavelength, although the imbalance caused by transpositions is less pronounced in the case of current injection. Figure 3.43 indicates that even 1% of harmonic current injection can produce 3 or 4% voltage harmonic content at the point of harmonic current injection, which is above the levels normally permitted by harmonic legislation.

As the harmonic order increases, the line experiences higher levels of voltage distortion. For example, the case of 5th harmonic current injection, illustrated in Figure 3.44 shows a peak voltage of about 4.5% in one of the phases. However, in this case the transposed line is seen to reduce considerably the harmonic peaks for the quarter wavelength distance line.

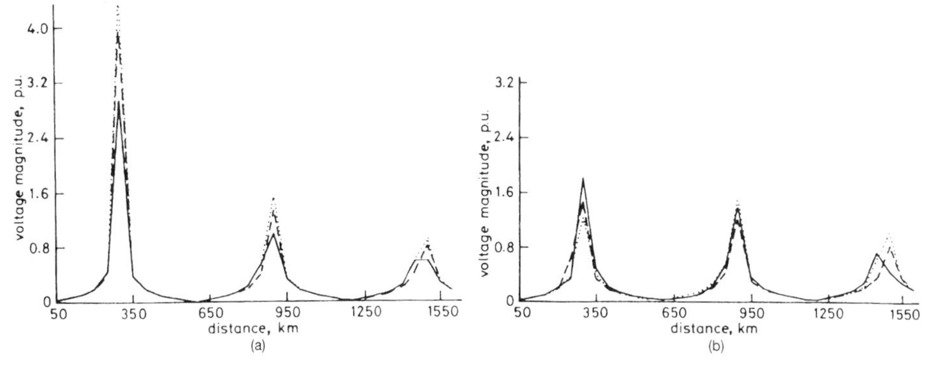

Figure 3.44 Three-phase fifth harmonic voltages caused by 1 p.u. third harmonic current at the point of harmonic current injection. For Key see Figure 3.36. Reproduced from [21] by permission of IEE

While the 5th harmonic current is normally eliminated by filters, this is not the case for non-characteristic orders like the 3rd, which will then distort the supply waveform and, in the absence of equidistant firing control, may increase further the production of 3rd harmonic current [26].

3.10.3 Harmonic Analysis of Transmission Line with Var Compensation [27]

Matrix Model of a Compensated Line Generally, long transmission lines are divided into two or three sections of equal length and VAR compensating equipment is connected between them in the form of series capacitors and shunt inductors or capacitors. An equivalent circuit of a typical long distance transmission line with conventional compensation elements is illustrated in Figure 3.45.

Each line section is represented by its harmonic admittance matrix

$$[Y] = \begin{bmatrix} [Y_{SS}] & [-Y_{SR}] \\ [-Y_{RS}] & [Y_{RR}] \end{bmatrix}_h \tag{3.116}$$

where

$$[Y_{SS}] = [Y_{RR}] = [Z]_{EPM}^{-1} + \tfrac{1}{2}[Y]_{EPM}$$
$$[Y_{SR}] = [Y_{RS}] = [Z]_{EPM}^{-1}$$

the suffix EPM indicating equivalent PI model.

With the assumption that the compensating inductances and capacitances are uncoupled and linearly dependent with frequency their corresponding harmonic matrix admittances are:

$$[Y_c] = \begin{bmatrix} jh_o\omega_o C_c & 0 & 0 \\ 0 & jh_o\omega_o C_c & 0 \\ 0 & 0 & jh_o\omega_o C_c \end{bmatrix} [Y_L] = \begin{bmatrix} \dfrac{-j}{h\omega_o L_c} & 0 & 0 \\ 0 & \dfrac{-j}{h\omega_o L_c} & 0 \\ 0 & 0 & \dfrac{-j}{h\omega_o L_c} \end{bmatrix} \tag{3.117}$$

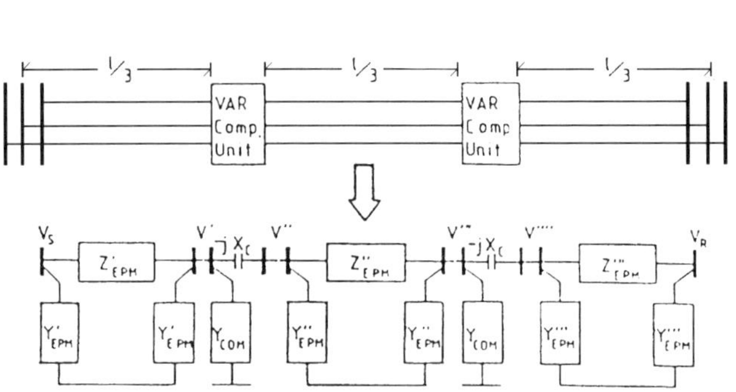

Figure 3.45 Equivalent circuit of VAR compensated transmission line

where L and C are the fundamental frequency values of the compensating inductance and capacitance derived from load flow studies,
and h is the harmonic order.

Finally, the admittance matrices of the individual subsystems, i.e. the line sections and VAR compensating units, must be combined into a single admittance matrix.

If the purpose is to observe harmonic voltages at the far end of the transmission system, the individual subsystems must be transformed to ABCD parameters and then cascaded to obtain an equivalent ABCD matrix equation. This in turn can be converted back to an equivalent admittance matrix, which relates the currents and voltages at the two ends of the line, as described by Equations (3.113), (3.114) and (3.115) and Figure 3.34 in the previous section.

Harmonic Voltage Excitation The test line is a 1000 km of the same configuration as in section 3.10 (see Figure 3.35).

The addition of shunt inductive compensation effectively increases the characteristic impedance and thus reduces the load that causes the optimum voltage profile.

For the positive sequence shunt admittance values of the test system, a standard load flow programme was used to derive the optimal discrete shunt inductances required to provide a practically constant voltage along the line at the fundamental frequency.

However, the addition of shunt inductance isolates the line from ground (reducing its ability to act as a low pass filter) and thus reduces its ability to dampen harmonics.

The results, plotted in Figure 3.46 correspond to an open-ended line and show that while the fundamental frequency voltage profile is good, the line performance at harmonic frequencies is worse than without compensation. In particular, the level of the receiving end voltage for second harmonic injection has increased dramatically.

In the absence of compensation, the natural load of the line under consideration is approximately 950 MW, but the maximum nominal loading planned is 1450 MW, i.e. 1.5 times the natural load.

For this loading condition Figure 3.47 shows the effect of a combined compensation scheme, consisting of shunt and series capacitors. It is noted that shunt capacitors tend to amplify harmonic distortion at the compensation points, while having the opposite effect elsewhere.

Harmonic Current Excitation In this case, a one percent harmonic current was injected at the receiving end of the line.

The effect of shunt inductive compensation in the harmonic behaviour of the unloaded line is shown in Figure 3.48; again the second harmonic shows the greater amplification.

The results of combining series and shunt capacitive compensation for the case of a heavily-loaded line are shown in Figure 3.49.

The magnitudes of the harmonic voltages for the loaded line are smaller than those of an open-ended line.

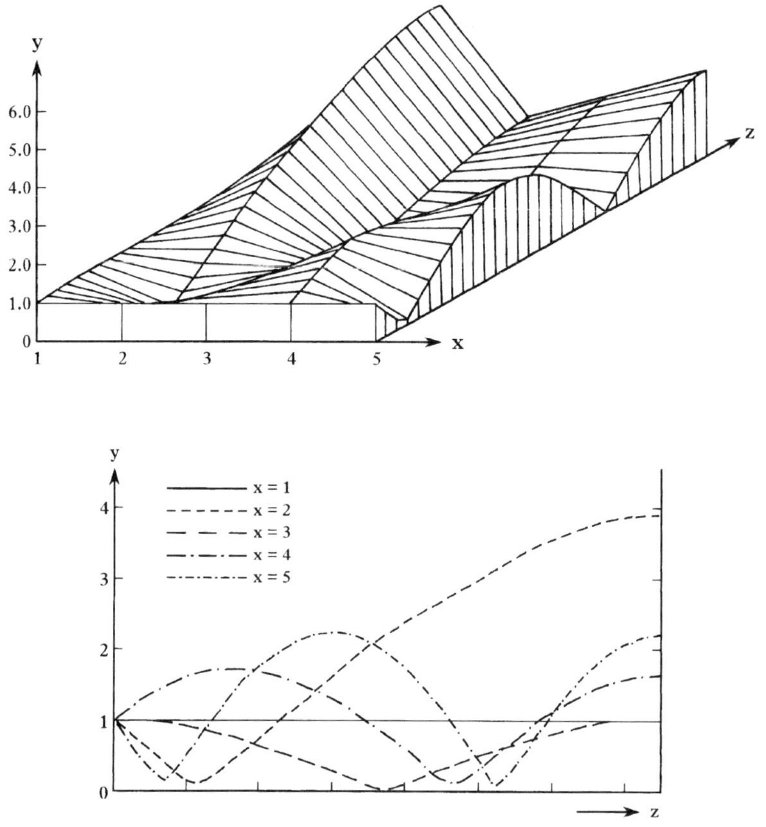

Figure 3.46 Fundamental and harmonic voltage levels along the unloaded line with shunt inductive compensation and 1 per unit voltage at the sending end. x—harmonic order; y—voltage magnitude; z—line position w.r.t. point of harmonic injection

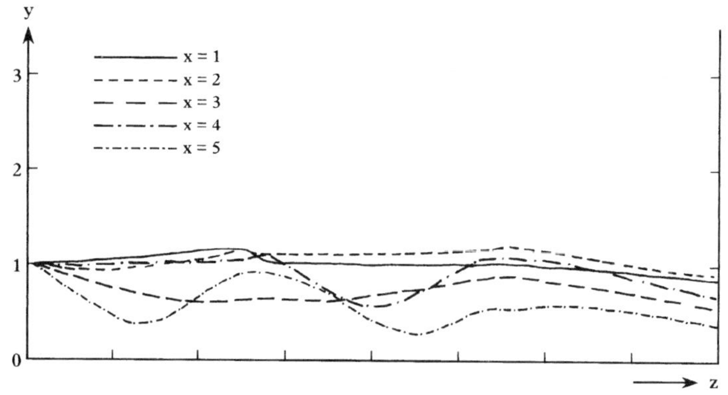

Figure 3.47 Standing waves along the heavy-loaded line with series as well as shunt capacitive compensation

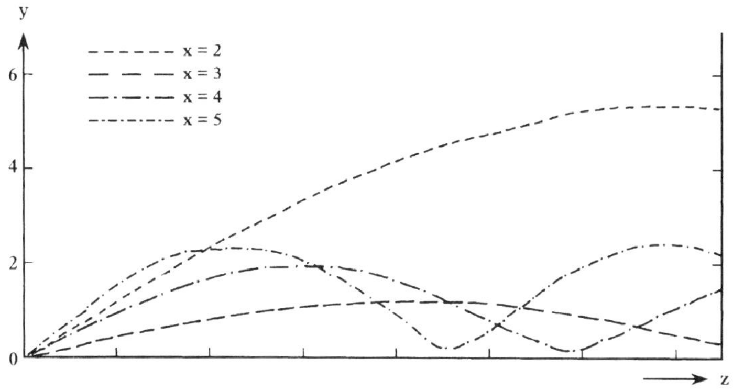

Figure 3.48 Harmonic voltage levels along the unloaded line with shunt inductive compensation

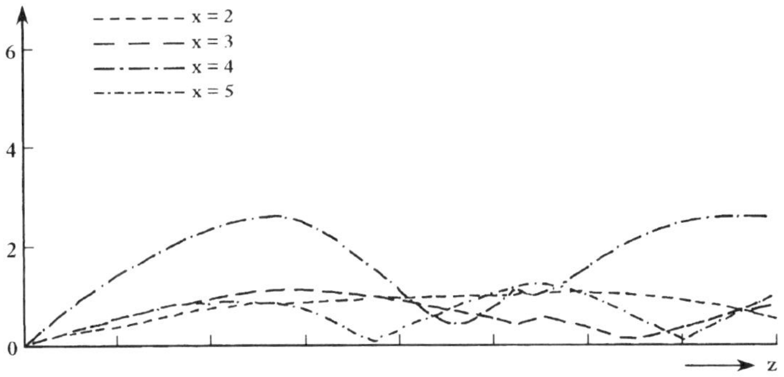

Figure 3.49 Harmonic voltage levels along the loaded line with series as well as shunt capacitive compensation

3.10.4 Harmonic Analysis in a Hybrid HVdc Transmission Link [28]

Due to the limited number of phases and switching devices, the dc output voltages at converter stations contain considerable ripple. Under perfectly symmetrical ac supply and switching conditions, the voltage ripple consists only of twelve pulse related harmonics. In practice, however, ac system imbalance and asymmetrical firing may lead to other frequencies being present in the dc voltage waveforms.

It is, therefore, necessary to derive the full spectrum of harmonic admittances of the dc link. For generality the test system, based on the New Zealand system, contains overhead lines and submarine cables and each of them must be represented by the frequency dependent models derived in earlier sections. The New Zealand HVdc link, illustrated in Figure 3.50, consists of six major subsystems, (ii), (v), (vi), (vii), (viii) and (ix) and three auxiliary components (iii), (iv) and (x). The distances of the main transmission components are:

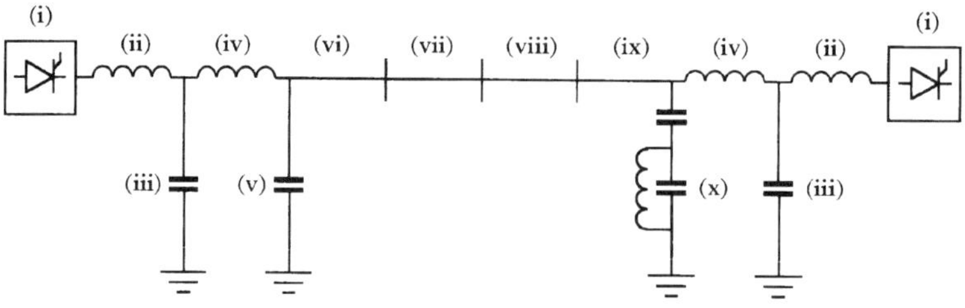

(i) Convertor terminal
(ii) Smoothing reactor
(iii) High frequency capacitor (0.1 μF)
(iv) Line trap (1.05 mH)
(v) Surge capacitor (0.9 μF)
(vi) SI inland line section
(vii) SI coastal line section
(viii) Cook Strait cable
(ix) NI coastal line section
(x) 600 Hz damping circuit

Figure 3.50 Layout of NZ dc link, including ancillary components

SI inland line	:	484 km
SI coastal line	:	49 km
Cook Strait cable	:	37 km
NI coastal line	:	34.5 km

Figure 3.51 shows a typical tower, with relevant data (for the inland line section) given in Table 3.9.

Towers in the coastal sections are identical to those in Figure 3.51 except that the spacing between conductors is increased from 8.23 to 12.8 m.

A cross-section of the submarine cable is shown in Figure 3.26, and relevant information for the cable is given in Table 3.10.

A relative permittivity of 3.5 and a relative permeability of 1.0 for the insulation are also assumed.

Derivation of Parameters [29] Considering the perfectly balanced self and mutual impedance of the line, the HVdc scheme is best analysed using sequence networks (of positive and zero sequence). With reference to the circuit diagram of Figure 3.52, the positive sequence current is defined as the average current flowing from node 1 to node 2, i.e.

$$I_+ = (I_1 - I_2)/2 \tag{3.118}$$

and the zero sequence component is the average current flowing into the network, and returning by some other path, i.e.

$$I_0 = (I_1 + I_2)/2. \tag{3.119}$$

The relationships between the phase and sequence components of current and voltage are:

$$\begin{bmatrix} I_0 \\ I_+ \end{bmatrix} = [T] \begin{bmatrix} I_1 \\ I_2 \end{bmatrix} \tag{3.120}$$

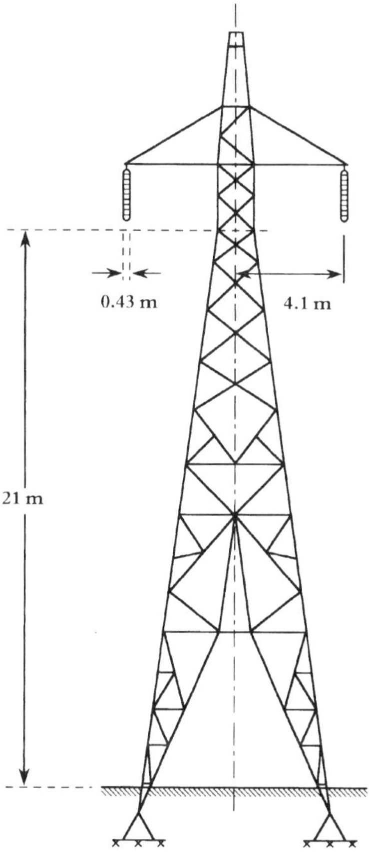

Figure 3.51 Typical dc tower

Table 3.9 Inland Line Section Data

	Main Conductors	Earth Wire
Resistance	0.351	3.1 Ohms/km
Diameter	38.4 mm	11.5 mm
Skin ratio T/D	0.3859	0.5
Bundle spacing	431.8 mm	—

Table 3.10 Data for the New Zealand Cable System

Conductor	Inside Diameter (mm)	Outside Diameter (mm)	Resistance
core	13.462	33.477	0.0331 Ohms/km
sheath	63.242	71.044	0.2865 Ohms/km
armour	88.189	98.958	0.1148 Ohms/km
sea	110.896	—	0.21 Ohms-metres

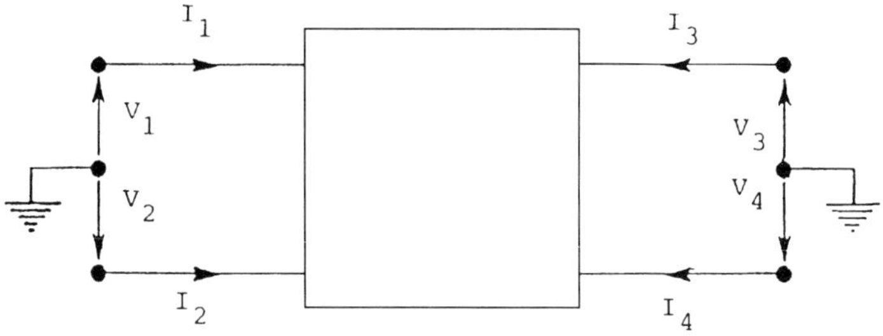

Figure 3.52 Components of a two phase system

$$\begin{bmatrix} V_0 \\ V_+ \end{bmatrix} = [T] \begin{bmatrix} V_1 \\ V_2 \end{bmatrix} \tag{3.121}$$

where

$$[T] = \frac{1}{2} * \begin{bmatrix} 1 & 1 \\ 1 & -1 \end{bmatrix}. \tag{3.122}$$

In terms of sequence components the impedances and admittances of the series and shunt elements of the equivalent PI circuits of the subsystems are:

$$[Z_{sequence}] = \begin{bmatrix} Z_o & 0 \\ 0 & Z_+ \end{bmatrix} \tag{3.123}$$

$$[Y_{sequence}] = \begin{bmatrix} Y_o & 0 \\ 0 & Y_+ \end{bmatrix}. \tag{3.124}$$

The above equation may be written as

$$\begin{bmatrix} Z_s & Z_m \\ Z_m & Z_s \end{bmatrix} = 0.5 \begin{bmatrix} 1 & 1 \\ 1 & -1 \end{bmatrix} \begin{bmatrix} Z_o & 0 \\ 0 & Z_p \end{bmatrix} \begin{bmatrix} 1 & 1 \\ 1 & -1 \end{bmatrix}. \tag{3.125}$$

Therefore

$$Z_s = 0.5[Z_o + Z_p] \text{ and } Z_m = 0.5[Z_o - Z_p]. \tag{3.126}$$

Similarly

$$Y_s = 0.5[Y_o + Y_p] \text{ and } Y_m = 0.5[Y_o - Y_p]. \tag{3.127}$$

The impedance and admittance matrices for each of the sections must be transformed in *ABCD* parameter matrices in order to cascade the section. For the circuit of Figure 3.53, the *ABCD* parameter transformation equations are as follows:

$$\begin{aligned} A &= 1 + Y_2 Z \\ B &= Z \\ C &= Y_1 + Y_2 + Z Y_1 Y_2 \\ D &= 1 + Y_1 Z. \end{aligned} \tag{3.128}$$

For the situation under consideration, the scalar quantities Z, Y_1 and Y_2 must be replaced with the appropriate matrices, i.e.

$$\begin{aligned} Y_1 = Y_2 &= 0.5[Y_{phase}] \\ Z &= [Z_{phase}]. \end{aligned} \tag{3.129}$$

Therefore, the $ABCD$ parameter matrix equations become:

$$\begin{aligned} [A] &= [U] + 0.5[Y_{phase}][Z_{phase}] \\ [B] &= [Z_{phase}] \\ [C] &= [Y_{phase}] + 0.5[Z_{phase}][Y_{phase}]^2 \\ [D] &= [A]. \end{aligned} \tag{3.130}$$

The final form of the $ABCD$ parameter matrix for a particular section is, therefore:

$$[ABCD] = \begin{bmatrix} [A] & [B] \\ [C] & [D] \end{bmatrix}. \tag{3.131}$$

The sections may then be cascaded by simply multiplying their respective $ABCD$ matrices together, i.e.

$$\begin{bmatrix} [A] & [B] \\ [C] & [D] \end{bmatrix} = \begin{bmatrix} [A^1] & [B^1] \\ [C^1] & [D^1] \end{bmatrix} \times \ldots \times \begin{bmatrix} [A^n] & [B^n] \\ [C^n] & [D^n] \end{bmatrix}. \tag{3.132}$$

The resulting 4×4 matrix must be converted back into a 4×4 admittance using the following inverse equations:

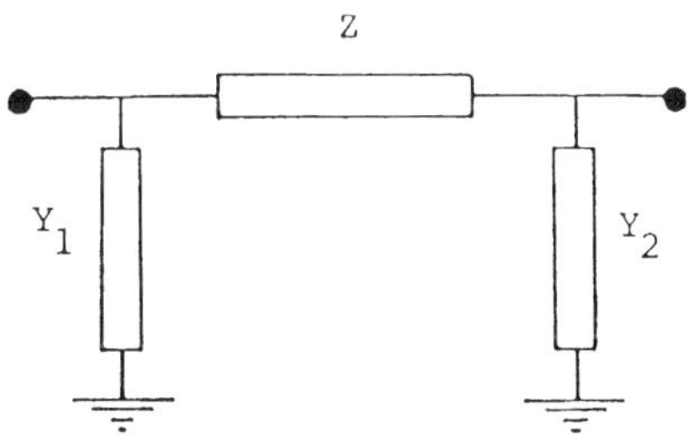

Figure 3.53 Circuit for calculating $ABCD$ parameters

$$
\begin{aligned}
[Y_{11}] &= [D][B]^{-1} \\
[Y_{12}] &= [C] - [D][B]^{-1}[A] \\
[Y_{21}] &= -[B]^{-1} \\
[Y_{22}] &= [B]^{-1}[A].
\end{aligned}
\tag{3.133}
$$

The final admittance matrix, in terms of phase components is

$$
[Y] = \begin{bmatrix} Y_{11} & Y_{12} & Y_{13} & Y_{14} \\ Y_{21} & Y_{22} & Y_{23} & Y_{24} \\ Y_{31} & Y_{32} & Y_{33} & Y_{34} \\ Y_{41} & Y_{42} & Y_{43} & Y_{44} \end{bmatrix} = \begin{bmatrix} [Y_{11}] & [Y_{12}] \\ [Y_{21}] & [Y_{22}] \end{bmatrix}. \tag{3.134}
$$

Thus the transmission network of Figure 3.52 may be represented by the following equation:

$$
[I] = [Y][V] \tag{3.135}
$$

where $[I] = [I_1, I_2, I_3, I_4]$

and $[V] = [V_1, V_2, V_3, V_4]$.

In order to calculate the impedance seen from the sending end terminals, the 4×4 admittance matrix must first be reduced to a 2×2 matrix by eliminating the receiving end voltages and currents. Since the receiving end converter can be approximated by a voltage source, it appears as a short circuit to harmonic frequencies. It may, therefore, be assumed that

$$
V_3 = V_4 \text{ and } I_3 = -I_4.
$$

This leads to the 2×2 admittance matrix in terms of I_1, V_1 and I_2, V_2 only:

$$
Y' = \begin{bmatrix} Y_{11} - Y_a & Y_{12} - Y_b \\ Y_{21} - Y_c & Y_{22} - Y_d \end{bmatrix} \tag{3.136}
$$

where

$$
Y_a = \frac{(Y_{13} + Y_{14})(Y_{31} + Y_{41})}{(Y_{33} + Y_{34} + Y_{43} + Y_{44})}
$$

$$
Y_b = \frac{(Y_{13} + Y_{14})(Y_{32} + Y_{42})}{(Y_{33} + Y_{34} + Y_{43} + Y_{44})}
$$

$$
Y_c = \frac{(Y_{23} + Y_{24})(Y_{31} + Y_{41})}{(Y_{33} + Y_{34} + Y_{43} + Y_{44})}
$$

$$
Y_d = \frac{(Y_{23} + Y_{24})(Y_{32} + Y_{42})}{(Y_{33} + Y_{34} + Y_{43} + Y_{44})}
$$

Moreover, if the midpoint between the two poles at the far end is earthed, then $V_3 = V_4 = 0$, and the above matrix simplifies to:

$$Y' = \begin{bmatrix} Y_{11} & Y_{12} \\ Y_{21} & Y_{22} \end{bmatrix}. \tag{3.137}$$

Finally, the positive sequence admittance is

$$Y_{pos} = 0.5[Y_{11} - Y_{12}] \tag{3.138}$$

Impedance Plots Figure 3.54 compares the impedance plots, as seen from the Benmore terminal, obtained both with and without ancillary components. As can be readily seen, the effect of including the ancillary components is quite dramatic at high frequencies, shifting all the resonances to the left, and markedly altering the magnitude of the very first peak.

Figure 3.55 compares the same impedance plots as seen from the Haywards end. Although the resonances are again shifted to the left, the effect of the cable, as discussed in the previous section, tends to mask out the standing wave effects occurring in the SI inland line section. Of note here, however, is the 600 Hz damping circuit, with the resonant point modified by the surrounding components. This resonant point may also be observed from the Benmore end, in Figure 3.54 although its magnitude is reduced by the masking effect of the cable.

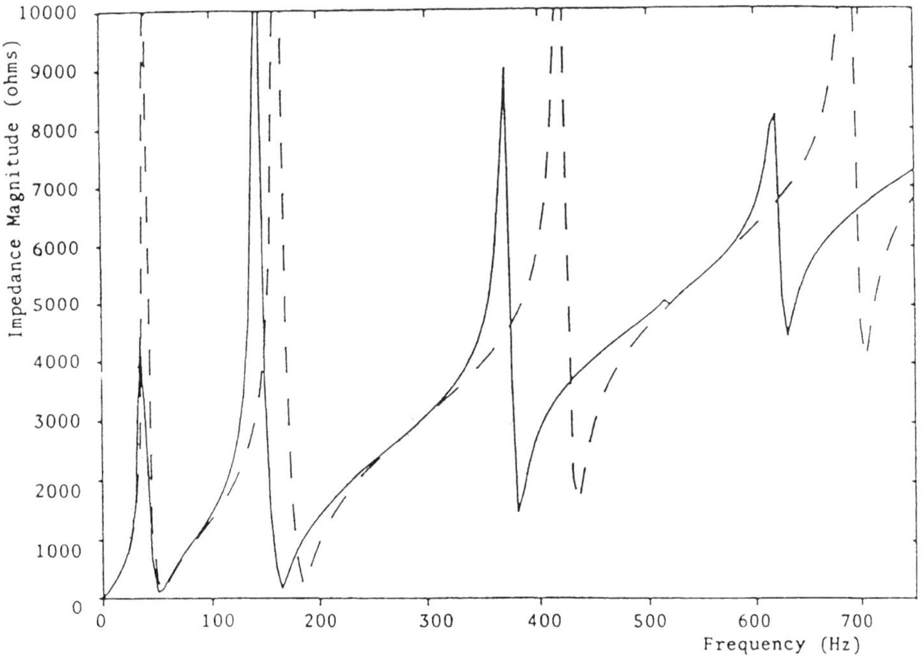

Figure 3.54 Harmonic impedances seen from Benmore
—— with ancillary components
------ without ancillary components

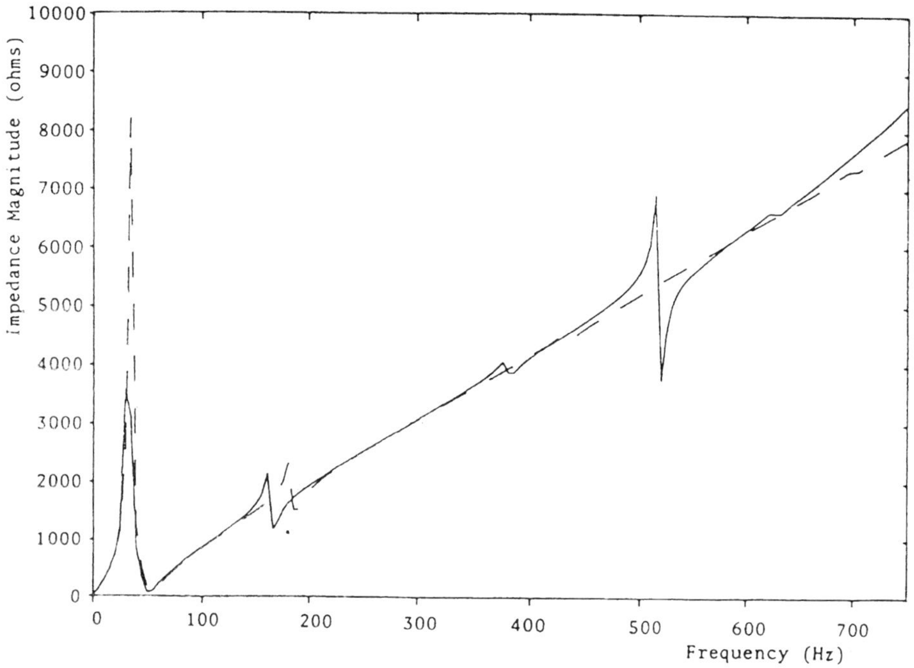

Figure 3.55 Harmonic impedances seen from Haywards
—— with ancillary components
------ without ancillary components

3.11 Summary

Most of the chapter has been devoted to the multiconductor transmission line, as the most influential component in harmonic analysis.

A frequency-dependent equivalent PI model has been described suitable for computer implementation. The formulation of earth path and conductor impedances taking into account skin effect has been carried out and then used to derive simpler solutions based on tabulated coefficients.

Frequency-dependent models have also been derived for the transformers and VAR compensating equipment. These models, combined with the line equivalent PIs are used to derive the network admittance matrix.

A detailed analysis of high voltage underground/submarine transmission systems has also been made, based on current cable technology.

Several practical examples of application of the models to ac and dc transmission have been included.

3.12 References

1. Fortescue, C L, (1918). Method of symmetrical co-ordinates applied to the solution of polyphase networks. *Trans. AIEE*, **37**(2), 1027–1140.

2. Carson, J R, (1926). Wave propagation in overhead wires with ground return. *Bell System Technical Journal*, **5**, 539–554.
3. Deri, A, *et al.* (1981). The complex ground return plane, a simplified model for homogeneous and multi-layer earth return. *IEEE Transactions on Power Apparatus and Systems*, PAS-100, 3686–3693.
4. Acha, E (1988). Modelling of power system transformers in the complex conjugate harmonic space. Ph.D. Thesis, University of Canterbury, New Zealand.
5. Semlyen, A and Deri, A, (1985). Time domain modelling of frequency dependent three-phase transmission line impedance. *IEEE Transactions on Power Apparatus and Systems*, PAS-104, 1549–1555.
6. Kimbark, E W, (1950). Electrical Transmission of Power and Signals, John Wiley, New York.
7. Elgerd, O, (1971). Electric Energy Systems Theory: An Introduction, McGraw Hill, New York.
8. Shultz, R D, Smith, R A and Hickey, G L, (1983). Calculation of maximum harmonic currents and voltages on transmission lines, *IEEE Trans*, PAS-102, 817–821.
9. Chen, M S and Dillon, W E, (1874). Power system modelling, *Proc. IEE*, **62**, 901.
10. Arrillaga, J, Arnold C P and Harker, B J, (1983). Computer Modelling of Electrical Power Systems, John Wiley & Sons Ltd, London.
11. Galloway, R H, Shorrocks, W N and Wedepohl, L M, (1964). Calculation of electrical parameters for short and long polyphase transmission lines, *Proc. IEE*, **111**, 2051–2059.
12. Bowman, W I and McNamee, J M, (1964). Development of equivalent PI and T matrix circuits for long untransposed transmission lines, *IEEE Trans*, PAS-84, 625–632.
13. Wilkinson, J H and Reinsch, C, (1971). Handbook for Automatic Computations, Vol II, Linear Algebra, Springer-Verlag, Berlin.
14. Dommel, H W (1978). Line constants of overhead lines and underground cables, Course E.E 553 notes, University of British Columbia.
15. Bianchi, G and Luoni, G, (1976). Induced currents and losses in single-core submarine cables, *IEEE Trans*, **PAS-95**, 49–58.
16. Densem, T J, (1983). Three phase power system harmonic penetration, Ph.D. Thesis, University of Canterbury, New Zealand.
17. Arrillaga, J, Densem, T J and Harker, B J, (1983). Zero sequence harmonic current generation in transmission lines connected to large converter plant, *IEEE Trans*, **PAS-102**, 2357–2363.
18. Hesse, M H, (1966). Circulating currents in parallel untransposed multicircuit lines. I Numerical evaluations. *IEEE Trans*, **PAS-85**, 802–811.
19. *EHV Transmission Line Reference Book*, Edison Electric Institute (1968), New York.
20. Robinson, G H, (1966). Harmonic phenomena associated with the Benmore-Haywards HVdc transmission scheme, *New Zealand Engineer*, **21**, 16–29.
21. Arrillaga, J, *et al.*, (1986). Ineffectiveness of transmission line transpositions at harmonic frequencies, *Proc. IEE*, **123C**(2), 99–104.
22. Semlyen, A, Eggleston, J F and Arrillaga, J, (1987). Admittance matrix model of a synchronous machine for harmonic analysis, *IEEE Trans*, **PWRS-2**(4), 833–840.
23. Arrillaga, J and Duke, R M, (1979). Thyristor-controlled quadrature boosting, *Proc IEE*, **126**(6), 493–498.
24. Stemmler, H and Guth, G, (1982). The thyristor-controlled static phase-shifter, *Brown Boveri Rev*, **69**(3), 73–78.
25. Arrillaga, J, (1983). *High Voltage Direct Current Transmission*, Peter Peregrinus Ltd, London.
26. Ainsworth, J D, (1967). Harmonic instability between controlled static converters and ac networks, *Proc. IEE*, **114**, 949–957.

27. Arrillaga, J, *et al.*, (1988). Ineffectiveness of transmission line VAR compensation at harmonic frequencies. *ICHPS III*, Nashville (Indiana), 233–238.
28. Eggleston, J F, *et al.*, (1988). Derivation of harmonic impedances of the inter-island HVdc link, *IPENZ* (New Zealand), **15**, 1/EMCh, 9–24.
29. Dommel, H W, (1982). *Line Constants Program Manual*, Part of the EMTP package, Bonneville Power Administration.

4

DIRECT HARMONIC SOLUTIONS

4.1 Introduction

When the calculation of the harmonic sources can be decoupled from the analysis of harmonic penetration a direct solution is possible. In such case, the expected voltage levels (or the results of a fundamental frequency load flow) are used to derive the current waveforms of the non-linear components. A Fourier analysis is then applied to obtain the harmonic currents injected by each non-linear component into the linear system.

The simplest model involves a single harmonic source and performs a single phase harmonic analysis. This model is commonly used to derive the system harmonic impedances at the point of common coupling as required in filter design.

In general, the network may contain several harmonic sources and may be unbalanced. The derivation of the harmonic voltages and currents will, therefore, require a three-phase harmonic flow solution.

Most power system non-linearities manifest themselves as harmonic current sources, but sometimes harmonic voltage sources are used to represent the distortion background present in the network prior to the installation of the new non-linear load; moreover, some power electronic applications apply voltage rather than current distortion. A comprehensive algorithm of general applicability should have the following capabilities:

- Model the steady-state phasor response of multi-phase networks to the presence of single or multiple current or voltage sources at harmonic, subharmonic or interharmonic frequencies.
- Represent the individual network elements, assumed linear, by accurate frequency dependent models.
- Provide graphical interfaces for the specification and display of the system to be analysed, and for the post-processing of the information obtained from the analysis.

4.2 Nodal Harmonic Analysis

The distribution of voltage and current harmonics throughout a linear power network containing one or more harmonic current sources is normally carried out using nodal analysis [1].

The nodal admittance matrix of the network at a frequency f is of the form:

$$[Yf] = \begin{bmatrix} Y_{11} & Y_{12} & \cdots & Y_{1i} & \cdots & Y_{1k} & \cdots & Y_{1n} \\ Y_{21} & Y_{22} & \cdots & Y_{2i} & \cdots & Y_{2k} & \cdots & Y_{2n} \\ \vdots & \vdots & \ddots & \vdots & \ddots & \vdots & \ddots & \vdots \\ Y_{i1} & Y_{i2} & \cdots & Y_{ii} & \cdots & Y_{ik} & \cdots & Y_{in} \\ \vdots & \vdots & \ddots & \vdots & \ddots & \vdots & \ddots & \vdots \\ Y_{kl} & Y_{k2} & \cdots & Y_{ki} & \cdots & Y_{kk} & \cdots & Y_{kn} \\ \vdots & \vdots & \ddots & \vdots & \ddots & \vdots & \ddots & \vdots \\ Y_{n1} & Y_{n2} & \cdots & Y_{ni} & \cdots & Y_{nk} & \cdots & Y_{nn} \end{bmatrix}. \quad (4.1)$$

where

Y_{ki} : mutual admittance between busbars k and i at frequency f.

Y_{ii} : self-admittance busbar i at frequency f.

A separate system admittance matrix is generated for each frequency of interest. The main difficulty is to determine which model best represents the various system components at the required frequency and obtain appropriate parameters for them. With this information, it is straight-forward to build up the system fundamental and harmonic frequency admittance matrices.

The three-phase nature of the power system always results in some load or transmission line asymmetry, as well as circuit coupling. These effects give rise to unbalanced self and mutual admittances of the network elements.

For the three-phase system, the elements of the admittance matrix are themselves 3×3 matrices consisting of self and transfer admittances between phases, i.e.

$$[Y_{ii}] = \begin{bmatrix} Y_{aa} & Y_{ab} & Y_{ac} \\ Y_{ba} & Y_{bb} & Y_{bc} \\ Y_{ca} & Y_{cb} & Y_{cc} \end{bmatrix}. \quad (4.2)$$

Figure 4.1 shows a case of two three-phase harmonic sources and an unbalanced ac system. The current injections, i.e. $I_{1h} - I_{3h}$ and $I_{4h} - I_{6h}$, can be unbalanced in magnitude and phase angle.

The system harmonic voltages are calculated by direct solution of the linear equation

$$[I_h] = [Y_h][V_h], \quad (4.3)$$

where $[Y_h]$ is the system admittance matrix.

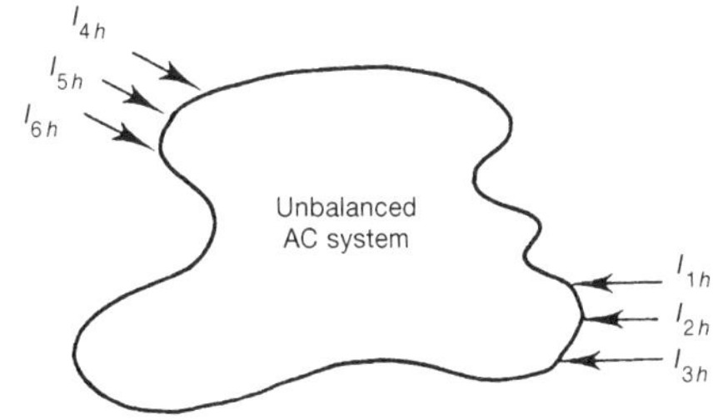

Figure 4.1 Unbalanced current injection into an unbalanced ac system

The direct solution involves $h-1$ independent sets of linear simultaneous equations, i.e.

$$\begin{aligned} [I_h] &= [Y_h] \quad [V_h] \\ &\ddots \quad \ddots \quad \ddots \\ [I_3] &= [Y_3] \quad [V_3] \\ [I_2] &= [Y_2] \quad [V_2]. \end{aligned} \tag{4.4}$$

The injected currents at most ac busbars will be zero, since the sources of the harmonic considered are generally from static converters. To calculate an admittance matrix for the reduced portion of a system comprising of just the injection busbars, the admittance matrix is formed with those buses, at which harmonic injection occurs, ordered last. Advantage is taken of the symmetry and sparsity of the admittance matrix [2], by using a row ordering technique to reduce the amount of off-diagonal element build-up. The matrix is triangulated using Gaussian elimination, down to but excluding the rows of the specified buses.

The resulting matrix equation for an n-node system with $n-j+1$ injection points is

$$\begin{bmatrix} 0 \\ \vdots \\ 0 \\ \hline I_j \\ \vdots \\ I_n \end{bmatrix} = \begin{bmatrix} \diagdown & & 0 & \\ & Y_{jj} & \cdots & Y_{jn} \\ 0 & \vdots & & \vdots \\ & Y_{nj} & \cdots & Y_{nn} \end{bmatrix} \cdot \begin{bmatrix} V_1 \\ \vdots \\ V_{j-1} \\ \hline V_j \\ \vdots \\ V_n \end{bmatrix} \tag{4.5}$$

As a consequence, $I_j \ldots I_n$ remain unchanged since the currents above these in the current vector are zero. The reduced matrix equation is

$$\begin{bmatrix} I_j \\ \vdots \\ I_n \end{bmatrix} = \begin{bmatrix} Y_{jj} & \cdots & Y_{jn} \\ \vdots & \cdots & \vdots \\ Y_{nj} & \cdots & Y_{nn} \end{bmatrix} \cdot \begin{bmatrix} V_j \\ \vdots \\ V_n \end{bmatrix} \tag{4.6}$$

and the order of the admittance matrix is three times the number of injection busbars. The elements are the self- and transfer-admittances of the reduced system as viewed from the injection busbars. Whenever required, the impedance matrix may be obtained for the reduced system by matrix inversion.

Reducing a system to provide an equivalent admittance matrix, as viewed from a specific bus, is an essential part of filter design.

By making $I_1 = \underline{|0^\circ}$ per unit, $I_2 = \underline{|-120^\circ}$ per unit, $I_3 = \underline{|120^\circ}$ per unit, the matrix equation

$$\begin{bmatrix} I_1 \\ I_2 \\ I_3 \end{bmatrix} = \begin{bmatrix} Y_{11} & Y_{12} & Y_{13} \\ Y_{21} & Y_{22} & Y_{23} \\ Y_{31} & Y_{32} & Y_{33} \end{bmatrix} \cdot \begin{bmatrix} V_1 \\ V_2 \\ V_3 \end{bmatrix} \tag{4.7}$$

can be solved for V_1, V_2 and V_3, yielding the following equivalent phase impedances:

$$Z_1 = \frac{V_1}{I_1}, \quad Z_2 = \frac{V_2}{I_2}, \quad Z_3 = \frac{V_3}{I_3}. \tag{4.8}$$

Restricted measurements on the physical network limit the ability to compare a three-phase model with test results. The data obtained from live three-phase systems only includes the phase voltages and currents of the coupled phases. To compare measured and simulated impedances at a current injection busbar it is thus necessary to derive equivalent phase impedances from the 3 x 3 admittance matrix.

4.2.1 Incorporation of Harmonic Voltage Sources

A system containing harmonic voltages at some busbars and harmonic current injections at other busbars is solved by partitioning the admittance matrix and performing a partial inversion. This then allows the unknown busbar voltages and unknown harmonic currents to be found. If V_2 represents the known voltage sources then I_2 are the unknown variables. The remaining busbars are represented as a harmonic current injection I_1 (which can be either zero or specified by harmonic current source) and the corresponding harmonic voltage vector V_1 represent the unknown variables.

Partitioning the matrix equation to separate the two types of nodes gives:-

$$\begin{bmatrix} Y_{11} & Y_{12} \\ Y_{21} & Y_{12} \end{bmatrix} \begin{bmatrix} V_1 \\ V_2 \end{bmatrix} = \begin{bmatrix} I_1 \\ I_2 \end{bmatrix} \tag{4.9}$$

The unknown voltage vector V_1 is found by solving

$$[Y_{11}][V_1] = [I_1] - [Y_{12}][V_2] \tag{4.10}$$

The harmonic currents injected by the harmonic voltage sources are then found by solving:

$$[Y_{21}][V_1] - [Y_{22}][V_2] = [I_2] \tag{4.11}$$

With this formulation some extra processing is required to obtain the reduced admittance matrix, which is not generated as part of the solution.

4.3 Harmonic Impedances

The nodal admittance matrix described in Section 4.2 requires modelling all the transmission lines, generators, transformers and loads connected to the individual nodes. The transmission system has the greatest influence on the parameters of the matrix and special attention has been given to their modelling in Chapter 3.

When calculating harmonic flows throughout the network each node has to be explicitly represented in the analysis. Besides the transmission system, such studies require information on the main plant components involved, in particular, the generators, transformers, distribution systems and loads. Such models are discussed in Sections 4.3.1 to 4.3.3.

Often the purpose of the studies is to determine the voltage harmonic distortion levels at points of common coupling (PCC) between a distorting component and other consumers. In this case, the harmonic equivalent impedances of the relevant individual components are combined into impedance loci or contours; because of their importance in filter design these are given special attention in Section 4.3.5.

4.3.1 Generator and Transformer Modelling

For the purpose of determining the network harmonic admittances the generators and transformer can be modelled as a series combination of resistance and inductive reactance, i.e.

$$Y_{gh} = \frac{1}{R\sqrt{h} + jX_d''h} \quad \text{for the generator} \tag{4.12}$$

$$Y_{th} = \frac{1}{R\sqrt{h} + jX_t h} \quad \text{for the transformer} \tag{4.13}$$

where R is derived from the machine power losses.

X_d'' is the generator subtransient reactance

and X_t is the transformer's short-circuit reactance.

A frequency dependent multiplying factor can be added to the reactance terms to account for skin effect. An example of a typical variation of the inductive coefficient of a transformer with frequency is shown in Figure 4.2.

The magnetising admittance is usually ignored since under normal operating conditions its contribution is not significant. If, however, the transformer is under severe saturation, appropriate current harmonic sources must be added at the transformer terminals; this matter is discussed in Chapter 7.

The modelling of multi-phase transformers as part of a transmission system is discussed in Chapter 3 (Section 3.7).

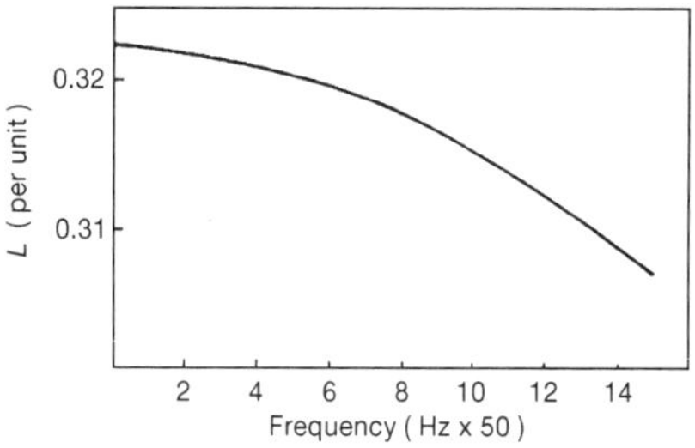

Figure 4.2 Transformer per unit inductance versus frequency

4.3.2 Distribution and Load System Modelling

The harmonic impedances seen from primary transmission system buses are greatly affected by the degree of representation of the distribution system and consumers' loads fed radially from each busbar.

A typical simplified *dominant* configuration of a distribution feeder is shown in Figure 4.3. Generally, the bulk of the load fed from distribution feeders is located behind two transformers downstream. Thus, to calculate the harmonic impedances seen from the high voltage primary transmission side it may be sufficient to use a discrete model of the composite effect of many loads and distribution system lines and transformers at the high voltage side of the main distribution transformers; typically the 110 kV in a system using 400 and 220 kV transmission.

The aggregate nature of the load makes it difficult to establish models based purely on theoretical analysis. Attempts to deduce models from measurements have been made [3, 4] but lack general applicability. Utilities should be encouraged to develop data base of their electrical *regions*, with as much information as possible to provide accurate equivalent harmonic impedances for future studies.

The following guidelines are recommended for the derivation of distribution feeder equivalents [5].

- Distribution lines and cables (e.g. 69–33 kV) should be represented by an equivalent-π model. For short lines, the total capacitance at each voltage level should be estimated and connected at that busbar.

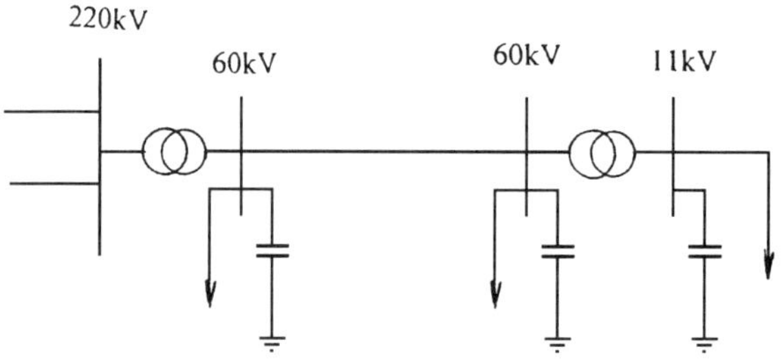

Figure 4.3 Typical distribution feeder

- Transformers between distribution voltage levels should be represented by an equivalent element.
- As the active power absorbed by rotating machines does not correspond to a damping value, the active and reactive power demand at the fundamental frequency may not be used in a straightforward manner. Alternative models for load representation should be used according to their composition and characteristics.
- Power factor correction (PFC) capacitance should be estimated as accurately as possible and allocated at the corresponding voltage level.
- Other elements, such as transmission line inductors, filters and generators, should be represented according to their actual configuration and composition.
- The representation should be more detailed nearer the points of interest. Simpler equivalents for the transmission and distribution systems should be used only for remote points.
- All elements should be uncoupled three-phase branches, including unbalanced phase parameters.

There are no *generally acceptable* load equivalents for harmonic analysis [6]. In each case the derivation of equivalent conductance and susceptance harmonic bandwidths from specified P (active) and Q (reactive) power flows will need extra information on the actual composition of the load. Power distribution companies will have a reasonable idea of the proportion of each type in their system depending on the time of day and should provide such information.

Consumers' loads constitute not only the main element of the damping component but may affect the resonance conditions, particularly at higher frequencies. Some early measurements [7] showed that maximum plant conditions can result in reduced impedances at lower frequencies and increased impedances at higher frequencies. Simulation studies [8] have also demonstrated that the addition of detailed load representation can result in either an increase or decrease in harmonic flow.

Consequently, an adequate representation of the system loads is needed. There are basically three types of loads – passive, motive and power electronic.

(1) Predominantly passive loads (typically domestic) can be represented approximately by a series R, X impedance i.e.

$$Z_r(\omega) = R_r\sqrt{h} + jX_r h \tag{4.14}$$

where R_r = Load resistance at the fundamental frequency.
X_r = Load reactance at the fundamental frequency.
h = Harmonic orders (ω/ω_1)

The weighting coefficient $\sqrt{h}$, used above for frequency dependence of the resistive component, is different in different models; for instance, reference [6] uses a factor of 0.6 $\sqrt{h}$ instead.

The equivalent resistance is estimated from the active power at fundamental frequencies, i.e. $R = V^2/P$ (V being the nominal voltage). The equivalent inductance represents the relatively small motor content when known.

(2) Various models of predominantly motive loads have been suggested using resistive-inductive equivalents, their differences being often due to the boundary of system representation. A detailed analysis of the induction motor response to harmonic frequencies, leading to a relatively simple model, is described in Section 4.3.3.

(3) Modelling the power electronic loads is a more difficult problem because, besides being harmonic sources, these loads do not present a constant *R*, *L*, *C* configuration and their non-linear characteristics cannot fit within the linear harmonic equivalent model. The presence of system non-linearities has been discussed earlier. In the absence of detailed information the power electronic loads are often left open-circuited when calculating harmonic impedances. However, their contribution to the harmonic current flow may have to be considered when the power ratings are relatively high, such as arc furnaces, railway locomotives, etc.

An alternative approach to explicit load representation based on detailed information, is the use of empirical models derived from measurements.

In studies concerning mainly the transmission network the loads are usually equivalent parts of the distribution network, specified by the consumption of active and reactive power. In this situation the load model A suggested by reference [6] can be used. The A model is a parallel connection of inductive reactance and resistance whose values are:

$$X = j\frac{V^2}{(0.1h + 0.9)Q} \qquad R = \frac{V^2}{(0.1h + 0.9)P} \tag{4.15}$$

where P and Q are fundamental frequency active and reactive powers.

When studying the transmission network it is strongly recommended to model at least part of the next lower voltage level and place the load equivalents there. To illustrate the importance of the loading level on the harmonic impedances, Figures 4.4 and 4.5 show the effect of halving the load level on the magnitude and phase of the individual harmonics at a converter bus connected to a 400 kV system [8].

4.3.3 Induction Motor Model

The circuit shown in Figure 4.6(a) is an approximate representation of the induction motor, with the magnetizing impedance ignored.

The motor impedance at any frequency can be expressed as:

$$Z_m(\omega) = R_{mh} + \mathrm{j}X_{mh}. \tag{4.16}$$

At the fundamental frequency ($h = 1$)

$$X_{m1} = X_1 + X_2 = X_B \tag{4.17}$$

$$R_{m1} = R_1 + \frac{R_2}{S} = R_B\left(a + \frac{b}{S}\right) \tag{4.18}$$

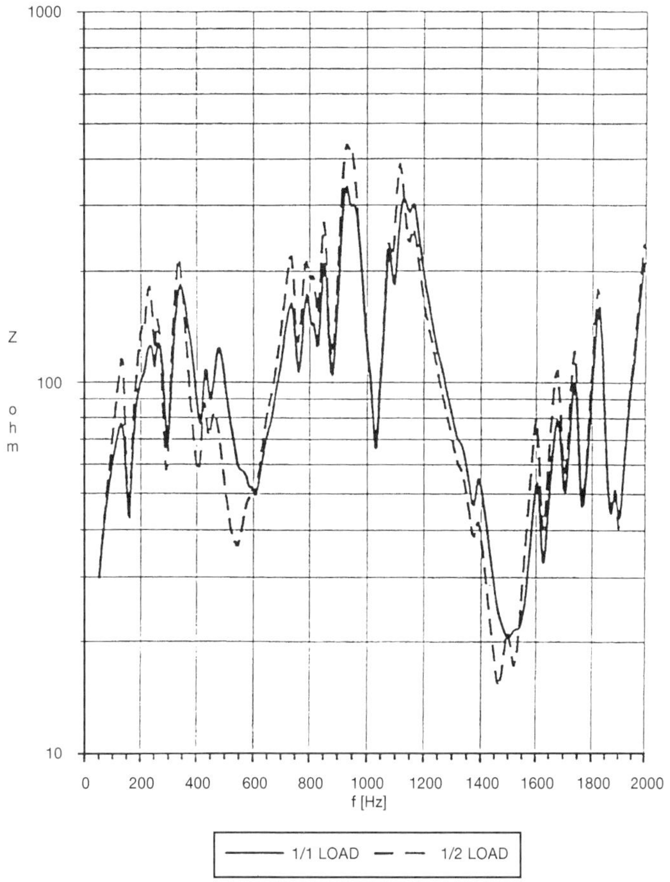

Figure 4.4 Load effect on network harmonic impedance magnitude. Reproduced from [9] by permission of CIGRE

where

R_B = total motor resistance with the rotor locked
R_1 = stator resistance related to R_B by coefficient a (which is typically 0.45)
R_2 = rotor resistance related to R_B by coeffi :ent b (which is typically 0.55)
X_B = total motor reactance with the rotor locked
$S = \text{Slip} = \frac{\omega_s - \omega_r}{\omega_s}$.

At harmonic frequencies:

$$X_{mh} = h.X_B \tag{4.19}$$

$$R_{mh} = R_B\left(a.k_a + \frac{b}{S_h}.k_b\right) \tag{4.20}$$

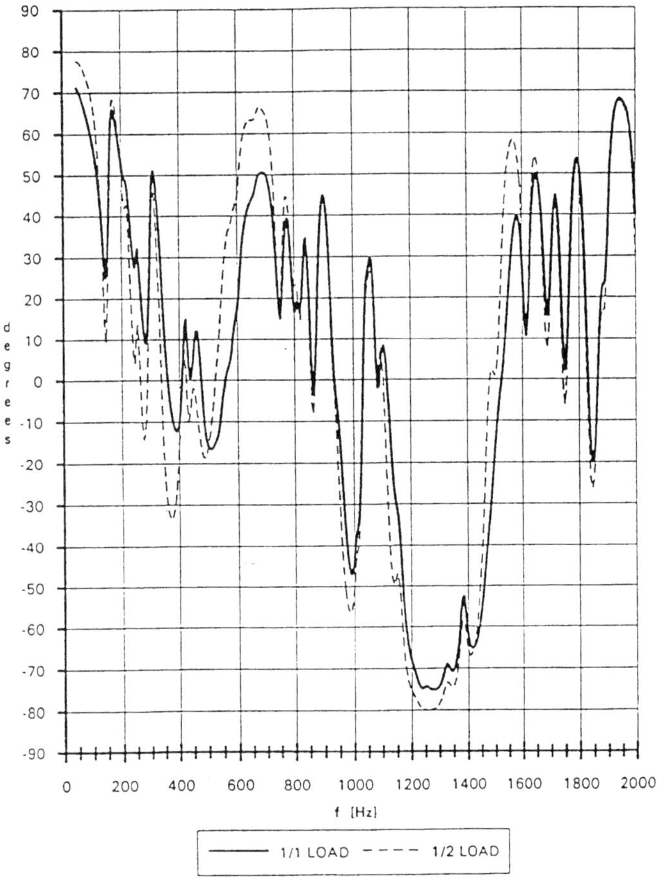

Figure 4.5 Load effect on network harmonic impedance phase. Reproduced from [9] by permission of CIGRE

where

k_a, k_b = correction factors to take into account skin effect in the stator and rotor, respectively.

S_h = apparent slip at the superimposed frequency, i.e.

$$S_h = \frac{\pm h\omega_s - \omega_r}{\pm h\omega_s} \tag{4.21}$$

i.e. $S_h \approx 1 - \dfrac{\omega_r}{h\omega_s}$ for the positive sequence harmonics

$S_h \approx 1 + \dfrac{\omega_r}{h\omega_s}$ for the negative sequence harmonics

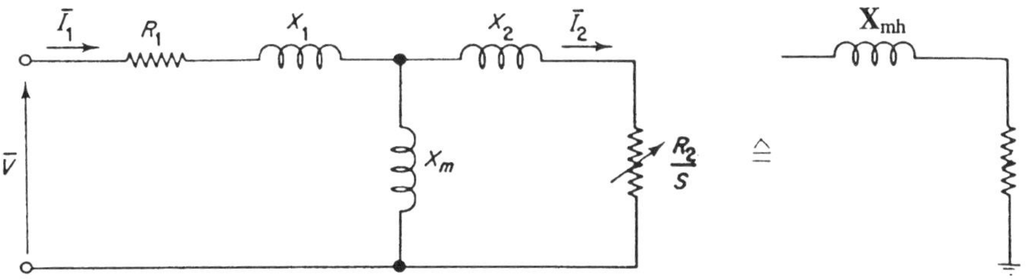

Figure 4.6(a) Approximate representation of the induction motor

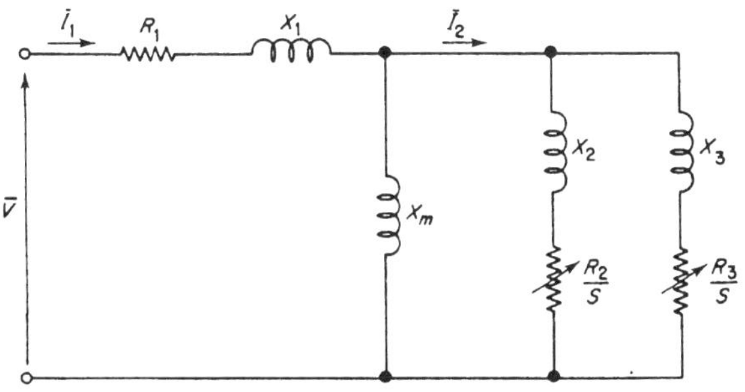

Figure 4.6(b) Accurate induction motor model

assuming an exponential variation of the resistances with frequency, i.e.

$$k_b = h^{\alpha}$$

$$k_b = (\pm h - 1)^{\alpha}$$

the motor equivalent resistance for $\alpha = 0.5$ becomes:

$$R_{mh} = R_B[a\sqrt{h} + (\pm h.b\sqrt{\pm h - 1})/(\pm h - 1)]. \tag{4.22}$$

An accurate model of a double cage induction motor is shown in Figure 4.6(b).

4.3.4 Detail of System Representation

As the system harmonic admittances vary with the network configuration and load patterns, large amounts of data are generated.

Considering the large number of studies involved in filter design, it is prohibitive to represent the whole system with the same degree of detail for every possible operating condition. The detail of components representation depends on their relative position with respect to the harmonic source, as well as their size in comparison with the harmonic source.

Any local plant components such as synchronous compensators, static capacitors and inductors etc., will need to be explicitly represented.

As the high voltage transmission system has relatively low losses, it is also necessary to consider the effect of plant components with large (electric) separation from the harmonic source. It would thus be appropriate to model accurately at least all the primary transmission network (i.e. using the models described in Chapter 3). Moreover, due to the standing wave effect on lines and cables, a very small load connected via a line or cable can have a dramatic influence on the system response at harmonic frequencies.

It is recommended to consider the loads on the secondary transmission network in order to decide whether these should be modelled explicitly or as an equivalent circuit. If these loads are placed directly on the secondary side of the transformer, their damping can be overestimated when using simple equivalents.

Increasing network complexity results in a greater number of resonant frequencies. By way of an illustration [9], Figure 4.7 shows the harmonic impedance at a converter bus of a primary (400 kV) system with either 25, 232 or 1682 buses included; the 25 bus case includes the nearest 400 kV lines terminated by equivalent circuits plus the transformers and large generators in this area. The continuous thick line shows the same information when the network representation consists of 1682 buses which include the complete 400 kV, 220 kV and 110 kV networks plus the generators down to the 1 MVA size; however, it must be emphasized that the number of buses is not the only relevant criterion for increased accuracy. The considerable differences observed are due to the hand made formation of the small network while the large network is produced automatically from the network data base without any equivalencing. Because modern computers can handle the larger network in reasonable times, the larger representation must be recommended as it gives accurate results at any point in the network and only one model for the whole network has to be maintained.

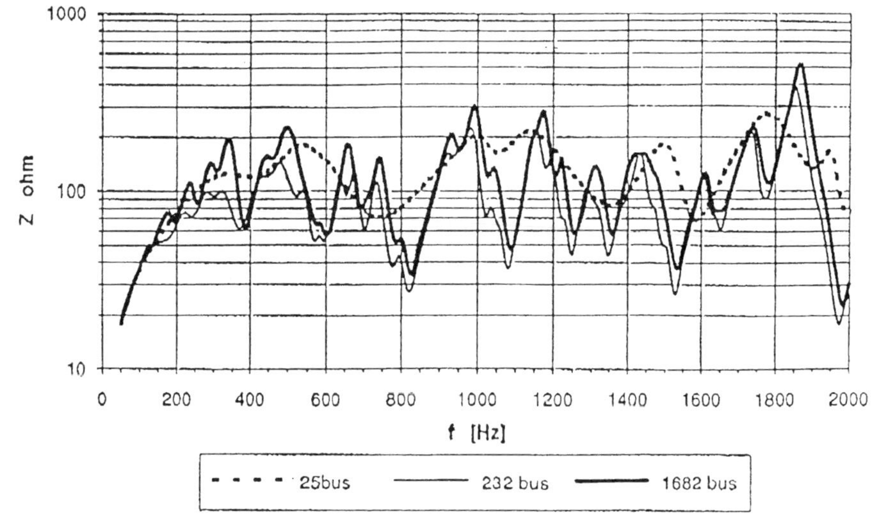

Figure 4.7 Effect of size of system representation. Reproduced from [9] by permission of CIGRE

Radial parts of the system or neighbouring interconnected systems that remain invariant when performing multiple case studies can be replaced by frequency-dependent equivalent circuits [10], or by their reduced harmonic admittance matrices at the point of connection.

4.3.5 System Impedances

The measured or calculated harmonic impedances of a given ac network configuration viewed from the location of a harmonic source are often displayed using a polar impedance/frequency plot or locus. Such loci assume spiral-like shapes which illustrate that the system, although normally inductive at fundamental frequency, changes from inductive to capacitive and back as the frequency increases, producing a number of resonant points at which the system is purely resistive.

With a large transmission system being so sensitive to frequency, calculations made at discrete frequencies could miss an important resonance. Moreover, subsequent fluctuations in frequency and changes in system configuration could shift this resonance on to a harmonic number, with important consequences for harmonic penetration throughout the system.

The computer solution, unlike harmonic measurements, can obtain accurate impedance information at any frequency without the need for interpolation, i.e. it is not restricted to multiples of the fundamental frequency. In practice, each configuration of a three-phase power system will have not one but three loci (if mutual effects are ignored) or a matrix of 3×3 (when mutual effects are included). As an illustration, the complete representation of the harmonic impedances for the system of Figure 4.8 is shown in Figure 4.9. An efficient way of deriving impedance loci is described in Appendix I.

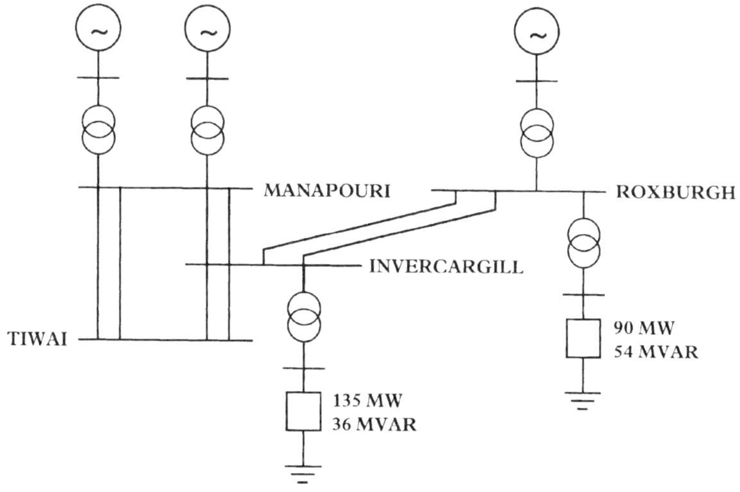

Figure 4.8 The lower South Island of New Zealand test system

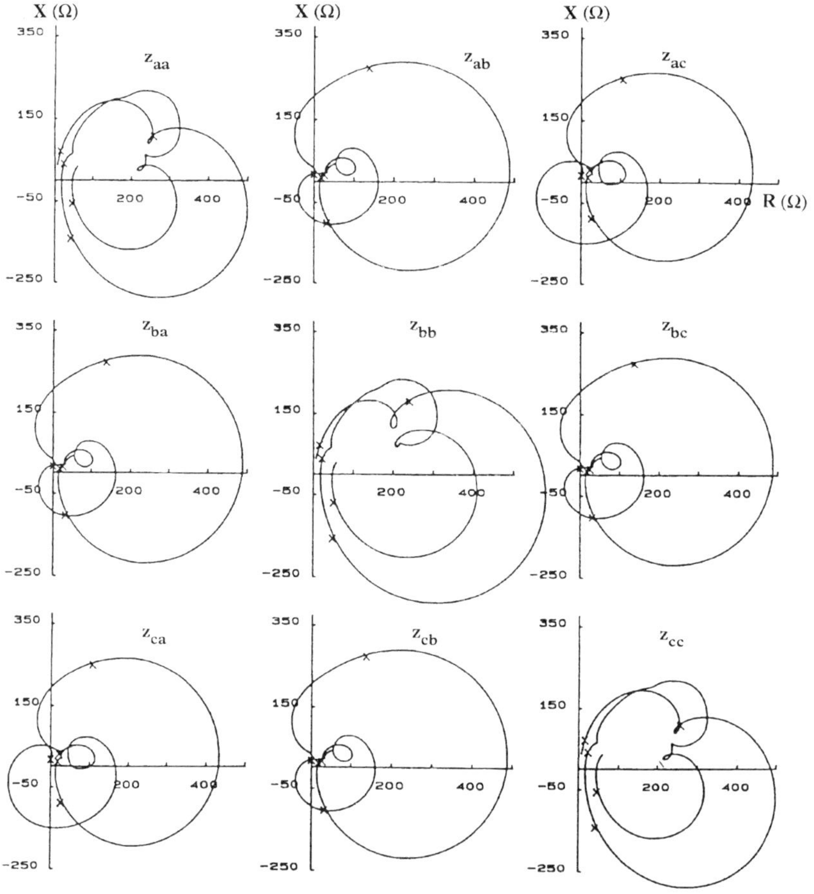

Figure 4.9 Impedance loci matrix of the test system

In the past, an impedance circle [11], as shown in Figure 4.10 encompassing all evaluated harmonic impedances, was used for all harmonics together with computer search techniques which maximized voltage distortion at a specified busbar. This approach leads to unduly pessimistic filter designs, particularly at low order harmonics. Besides, such an approach requires considerable computing and engineering time which is often not available at the tendering stage.

The Annular Sector Concept The annular sector approach, illustrated in Figure 4.11, restricts the geometric area applicable to each harmonic by setting upper and lower limits to the magnitude and phase of the harmonic impedance. Taking into account all the relevant operating conditions, a comprehensive scatter plot is produced for each harmonic on the impedance plane; all these points are then encompassed by two circles and a sector and the resulting values of Z_1, Z_2, θ_1 and θ_2 are tabulated.

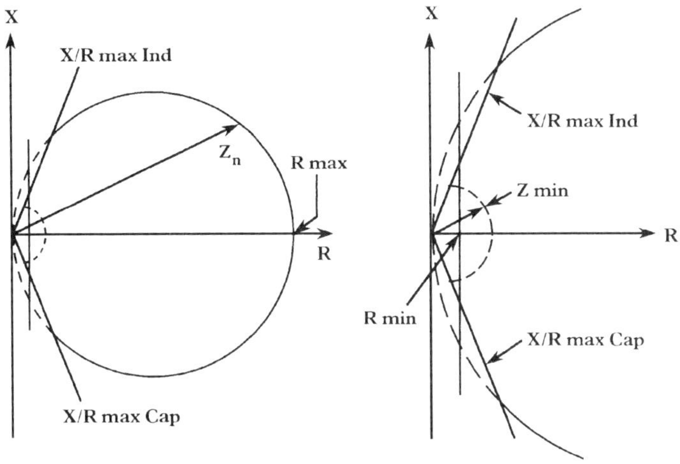

Figure 4.10 Traditional boundary for ac network impedance. Reproduced from [9] by permission of CIGRE

This approach was used in the design of the filters attached to the expansion of the New Zealand HVdc link and the information obtained is shown in Table 4.1.

The Discrete Polygon Concept In this case a distinction is made between low and high harmonic orders. At the lower harmonics discrete points are obtained for the different operating conditions as for the annular sector. Encompassing these points by a polygon results in a set of polygons for each harmonic of interest.

At higher harmonic frequencies, e.g. 14th to 49th, the scatter of the $R \pm jX$ values and hence the boundary of the encompassing polygons would become increasingly large. Additionally, the impedances would begin to extend into the capacitive region of the impedance plane. From detailed information of the particular system involved it is possible to decide on the use of a realistic outer boundary with a single geometrical shape without introducing an unacceptable degree of pessimism into the filter design studies.

A computer technique is then used to search each polygon in turn to evaluate the system impedance which maximizes voltage distortion at, or current injection into, the point of common coupling.

This approach was used in the design of the ac harmonic filters on the 2000 MW Cross-Channel HVdc scheme [4]. Individual search areas were defined for harmonics 1–13 as shown in Figures 4.12, 4.13 and 4.14 for 24 defined operating conditions as follows:

- bus reactor in and out
- filters at nearby busbars in and out
- single circuit outages

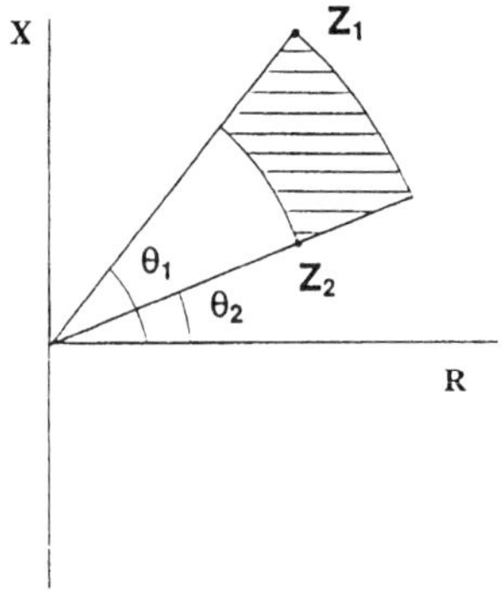

Figure 4.11 The annular sector concept

- double circuit outages
- minimum generating plant
- maximum generating plant.

In the case of the Cross-Channel scheme, a circle of centre $750 + jo\Omega$ and radius 750Ω (as shown in Figure 4.15) was considered sufficient to encompass all possible impedance loci derived from the 24 operating conditions considered. These figures indicate that the first harmonic to exhibit a resonance condition is the 13th, whereas a generalized impedance circle approach would have allowed even low order harmonics (2nd, 3rd) to exhibit resonance.

In this particular application a further refinement was introduced. Having chosen the particular worst (resonance) condition from the polygon search areas, the remaining system impedances for harmonic numbers 2 to 25 were chosen from a number of tables of harmonic impedance, from the column which included the impedance closest to the resonant impedance. For harmonic numbers greater than 25, the network impedance was chosen from the impedance circle of Figure 4.15 to maximise the voltage distortion at each harmonic.

The calculated $R \pm jX$ values used in the polygons are the equivalent Thevenin impedances of the entire network reduced to the Sellindge 400 kV busbar. These include the harmonic impedances of individual plant items such as transmission lines, generators, transformers, etc.

It must be understood that the quantitative impedance plots used in this scheme cannot be taken as typical and used as a default option in other schemes. For instance, in cases of ac networks with long EHV or UHV lines the first resonant frequency may even occur below the second harmonic.

The discrete polygon approach provides a realistic way of representing the ac network for the purposes of ac filter design. It avoids the pessimism of a generalized approach using a single search area, and provides a technique which provides acceptably quick solution times for the highly iterative task of filter design.

Table 4.1 Boundaries of Benmore 220 kV ac system harmonic impedance sectors

Harmonic Order	Z_1 (ohms)	θ_1 (degrees)	Z_2 (ohms)	θ_2 (degrees)
2	47.6	87	19.7	82
3	37.2	77	26.6	30
4	85.7	78	45.5	49
5	71.2	61	29.2	17
6	70.1	58	30.8	29
7	66.60	78	47.6	40
8	114.9	78	55.7	54
9	97.3	63	70.9	24
10	156.9	63	81.1	26
11	168.9	55	109.6	−38
12	93.6	48	24.1	−10
13	121.8	79	51.8	56
14	198.38	55	129.7	−19
15	117.7	43	35.5	−10
16	99.7	51	43.1	24
17	97.7	75	54.7	53
18	140.9	78	84.6	68
19	304.4	82	146.0	68
20	604.2	76	236.4	10
21	657.9	52	163.0	−69
22	291.2	0	65.9	−39
23	128.3	86	37.4	−35
24	146.5	23	72.6	0
25	204.8	71	62.0	58
26	341.9	59	157.8	51
27	525.1	52	200.3	1
28	1319.6	46	381.7	−61
29	460.9	−49	179.0	−72
30	97.0	0	56.0	−34
31	431.3	43	162.8	−34
32	449.8	−14	118.2	−53
33	372.6	5	107.6	−54
34	333.0	−22	82.4	−69
35	130.5	25	41.6	−5
36	238.2	8	102.8	−37
37	368.5	−13	210.3	−65
38	209.8	−0	136.8	−33
39	172.2	−40	114.4	−52
40	169.9	−73	88.7	−76
41	143.1	6	66.2	−12
42	149.6	−69	70.5	−79
43	145.0	0	69.2	−70
44	105.9	−53	52.0	−70
45	73.3	−58	37.5	−70
46	74.9	29	37.1	11
47	136.5	−55	90.6	−63
48	102.2	−65	41.8	−71
49	55.1	−60	28.4	−69
50	43.1	−2	35.6	−22

4.3.6 Existing Non-linearities

When calculating the system harmonic impedances a distinction must be made between linear and non-linear ac plant components. To a greater or lesser extent every electrical component is affected by frequency non-linearities due to skin effect. Such an effect, however, can be accounted for in the equivalent circuit, and for each harmonic study the resulting equivalent circuit is linear.

It will be shown in later chapters that for a given operating condition it is possible to derive harmonic Norton equivalents and, therefore, appropriate harmonic impedances for the non-linear components. Also, the cumulative effect of the existing non-linearities will impose an extra burden on the converter plant (filters in particular) at the bus under investigation.

Traditionally the harmonic currents injected by the converter are increased by some percentage (typically 10%) to take into account the presence of existing harmonics. When the ac system model is based on the impedance/frequency loci it is difficult to do otherwise.

4.4 Computer Implementation

4.4.1 Structure of the Algorithm

The main tasks of a direct solution are shown in Figure 4.16. Blocks (i) to (iii) describe the input data requirements, which will be single or multiphase depending on the information required. Block (iv) is commonly used to assess the voltage harmonic content at the point of common coupling (PCC) of a new non-linear load,

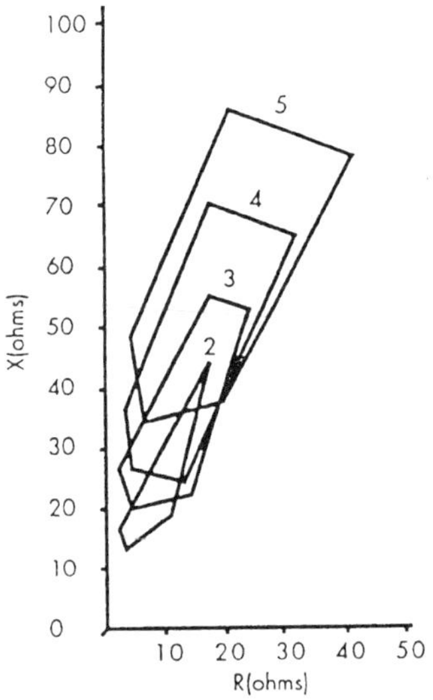

Figure 4.12 Harmonic impedances for harmonic order 2 to 5. Reproduced from [9] by permission of CIGRE

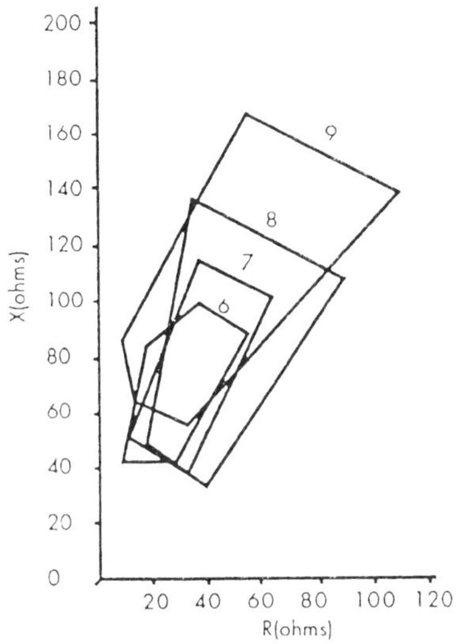

Figure 4.13 Harmonic impedances for harmonic order 6 to 9. Reproduced from [9] by permission of CIGRE

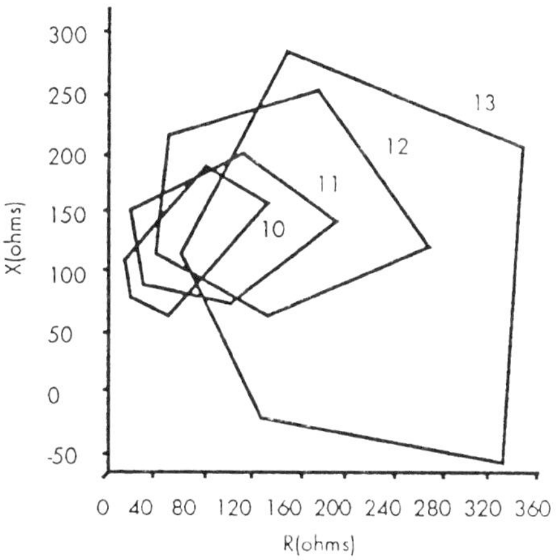

Figure 4.14 Harmonic impedances for harmonic order 10 to 13. Reproduced from [9] by permission of CIGRE

as required for filter design, on the assumption that there are no other harmonic injections in the system.

In single phase studies data entry is normally manual and the harmonic parameters for each component are scaled from the fundamental frequency parameters; therefore only one admittance matrix is constructed and solved at a time. Since the structure of the harmonic admittance matrices is the same at all frequencies, the elements only varying in size, the sparse techniques only require updating the array containing the non-zero elements (but not the pointers), to form the next harmonic admittance.

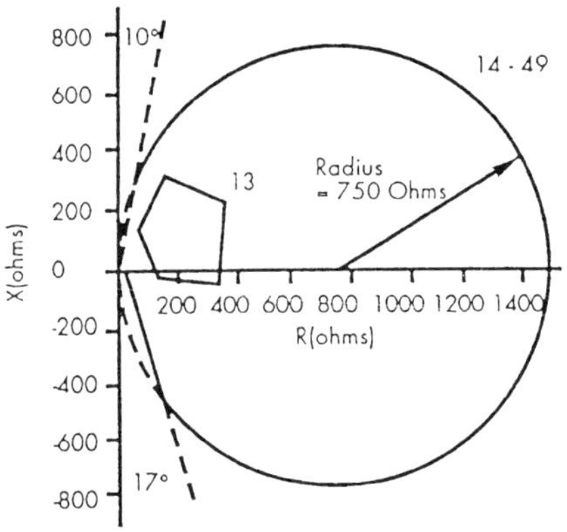

Figure 4.15 Harmonic impedances for harmonic order 13, and envelope of harmonic impedance locii for harmonic order 14 to 49. Reproduced from [9] by permission of CIGRE

When the effect of unbalance, and particularly the presence of zero sequence current, needs to be determined, there is no alternative to three-phase modelling with accurate representation of the frequency dependence of the transmission lines; however, scaled versions of the fundamental frequency parameters are still used for loads, generators and transformers.

Figure 4.17 shows the data files and data flow between a typical set of component programs used in three-phase harmonic analysis. Before describing these programs in detail it will be helpful to make some general observations.

The calculation of the electrical parameters of lines and cables takes up a significant amount of computation. Moreover, the harmonic impedances and levels alter with varying loading and configuration and many studies are normally required. Therefore, the frequency dependent parameters of all components are stored in a file so that they can be used as required for many simulations.

The electrical parameters of branches are calculated for all frequencies and stored in a data file, and all the harmonic admittance matrices are constructed simultaneously. This data is read sequentially once and the electrical parameters are added to the appropriate admittance matrix as it is read. As all admittance matrices are constructed simultaneously, memory constraints become important. The amount of memory required limits the size of the arrays and this, in turn, constrains either the size of the system or the number of harmonics to be modelled. This problem could be avoided if each admittance matrix was read and solved individually. However, in the three-phase case, this would require complicated file handling to allow only the correct frequency data to be incorporated, and file rewind and reread for each harmonic considered.

4.4.2 Data Programs

GIPS Gathering system data used to be a time consuming process, which involved entering numbers in specified columns of a text file. This process is simplified by a graphical entry program such as GIPS (graphical interface for power systems). This

Figure 4.16 Structure diagram of the direct solution

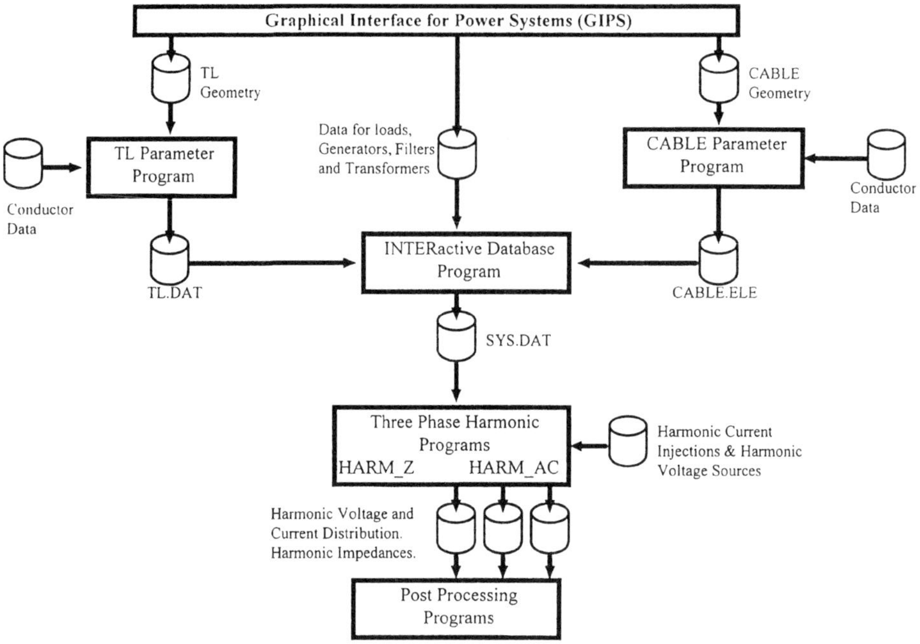

Figure 4.17 Flow diagram of the direct solution components

program collects the physical geometry of transmission lines and cables, as well as the electrical parameters of the various components involved, and writes out files for other data processing programs.

The use of windows facilities makes the graphical interface easy and intuitive to use, due to the users familiarity with the windows environment. Menus are activated with the help of push buttons across the top of the screen, and the various power system components are displayed by icons on the left-hand side of the screen, with the active icon in a depressed state.

Due to the limited amount of information that can be displayed on the screen, it is essential that the drawing area is not limited by the bounds of the screen. This is achieved by providing panning and zooming features. Moreover, due to the size and complexity of a modern power system, it is impractical to have all the system permanently on display, as this hinders visual interpretation of the information. Instead, it is possible to view a part of the system, the selection terms made by voltage level, area, or other user specified criteria.

It is impossible to draw a straight line with the mouse without moving slightly off track; instead, the use of auto-routing and line straightening features allows the lines to be entered quickly and perfectly straight.

Writing drivers for hardware is a tedious and time consuming task. Fortunately, there are now commercially available graphics libraries, and a simple call to a subroutine in the graphics library allows selection of the appropriate built-in drivers. Moreover, new hardware devices can be accommodated quickly and easily. The purchase of an updated version of the graphics library allows immediate use of new printers, graphics cards and plotters, with no changes required by the graphical interface program. Another advantage is that the graphics library is available for different platforms, hence allowing portability.

Easy entry of power system data is achieved by using pop-up windows, called FORMs, which appear when a component is selected for editing. The FORM is a table with the left column being a protected field identifying the data required, and the right column being available for the user to enter the data. A help field in the FORM also gives useful information to the user. In a power system there are normally many components that are identical, and entering their parameters individually would be laborious, therefore the ability to copy data from one to another with a few key pushes is an important feature.

The modelling of multiple transmission lines with different terminations is achieved by using multi-node junctions. Without these the graphical interface program assumes that all circuits are terminated on the same busbar or junction. With multi-node junctions the FORMs for all branches will have additional pop-up FORMs allowing specification of internal node connections. It is also necessary to modify the system diagram and data to carry out repetitive studies with individual components taken out of service or reinstated, without re-entering the data. This is simply achieved by selecting an out of service command, which is indicated by a colour change in the selected component. While the component FORMs allow system data modifications with great ease, modifying the system diagram is slightly more complicated. Any component can be deleted at any time and the connectors linking the deleted component to a busbar or junction will be deleted also. Components can be picked up and moved around. However, connectors will not move with them, they must be deleted and re-entered. As there is no data associated with connectors this is a trivial task. *Dragging* and *rubber-banding* are only used in drawing transmission lines, cables and connectors, as the extra code needed to extend their use to other components is not warranted.

Although the graphical interface could have been added directly to the harmonic analysis software, it is desirable to keep it separate, as shown in Figure 4.17. This is because the software code required to draw the system and handle all the data requires a large amount of memory by itself. If incorporated as part of the analysis program it would further restrict the memory available and hence the size of system that can be analysed. Another benefit is that it is easier to maintain groups of smaller programs than one large monolithic program.

GIPS stores all the data and drawing instructions collected from the User in one file under a name specified by the User. A typical screen is shown in Figure 4.18. When data entry is completed GIPS creates the three files required for harmonic analysis (default names are LINE.DAT, CABLE.DAT and COMP.DAT).

The geometry and conductor codes of the transmission lines are stored in the LINE.DAT files, an example of which for the system drawn in Figure 4.18 is:

```
MANAPOURI220 INVERCARG220
 0 0 1 1 1 1 1 1 152.90  50  5 1   0.00  12.50   31 220  1 2 1    10.1967
 6.47  0.00  0.00  0.00 -6.47  0.00  4.61  5.91    RYB    RYB    ASS
```

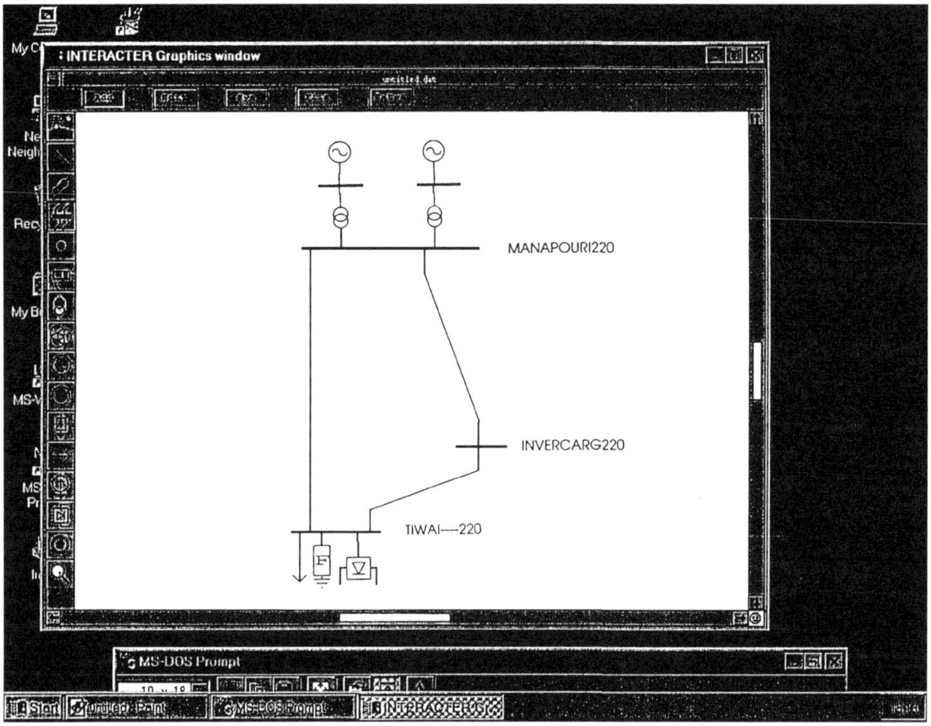

Figure 4.18 Sample GIPS screen display

```
INVERCARG220 TIWAI——220
 0 0 1 1 1 1 1 1 24.30   50 5 1   0.00 12.50   0  220  1 9 1   1.1973
 7.20 0.00 0.00 0.00 −7.20 0.00 0.00 0.00     RYB     RYB       ASS

TIWAI——220 MANAPOURI220
 0 0 1 1 1 1 1 1 175.60  50 5 1   0.00 12.50   0  220  1 9 1   5.1972
 7.20 0.00 0.00 0.00 −7.20 0.00 0.00 0.00     RYB     RYB       ASS
```

Similarly, the geometry and conductor loads of the cables are stored in the CABLE.DAT files, an example of which is:

```
C          Cable parameter file accessed from Cable.exe

2 left @        bottom      right $        bottom
   1     0      0     1     1     2
   1.00000      1.00000      1.00000      2.00000      1.00000      3.00000
   1.00000      4.00000      1.00000      5.00000      1.00000      6.00000
   100.0000     1.0000       1.0000       110.0000     2.1996

1 left          bottom
   0     1      0     0     3     1
   2.00000      1.00000      2.00000      2.00000      2.00000      3.00000
   0.0100       1.0000       10.0000      220.0000     2.1996

File End
```

Table 4.2 Sample Conductor data file (CONDUCT.DAT)

Conductor stranding	Name	Diameter (mm)	Aluminium equiv. cross section (mm^2)	GMR (mm)	AC resistance at 20°C (Ω/km)	Current ratings (A) at (air/conductor) temperatures (°C)					
						30/50	20/50	30/70	20/70	30/75	20/75
84/3.70 + 19/2.22 ACSR	CHUKAR	40.69	875	16.34	0.03134	1020	1250	1410	1570	1490	1640
54/4.36 + 19/2.62 ACSR	SPECIAL	39.24	785	15.76	0.03819	955	1165	1320	1470	1395	1530
76/3.72 + 7/2.89 ACSR	SPECIAL	38.40	794	15.39	0.03511	960	1175	1330	1480	1400	1540
54/3.90 + 19/2.34 ACSR	PHEASANT	35.10	628	14.20	0.04491	825	1000	1140	1270	1200	1320
54/3.18 + 7/3.18 ACSR	ZEBRA	28.58	400	11.55	0.07009	610	750	845	940	890	980
30/3.71 + 7/3.71 ACSR	GOAT	25.96	300	10.70	0.09010	525	640	725	805	765	840
30/3.00 + 7/3.00 ACSR	PANTHER	20.98	200	8.76	0.13790	405	495	560	620	590	650
30/2.59 + 7/2.59 ACSR	WOLF	18.14	150	7.47	0.18450	330	400	455	505	475	525
26/2.57 + 7/2.00 ACSR	PARTRIDGE	16.31	125	6.61	0.21320	300	365	410	460	435	480
26/2.54 + 7/1.91 ACSR	COYOTE	15.88	125	6.40	0.22120	295	360	405	450	425	470
16/2.86 + 19/2.48 ACSR	BRAHMA	18.14	100	3.17	0.29170	270	330	370	415	390	430
7/4.39 + 7/1.93 ACSR	HYENA	14.58	100	2.48	0.30700	255	315	355	395	370	410
6/4.72 + 1/4.72 ACSR	HARE	14.17	100	2.44	0.30800	255	310	350	390	370	405
6/4.72 + 7/1.57 ACSR	DOG	14.17	100	2.44	0.30570	255	310	350	390	370	405
6/4.25 + 1/4.25 ACSR	PIGEON	12.75	80	1.83	0.33800	220	270	300	340	320	355
12/2.59 + 7/2.59 ACSR	SKUNK	12.95	60	1.83	0.45690	190	235	265	295	275	305
6/3.66 + 1/3.66 ACSR	MINK	10.97	60	1.49	0.47290	180	220	250	280	265	290
37/3.66 COPPER	COPPER	25.60	600	9.83	0.04600	720	880	995	1110	1050	1160
61/2.62 COPPER	COPPER	23.55	500	9.09	0.05450	650	790	895	995	940	1040
37/2.62 COPPER	COPPER	18.31	300	7.03	0.09000	470	570	645	720	680	750

continued

Table 4.2 *continued*

Conductor stranding	Name	Diameter (mm)	Aluminium equiv. cross section (mm^2)	GMR (mm)	AC resistance at 20°C (Ω/km)	Current ratings (A) at (air/conductor) temperatures (°C)					
						30/50	20/50	30/70	20/70	30/75	20/75
7/4.72 H.D. ALUMINIUM	WEKE	14.16	100	5.14	0.23280	275	335	380	420	400	440
19/4.22 H.D. ALUMINIUM	COCKROACH	21.10	250	7.99	0.10830	455	555	625	700	660	730
72/4.41 + 7/2.94 ACSR	KIWI	44.07	966	17.37	0.02956	1160	1410	1600	1780	1690	1860
6/3.35 + 1/3.35 ACSR	RABBIT	10.06	52	1.37	0.54040	165	200	225	240	235	250
6/3.00 + 1/3.00 ACSR	ROBIN	9.02	40		0.67400	140	175	195	215	205	225

Table 4.3 Sample cable data file (CABLE.INF)

```
C  This is the data input file for the physical and electrical parameters of the different cable
types which could be called when the cable parameter program is run.
TIDDLY CABLE 1                    1
0.01            0.02              0.04       0.045            0.05       0.058
        0.07
1.68E-081.               2.0E-07            1.            1.8E-07          1.
4.1             0.                1.         2.3              0.         1.
        1.               0.                 1.

MEGA CABLE   5                    2
0.001           0.022             0.0395     0.044            0.0475     0.0583
        0.0653
1.68E-081.               2.20E-071.          1.8E-07          1.
4.1             0.                1.         2.3              0.         1.
        1.               0.                 1.

SUPER CABLE  5                    3
0.001           0.022             0.043      0.044            0.0475     0.0583
        0.0653
1.68E-081.               2.20E-071.          1.8E-07          1.
4.1             .5                1.         2.3              0.     1.
        1.               0.                 1.

SUPER CABLE2 5                    4
0.001           0.022             0.043      0.044            0.0475     0.0583
        0.0653
1.68E-081.               2.20E-071.          1.8E-07          1.
4.1             .5                1.         2.3              0.         1.
        1.               0.                 1.

SUPER CABLE2 5                    5
0.001           0.022             0.043      0.044            0.0475     0.0583
        0.0653
1.68E-081.               2.20E-071.          1.8E-07          1.
4.1             .5                1.         2.3              0.         1.
        1.               0.                 1.
```

FILE END
The cable entry number is to enable cable.exe to use a data input file created by GIPS as GIPS cannot at present pass the cable type strings through to the output files. The cable entry numbers do not need to be in order as Cable.exe scans the entire file.

```
Cable Type      Case Number      Cable Entry Number
Cond-Inner(m)   Cond-Outer       Sheath-Inner       Sheath-Outer     Armour-Inner      Armour-Outer
        Cable-Radius
Con-res(ohm-m)Con-permeabil She-resistivity     She-permeabil    Arm-resistivity   She-permeabil
c2s dielectr    c2s loss angle   c2s permeabil      s2a dielectric   s2a loss angle    s2a permeabil
        a2ext dielec        a2ext loss ang        a2ext permea
```

Case 1 = Conductor and insulator
Case 2 = Conductor, insulator and sheath
Case 3 = Conductor, insulator, sheath and insulator
Case 4 = Conductor, insulator, sheath, insulator and armour
Case 5 = Conductor, insulator, sheath, insulator, armour and insulator

TL The Transmission Line program uses data contained in two files; the line geometry and conductor code file created by GIPS (default name LINE.DAT) and a look-up table giving details of the conductors (called CONDUCT.DAT). The CONDUCT.DAT file is shown in Table 4.2.

Taking into account line geometry and conductor types, this program calculates the electrical parameters of the transmission lines at all relevant frequencies using the equivalent PI model. The results are used to create an output file called TL.DAT. As described in Chapter 3, the TL program models skin effect and earth return as well as different three-phase line geometries with one or more earth wires.

Where a transmission line consists of several sections with different geometry, TL calculates the electrical parameters for each section independently. The harmonic analysis programs read each section data and convert it to ABCD parameters; then the sections are cascaded by multiplying them together to obtain one equivalent PI for the complete length. To save space TL.DAT is an unformatted file and hence a sample of it cannot be shown.

CABLE The CABLE program uses data contained in two files;

- The position of the cables in the ground is contained in a file created by GIPS (default name CABLE.DAT).
- Cable data such as permittivity of insulation and radius's of layers in cable is contained in a look-up table (called CABLE.INF). A sample file is shown in Table 4.3.

Taking into account cable geometry and conductor types, this program calculates the electrical parameters of the cable at all relevant frequencies in a form applicable to equivalent PI model, as described in Chapter 3. The CABLE program models skin effect and earth return. The electrical parameters are written to an output file (called CABLE.ELE).

INTER Considering the large amount of data required to perform three-phase studies it is necessary to use an auxiliary program for adequate preparation of the data in order to minimise errors. This is the purpose of INTER, a program used for the interactive data preparation.

INTER reads the file TL.DAT, previously generated by the TL program, the file CABLE.DAT, previously generated by the CABLE program, and the components data contained in a COMP.DAT file and creates interactively a large file called SYS.DAT containing all the system data. INTER also permits the User to add or eliminate components or eliminate lines and cables. This permits carrying out small modifications without the need to recalculate all the electrical parameters of the lines and cables.

Examples of the SYS.DAT file for the first three harmonic orders of the system of Figure 4.18 follow:

```
TITLE LINE 1
TITLE LINE 2
33.333000 0.000000 40 1 10000000000 025
TIWAI——220 BB 1.000 0.00 166.60 50.00 166.60 50.00 166.60 50.00 1 1
MANAPOURI220 BB 1.000 0.00 0.00 0.00 0.00 0.00 0.00 0.00 0 1
```

```
MANAPOUR1014 BB 1.000 0.00 0.00 0.00 0.00 0.00 0.00 0.00 0 1
INVERCARG220 BB 1.000 0.00 0.00 0.00 0.00 0.00 0.00 0.00 0 1
END BUS
MANAPOUR1014 GG MANAPOUR1014 0.0000 0.0200 0.0000 0.0400 0.0000 0.0370 0.0000
0.0000 0.00 END GEN
TIWAI——220 FF 33.0000
1 1 0.2490 3.6000 23.2000 0.0000 0.0000 11 0
TIWAI——220 FF 33.0000
1 1 0.3050 3.7500 16.1000 0.0000 0.0000 13 0
TIWAI——220 FF 33.0000
2 1 4.5000 0.6000 32.1000 45.0000 0.0000 15 0
END FIL
MANAPOURI220 MANAPOUR1014
                      2 1 0
1TT 2 0.
0 0 0.00060000 0.02690000
0.02500.0000 4
MANAPOURI220 INVERCARG220
                      2 1 25
1LL 2 50.
1 1 0.04578067 0.21457702 1 2 0.02389826 0.08929160 1 3 0.02348670 0.07587518
2 1 0.02389826 0.08929160 2 2 0.04641612 0.21410112 2 3 0.02389826 0.08929160
3 1 0.02348670 0.07587518 3 2 0.02389826 0.08929160 3 3 0.04578067 0.21457702
1 1 0.00011238 0.18626426 1 2 0.00002561 -0.02567649 1 3 0.00004093 -0.00885933
2 1 0.00002561 -0.02567650 2 2 0.00011081 0.19061552 2 3 0.00002561 -0.02567650
3 1 0.00004093 -0.00885934 3 2 0.00002561 -0.02567649 3 3 0.00011238 0.18626428
1LL 2 100.
1 1 0.06514066 0.39519861 1 2 0.04349704 0.14754842 1 3 0.04256839 0.12173259
2 1 0.04349704 0.14754842 2 2 0.06644361 0.39331022 2 3 0.04349704 0.14754842
3 1 0.04256839 0.12173259 3 2 0.04349704 0.14754842 3 3 0.06514066 0.39519861
1 1 0.00064294 0.37565178 1 2 0.00026900 -0.05114302 1 3 0.00034065 -0.01726904
2 1 0.00026900 -0.05114302 2 2 0.00061635 0.38431168 2 3 0.00026900 -0.05114302
3 1 0.00034065 -0.01726904 3 2 0.00026900 -0.05114302 3 3 0.00064294 0.37565181
1LL 2 150.
1 1 0.07431545 0.55574954 1 2 0.05277996 0.19225056 1 3 0.05161515 0.15505637
2 1 0.05277997 0.19225056 2 2 0.07580727 0.55224729 2 3 0.05277997 0.19225056
3 1 0.05161515 0.15505637 3 2 0.05277996 0.19225056 3 3 0.07431545 0.55574954
1 1 0.00181339 0.57127279 1 2 0.00089105 -0.07635127 1 3 0.00107668 -0.02493453
2 1 0.00089105 -0.07635127 2 2 0.00171399 0.58417124 2 3 0.00089105 -0.07635128
3 1 0.00107668 -0.02493452 3 2 0.00089105 -0.07635127 3 3 0.00181339 0.57127279
INVERCARG220 TIWAI——220
2 1 25
1LL 2 50.
1 1 0.00595909 0.03570833 1 2 0.00240662 0.01541142 1 3 0.00240666 0.01322512
2 1 0.00240662 0.01541142 2 2 0.00595912 0.03570830 2 3 0.00240662 0.01541142
3 1 0.00240666 0.01322512 3 2 0.00240662 0.01541142 3 3 0.00595909 0.03570833
1 1 0.00000034 0.02845132 1 2 0.00000001 -0.00457417 1 3 0.00000007 -0.00186026
2 1 0.00000001 -0.00457417 2 2 0.00000034 0.02906509 2 3 0.00000001 -0.00457417
3 1 0.00000007 -0.00186026 3 2 0.00000001 -0.00457417 3 3 0.00000034 0.02845132
1LL 2 100.
1 1 0.00840285 0.06925459 1 2 0.00475259 0.02867422 1 3 0.00475277 0.02430333
2 1 0.00475259 0.02867422 2 2 0.00840292 0.06925442 2 3 0.00475259 0.02867422
3 1 0.00475277 0.02430333 3 2 0.00475259 0.02867422 3 3 0.00840285 0.06925459
1 1 0.00000179 0.05691461 1 2 0.00000039 -0.00914771 1 3 0.00000067 -0.00371888
2 1 0.00000039 -0.00914771 2 2 0.00000171 0.05814191 2 3 0.00000039 -0.00914771
3 1 0.00000067 -0.00371888 3 2 0.00000039 -0.00914771 3 3 0.00000179 0.05691462
```

```
1LL 2 150.
1 1 0.01085802 0.10194825 1 2 0.00705311 0.04111161 1 3 0.00705355 0.03455950
2 1 0.00705311 0.04111161 2 2 0.01085814 0.10194774 2 3 0.00705311 0.04111161
3 1 0.00705355 0.03455950 3 2 0.00705311 0.04111161 3 3 0.01085802 0.10194825
1 1 0.00000497 0.08540143 1 2 0.00000161 -0.01372043 1 3 0.00000234 -0.00557466
2 1 0.00000161 -0.01372043 2 2 0.00000465 0.08724183 2 3 0.00000161 -0.01372043
3 1 0.00000234 -0.00557466 3 2 0.00000161 -0.01372043 3 3 0.00000497 0.08540142
TIWAI——220 MANAPOURI220
2 1 25
1LL 2 50.
1 1 0.04225127 0.25567842 1 2 0.01687885 0.10981817 1 3 0.01687820 0.09412731
2 1 0.01687885 0.10981817 2 2 0.04226027 0.25566909 2 3 0.01687885 0.10981817
3 1 0.01687820 0.09412731 3 2 0.01687885 0.10981817 3 3 0.04225127 0.25567842
1 1 0.00013120 0.20636229 1 2 0.00000516 -0.03299590 1 3 0.00002781 -0.01331885
2 1 0.00000516 -0.03299590 2 2 0.00013033 0.21077853 2 3 0.00000516 -0.03299590
3 1 0.00002781 -0.01331885 3 2 0.00000516 -0.03299589 3 3 0.00013120 0.20636232
1LL 2 100.
1 1 0.05609922 0.48290238 1 2 0.03093567 0.19627479 1 3 0.03093806 0.16552891
2 1 0.03093567 0.19627479 2 2 0.05612547 0.48284081 2 3 0.03093567 0.19627479
3 1 0.03093806 0.16552891 3 2 0.03093567 0.19627479 3 3 0.05609922 0.48290238
1 1 0.00070339 0.41735372 1 2 0.00015925 -0.06572535 1 3 0.00026892 -0.02597622
2 1 0.00015925 -0.06572535 2 2 0.00067195 0.42608768 2 3 0.00015925 -0.06572535
3 1 0.00026892 -0.02597622 3 2 0.00015925 -0.06572534 3 3 0.00070339 0.41735381
1LL 2 150.
1 1 0.06525607 0.68069899 1 2 0.04056496 0.26332098 1 3 0.04057887 0.21875937
2 1 0.04056496 0.26332098 2 2 0.06529872 0.68051624 2 3 0.04056496 0.26332098
3 1 0.04057887 0.21875937 3 2 0.04056496 0.26332098 3 3 0.06525607 0.68069899
1 1 0.00207110 0.63787633 1 2 0.00071223 -0.09795913 1 3 0.00101316 -0.03732961
2 1 0.00071223 -0.09795913 2 2 0.00192903 0.65073633 2 3 0.00071223 -0.09795912
3 1 0.00101316 -0.03732961 3 2 0.00071223 -0.09795912 3 3 0.00207110 0.63787627
END DATA
```

4.4.3 Applications Programs

HARM_Z This program reads the SYS.DAT file produced by INTER and creates a series of files with the information required for the analysis of the system harmonic impedances. The specific files produced are:

- ZFULL.DAT, with the values of the system harmonic impedances seen from a specified bus.
- ZSEQ.DAT, with the values of the sequence components of the system harmonic impedances seen from a specified bus.
- ZRED.DAT, with the values of the system harmonic impedances seen from a specified bus and reduced to only phase values assuming balanced currents.
- HYCONV.DAT, with the values of the system admittance matrix seen from a specified bus.

HARM_AC This program also reads the SYS.DAT file, and creates a series of files with the information required to carry out current and voltage harmonic penetration studies. The specific files produced are:

- HVC.DAT, with the values of the harmonic voltages (in phases *R*, *Y*, *B* as well as their sequence components) present at each of the system buses.

- HLC.DAT, with the values of the harmonic currents (in phases *R*, *Y*, *B* as well as their sequence components) present in the system

- BRNCURR.DAT, with an ordered list of branches for indicating the sensitivity of the result to branch outage.

HARMAC HARMAC creates the same files that both HARM_Z and HARM_AC create. Although HARMAC only allows harmonic current injections, and cannot handle as large a system as HARM_Z and HARM_AC, it is ideal for implementing more sophisticated models requiring the calculation of the system harmonic admittance and harmonic penetration in the same study.

An example is the accurate assessment of HVdc system harmonics, with the converter represented as a harmonic Norton equivalent, the Norton admittance being itself a function of the ac system harmonic impedance. In this case, a two-pass approach is adopted, the first pass being the derivation of the system harmonic impedances seen from the converter terminal bus, the second pass is a harmonic penetration study incorporating the converter's harmonic admittances (derived in Chapter 10).

4.4.4 Post-Processing

As mentioned previously, the large amount of information generated, particularly in three phase harmonic analysis, necessitates interpretation using graphics, as there are too many tables of numbers to be displayed. Maintaining graphics software is by nature difficult, as the rapid advances in computer technology quickly makes graphics software obsolete. Added to this is the difficulty in maintaining and updating complex software. It is thus recommended to make use of commercially available programs where possible, as well as to make use of commercially available graphics and numerical libraries when developing code.

Some post-processing programs use calls to commercially available graphics libraries for FORTRAN, others use C and WINDOWS.

In another approach, data formed by the HARMAC, HARM_Z and HARM_AC programs is converted into a MATLAB format. This creates the following files:

z-mat For a specified branch obtains the values of the harmonic impedances, resistances and reactances, as well as the impedance locus.

v-mat Contains the phase and sequence values of the harmonic voltages at all the network busbars.

i-mat Contains the phase and sequence values of the harmonic currents of each line.

MATLAB is then run, incorporating the output GIPS software such that a menu of options is displayed. The post-processing software should have the following features:

- Perform a variety of display formats
  - 2-D plots (X–Y plots)
  - 3-D plots
  - Loci plots
  - Scatter plots.
- View harmonics of selected parts of the system with selectable ranges.
- Plot and compare data from different files (simulations).
- View both the phase and sequence quantities.
- Calculate the profile of harmonics along the branches.
- Apply curve fitting and interpolation techniques.
- Plot both equally spaced data and X–Y pairs.
- Perform simple calculations on the harmonic data.
- Have flexibility to drive various printers/plotters.
- Be easy to use, maintain and update.

Although the 3-D plots are useful in getting an overview of the harmonic levels throughout the system it is difficult to get quantitative information from them. The 2-D plots are more informative in this respect. Harmonic impedance information can be displayed in the form of impedance loci, as explained in Section 4.3.5.

Figures 4.19 to 4.22 offer a selection of graphic displays illustrating the post processing facilities described above.

4.5 Summary

Assuming perfectly linear network components, other than those specified as the distortion sources, this chapter has described the state of the art for the analysis of currents and voltages in the network at the frequencies injected by the distorting source or sources. Although the purpose of the book is harmonic analysis, the direct solution described is equally applicable to sub-harmonic and inter-harmonic frequencies.

Because of their greater influence at frequencies other than the fundamental, the modelling of transmission lines and cables has been given special consideration in Chapter 3.

The computer implementation has been divided in two parts, to achieve more efficient solutions for their respective applications. These are HARM_Z, the calculation of harmonic impedances from any selected bus and HARM_AC, the

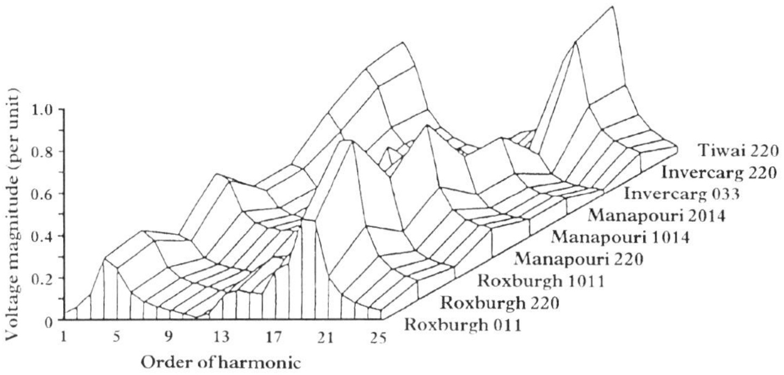

Figure 4.19 Three-dimensional plot of the positive sequence harmonic voltage magnitudes at each busbar of the system of Figure 4.8

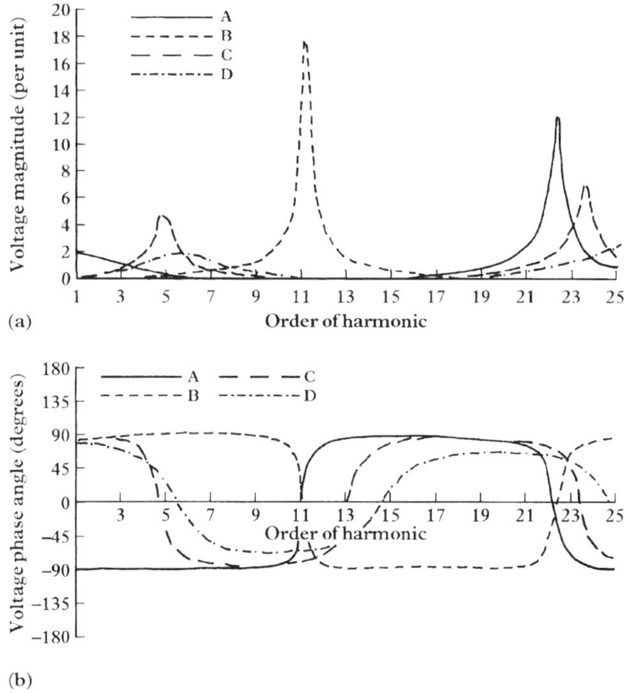

Figure 4.20 Two-dimensional plots of harmonic voltage magnitude (a) and phase (b) for the system of Figure 4.8 at the Invercargill busbar.
A—with Roxborough open circuited. B—with Roxborough short-circuited.
C—with Roxborough with load and generation

penetration of harmonic currents and voltages, also known as harmonic flow. The former constitute the basis of filter design while the latter are needed to assess the harmonic levels throughout the system. Both algorithms can be used by themselves, as described by the direct solution in this chapter, or as part of more elaborate iterative solutions as described in later chapters.

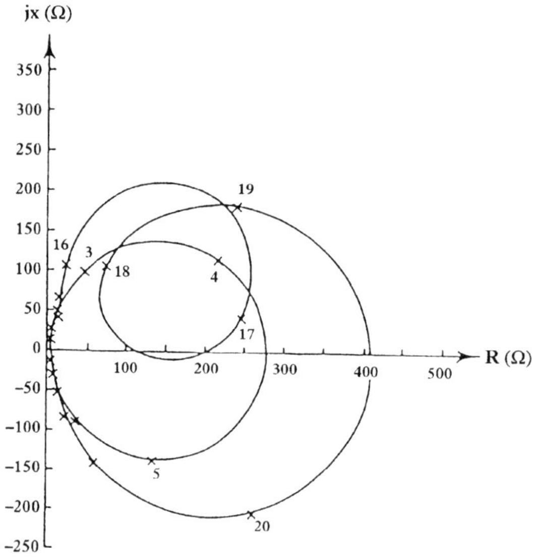

Figure 4.21 Impedance locus plot of the test system of Figure 4.8 seen from Tiwai

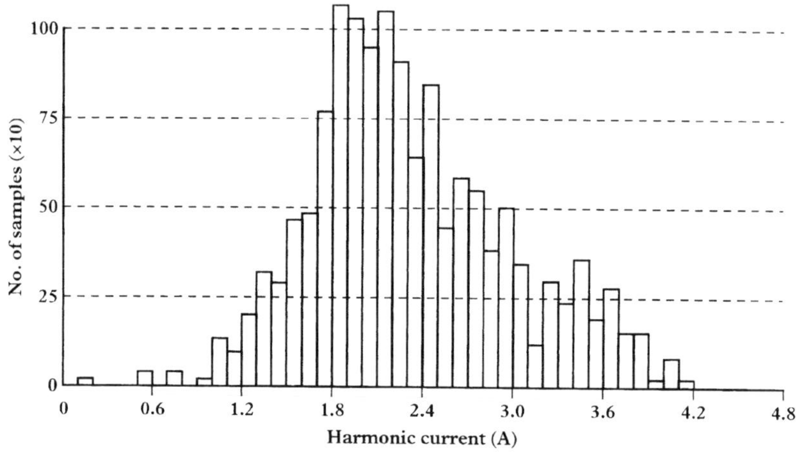

Figure 4.22 A bar chart (Histogram) of harmonic current and magnitude

4.6 References

1. Arrillaga, J, Bradley, D A, and Bodger, P S, (1985). *Power System Harmonics*, John Wiley & Sons, London.
2. Zollenkopf, K (1960). Bifactorization-basic computational algorithm and programming techniques, *Conference on large sets of sparse linear equations*, Oxford.

3. Bergeal, J, and Moller, L, (1980). Influence des charges sur la propagation de perturbations de type harmoniques — principales consequences, *Internal Report* EDF **HR/22-1034**.
4. Baker, W P, (1981). Measured impedances of power systems. *International Conference on Harmonics in Power Systems*, UMIST, Manchester, England.
5. Ribeiro, P F, (1985). Investigations of harmonic penetration in transmission systems. PhD. Thesis, UMIST (Manchester), England.
6. Pesonen, J A, (1981). Harmonics, characteristic parameters, method of study, estimates of existing values in the network, *Electra*, **77**, 35–54.
7. Huddart, K W and Brewer, G L, (1966). Factors influencing the harmonic impedance of a power system, *Conference on High Voltage d.c. Transmission, IEE*, No 22, 450–452.
8. Mahmoud, A A and Shultz, R D, (1982). A method for analyzing harmonic distribution in a.c. power systems, *IEEE Trans*, **PAS–101**(6), 1815–1824.
9. Arrillaga, J, *et al.*, (1996). A.c. System Modelling For a.c. Filter Design, An Overview Of Impedance Modelling, *ELECTRA*, **164**, 133–151.
10. Watson, N R, (1987). Frequency-dependent a.c. system equivalents for harmonic studies and transient converter simulation. PhD. Thesis, University of Canterbury, New Zealand.
11. Laurent, P G, Gary, C and Clade, J, (1962). D.c. interconnection between France and Great Britain by submarine cables. *CIGRE*, **Part III**, Paper 331.

5

AC–DC CONVERSION—FREQUENCY DOMAIN

5.1 Introduction

The direct harmonic analysis described in Chapter 4 requires information about the harmonic sources; this is obtained either from field measurements or from Fourier analysis of the expected current waveform of the various non-linear components. As the main individual contributor to power system harmonic distortion, the three-phase bridge converter requires special consideration in this respect.

Under balanced and undistorted ac terminal voltage and perfect dc current conditions, a p-pulse converter can be regarded as a linear frequency modulator interconnecting the ac and dc systems. Three main sets of frequencies are involved, one equal to npf (f being the fundamental frequency of the ac system and n an integer) on the dc voltage waveform and two equal to $npf \pm 1$ on the ac current waveforms.

In practice, the above ideal conditions never exist. There is always some asymmetry in the parameters of the plant components and in the operating conditions. The delay and commutation angles may all be different and as a result additional harmonics will appear on the ac and dc sides of the converter. Also, subharmonic and inter-harmonic frequencies often occur in non-synchronous interconnections.

Thus, a more general converter model is needed to simulate the variety of characteristic and non-characteristic frequencies involved.

After discussing the harmonic behaviour of the ideal converter, this chapter analyses the ac–dc converter in the frequency domain taking into account ac voltage distortion, dc current distortion and the effect of the converter controller.

5.2 Characteristic Converter Harmonics [1]

For medium and large power applications, such as industrial drives, smelters and HVdc transmission, the six-pulse bridge, shown in Figure 5.1, constitutes the basic

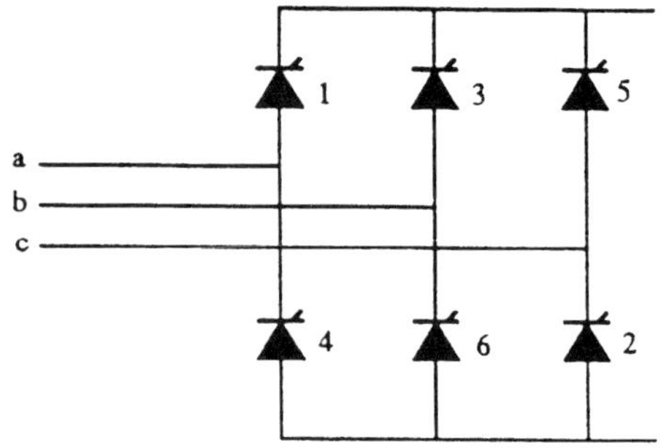

Figure 5.1 Three-phase bridge configuration

converter unit. A number of six-pulse bridges with their supply voltage phase-shifted by appropriate transformer connections are joined either in series or parallel on the dc side, depending on whether high voltage or high current is required.

Generally, these ac–dc converters have considerably more inductance on the dc side than on the ac side and the converter acts like a source of harmonic voltage on the dc side and of harmonic current on the ac side.

For the three-phase bridge configuration the orders of the harmonic voltages are $k = 6n$ and the corresponding dc voltage waveforms are illustrated in Figure 5.2.

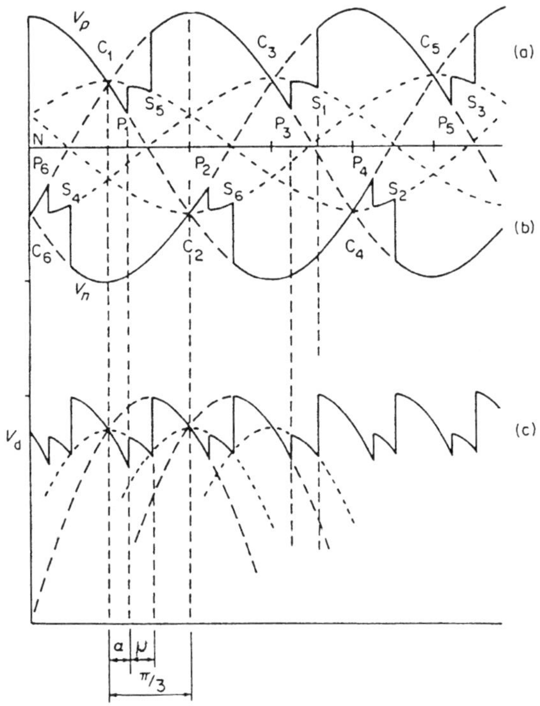

Figure 5.2 Six-pulse converter dc voltage waveforms: (a) at the positive terminal; (b) at the negative terminal; (c) between output terminals

The repetition interval of the waveform shown in Figure 5.2(c) is $\pi/3$, and it contains the following three different functions with reference to voltage crossing C_1:

$$v_d = \sqrt{2}V_c \cos\left[\omega t + \frac{\pi}{6}\right] \quad \text{for} \quad 0 < \omega t < \alpha \tag{5.1}$$

$$v_d = \sqrt{2}V_c \cos\left[\omega t + \frac{\pi}{6}\right] + \frac{1}{2}\sqrt{2}V_c \sin[\omega t] = \frac{\sqrt{6}}{2}V_c \cos[\omega t] \quad \text{for} \quad \alpha < \omega t < \alpha + \mu, \tag{5.2}$$

$$v_d = \sqrt{2}V_c \cos\left[\omega t - \frac{\pi}{6}\right] \quad \text{for} \quad \alpha + \mu < \omega t < \frac{\pi}{3}. \tag{5.3}$$

where V_c is the (commutating) phase to phase r.m.s. voltage, and α and μ the firing and commutation angles respectively.

From Equations 5.1, 5.2 and 5.3, the following expression is obtained for the r.m.s. magnitudes of the harmonic voltages of the dc voltage waveform:

$$V_k = \frac{V_{c0}}{\sqrt{2}(k^2 - 1)}\left\{(k-1)^2\cos^2\left[(k+1)\frac{\mu}{2}\right] + (k+1)^2\cos^2\left[(k-1)\frac{\mu}{2}\right] - 2(k-1)(k+1)\cos\left[(k+1)\frac{\mu}{2}\right]\cos\left[(k-1)\frac{\mu}{2}\right]\cos(2\alpha+\mu)\right\}^{1/2} \tag{5.4}$$

Figure 5.3 illustrates the use of Equation 5.4 to derive the variation of the sixth harmonic as a percentage of V_{c0}, the maximum average rectified voltage, which for

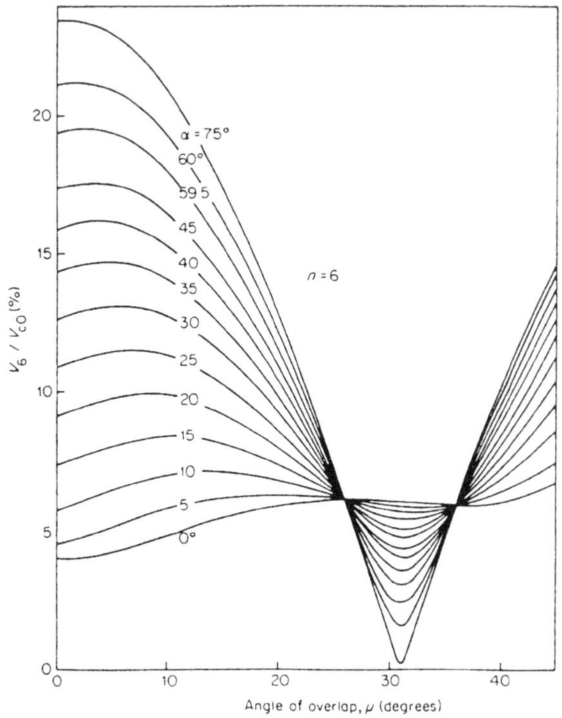

Figure 5.3 Variation of the sixth harmonic voltage

the six-pulse bridge converter is $3\sqrt{2}V_c/\pi$. These curves and equations show some interesting facts. Firstly, for $\alpha = 0$ and $\mu = 0$, Equation 5.4 reduces to

$$V_{k0} = \sqrt{2}\frac{V_{c0}}{k^2 - 1} \tag{5.5}$$

or

$$\frac{V_{k0}}{V_{c0}} = \frac{\sqrt{2}}{k^2 - 1} \approx \frac{\sqrt{2}}{k^2}, \tag{5.6}$$

giving 4.04, 0.99 and 0.44% voltage distortion for the sixth, 12th and 18th harmonics, respectively.

Generally, as α increases, harmonics increase as well, and for $\alpha = \pi/2$ and $\mu = 0$,

$$\frac{V_k}{V_{c0}} = \sqrt{2}\frac{k}{k^2 - 1} \approx \frac{\sqrt{2}}{k}, \tag{5.7}$$

which produces k times the harmonics content corresponding to $\alpha = 0$. This means that the higher harmonics increase faster with α. Equation 5.7 is of some importance as it represents the maximum proportion of harmonics in the system, particularly when it is considered that at $\alpha = 90^\circ$, μ is likely to be very small.

Neglecting commutation, the frequency domain representation of the ac current in phase 'a' is

$$i_a = \frac{2\sqrt{3}}{\pi} I_d \left(\cos \omega t - \frac{1}{5}\cos 5\omega t + \frac{1}{7}\cos 7\omega t - \frac{1}{11}\cos 11\omega t + \frac{1}{13}\cos 13\omega t - \frac{1}{17}\cos 17\omega t + \frac{1}{19}\cos 19\omega t - \ldots \right) \tag{5.8}$$

The three-phase currents are shown in Figure 5.4(b), (c) and (d), respectively. Some useful observations can now be made from Equation 5.8:

- the absence of triplen harmonics;
- the presence of harmonics of orders $6n \pm 1$ for integer values of n;
- the r.m.s. magnitude of the fundamental frequency is

$$I_1 = \frac{1}{\sqrt{2}}\frac{2\sqrt{3}}{\pi} I_d = \frac{\sqrt{6}}{\pi} I_d; \tag{5.9}$$

- the r.m.s. magnitude of the kth harmonic is

$$I_k = \frac{I_1}{k} \tag{5.10}$$

Moreover, those harmonics of orders $6n + 1$ are of positive sequence and those harmonics of orders $6n - 1$ are of negative sequence.

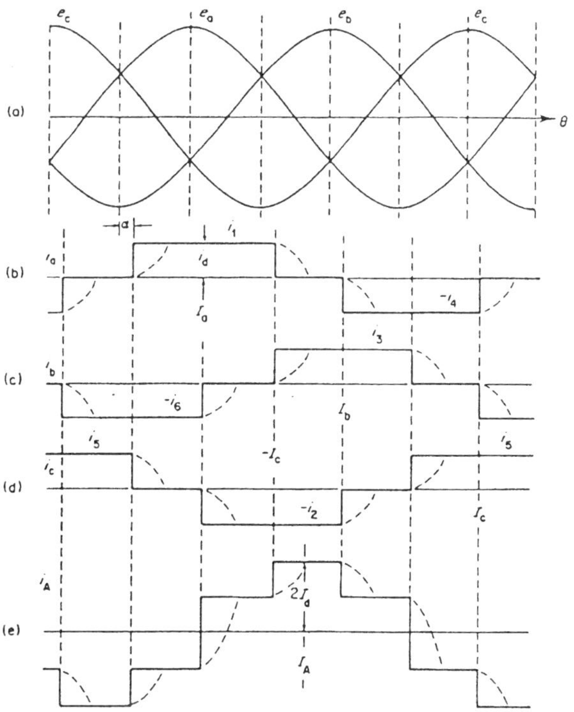

Figure 5.4 Six-pulse bridge waveforms: (a) phase to neutral voltages; (b)–(d) phase currents on the converter side; (e) phase current on the system side with *Delta*-Y transformer

5.2.1 Effect of Transformer Connection

If either the primary or secondary three-phase windings of the converter transformer are connected in delta, the ac side current waveforms consist of the instantaneous differences between two rectangular secondary currents 120° apart, as shown in Figure 5.4(e).

To maintain the same primary and secondary voltages as for the star-star connection, a factor of $\sqrt{3}$ is introduced in the transformer ratio, and the current waveform is as shown in Figure 5.5.

The resulting Fourier series for the current in phase 'a' on the primary side is

$$
\begin{aligned}
i_a = \frac{2\sqrt{3}}{\pi} I_d \Big(&\cos \omega t + \frac{1}{5}\cos 5\omega t - \frac{1}{7}\cos 7\omega t - \frac{1}{11}\cos 11\omega t \\
&+ \frac{1}{13}\cos 13\omega t + \frac{1}{17}\cos 17\omega t - \frac{1}{19}\cos 19\omega t - \ldots \Big)
\end{aligned}
\quad (5.11)
$$

This series only differs from that of a star-star connected transformer by the sign of harmonic orders $6n \pm 1$ for odd values of n, i.e. the fifth, seventh, 17th, 19th, etc.

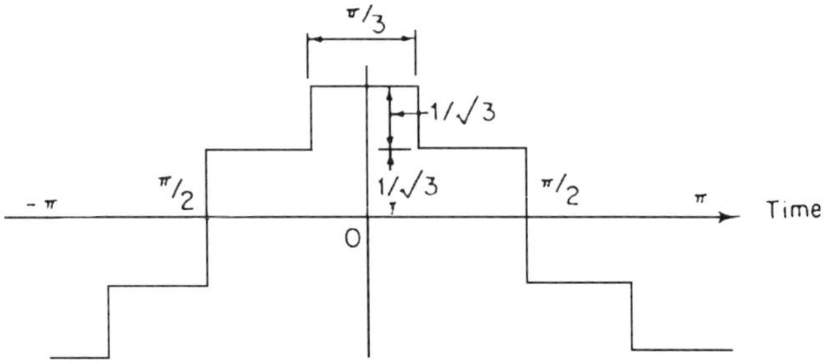

Figure 5.5 Time domain representation of a six-pulse waveform with delta-star transformer connection

5.2.2 Twelve-pulse Related Harmonics

Twelve-pulse configurations consist of two six-pulse groups fed from two sets of three-phase transformers in parallel, with their fundamental voltage equal and phase-shifted by 30°; a common 12-pulse configuration is shown in Figure 5.6.

Moreover, to maintain 12-pulse operation the two six-pulse groups must operate with the same control angle and, therefore, the fundamental frequency currents on the ac side of the two transformers are in phase with one another.

The resultant ac current is given by the sum of the two Fourier series of the star-star (Equation 5.8) and delta-star (Equation 5.11) transformers, i.e.

$$(i_{12}) = 2\frac{2\sqrt{3}}{\pi} I_d \left(\cos \omega t - \frac{1}{11} \cos 11\omega t + \frac{1}{13} \cos 13\omega t \right.$$
$$\left. - \frac{1}{23} \cos 23\omega t + \frac{1}{25} \cos 25\omega t - \ldots \right) \tag{5.12}$$

This series only contains harmonics of order $12n \pm 1$. The harmonic currents of orders $6n \pm 1$ (with n odd), i.e. $n = 5, 7, 17, 19$, etc, circulate between the two converter transformers but do not penetrate the ac network. The time domain representation of the 12-pulse waveform is shown in Figure 5.7(a) and the corresponding frequency domain representation in Figure 5.7(b).

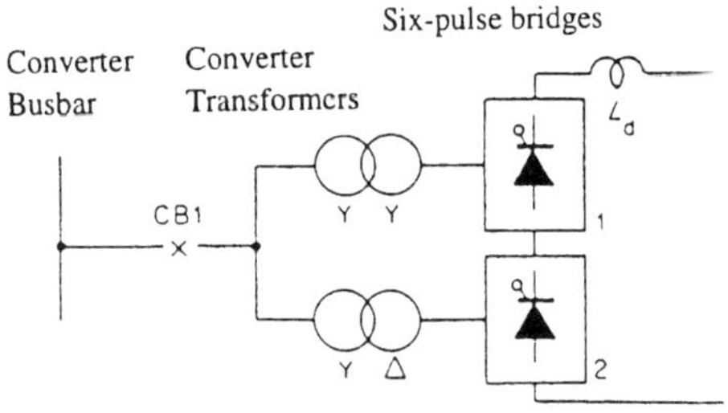

Figure 5.6 Twelve-pulse converter configuration

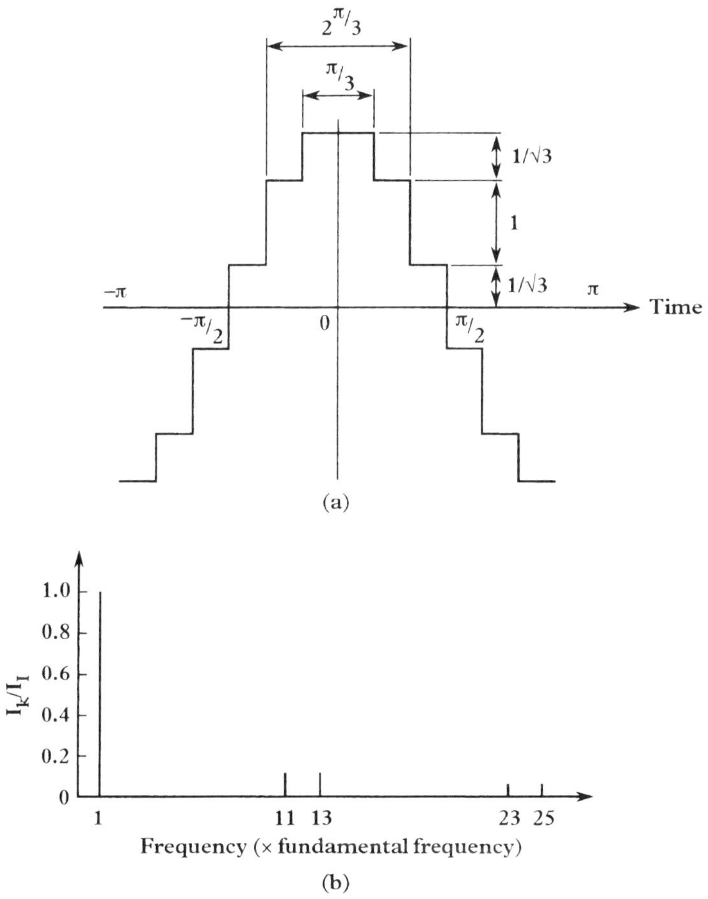

Figure 5.7 (a) Time domain representation of the 12-pulse phase current and (b) frequency domain representation of 12-pulse operation

On the dc side of the 12-pulse converter 30° of mains frequency correspond to a half-cycle of the sixth harmonic voltage and, therefore, this harmonic will be in phase opposition in the two bridges. On the other hand, for the 12th harmonic, 30° corresponds to one cycle, giving harmonics in phase; for the 18th harmonic, 30° corresponds to one and a half cycles, giving harmonics in opposition and so on.

5.2.3 Higher Pulse Configurations

In the last section, the use of two transformers with a 30° phase-shift has been shown to produce 12-pulse operation. The addition of further appropriately shifted transformers in parallel provides the basis for increasing pulse configurations.

For instance, 24-pulse operation is achieved by means of four transformers with 15° phase shifts and 48-pulse operation requires eight transformers with 7.5° phase-shifts. Although theoretically possible, pulse numbers above 48 are rarely justified due to the practical levels of distortion found in the supply voltage waveforms, which can have as much influence on harmonic generation as the theoretical phase-shifts.

Similarly to the case of the 12-pulse connection, the alternative phase-shifts involved in higher pulse configurations require the use of appropriate factors in the parallel transformer ratios to achieve common fundamental frequency voltages on their primary and secondary sides.

The theoretical harmonic currents are related to the pulse number (p) by the general expression $pn \pm 1$ and their magnitudes decrease in inverse proportion to the harmonic order. Generally, harmonics above the 49th can be neglected as their amplitude is small.

5.2.4 Insufficient Smoothing Reactance

Considering the limited inductance of the motor armature winding and the larger variation of firing angle, the constant dc current assumption of the large size converters cannot be justified in the case of dc drives.

The dc load can be represented as an equivalent circuit which in its simplest form includes resistance, inductance and back e.m.f. With sinusoidal supply voltage $V_m \sin(\omega t)$, the following equation applies:

$$V_m \sin(\omega t) = Ri + L\frac{\mathrm{d}i}{\mathrm{d}t} + E \tag{5.13}$$

and the load current has the piecewise form

$$i = K\mathrm{e}^{-Rt/L} + \frac{V_m}{\sqrt{R^2 + (\omega L)^2}} \sin(\omega t - \phi) - \frac{E}{R}, \tag{5.14}$$

where

$$\phi = \arctan\frac{\omega L}{R} \tag{5.15}$$

and the constant K is derived from the particular initial conditions.

Under nominal loading the firing delay is kept low, but during motor start or light load conditions the delay increases substantially and the current may even be discontinuous. This extreme operating condition is illustrated in Figure 5.8 for a six-pulse rectifier. Each phase current consists of two positive and two negative current pulses per cycle, which are derived from the general Expression 5.14 using the appropriate voltage phase relationships with a common reference.

The current in phase A with reference to the instant when V_{ab} is maximum in Figure 5.8 has the following components:

1. Over the range $\theta_1 < \omega t < \theta_2$

$$i = \frac{V_m}{R}\left\{\cos\phi\cos(\omega t - \phi) - \frac{E}{V_m} + \left[\frac{E}{V_m} - \cos\phi\cos(\theta_1 - \phi)\right]\mathrm{e}^{-R(\omega t - \theta_1)/\omega L}\right\} \tag{5.16}$$

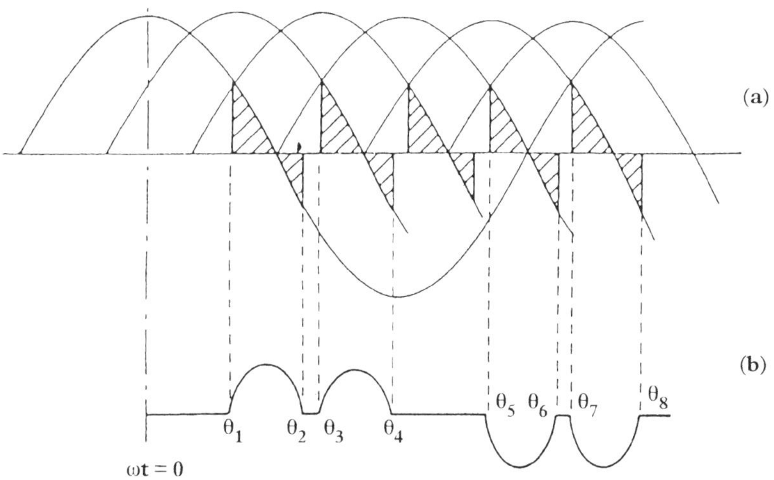

Figure 5.8 Discontinuous waveforms: (a) dc voltage; (b) ac current in phase a

2. When $\theta_3 < \omega t < \theta_4$, where $\theta_3 = \theta_1 + \pi/3$

$$i = \frac{V_m}{R}\left\{\cos\phi\cos(\omega t - \frac{\pi}{3} - \phi) - \frac{E}{V_m} + \left[\frac{E}{V_m} - \cos\phi\cos(\theta_1 - \phi)\right]e^{-R(\omega t - \pi/3 - \theta_1)/\omega L}\right\} \tag{5.17}$$

3. When $\theta_5 < \omega t < \theta_6$, where $\theta_5 = (\theta_1 + \pi)$

$$i = -\frac{V_m}{R}\left\{\cos\phi\cos(\omega t - \pi - \phi) - \frac{E}{V_m} + \left[\frac{E}{V_m} - \cos\phi\cos(\theta_1 - \phi)\right]e^{-R(\omega t - \pi/3 - \theta_1)/\omega L}\right\} \tag{5.18}$$

4. When $\theta_7 < \omega t < \theta_8$, where $\theta_7 = (\theta_1 + 2\pi/3)$

$$i = -\frac{V_m}{R}\left\{\cos\phi\cos\left(\omega t - \frac{2\pi}{3} - \phi\right) - \frac{E}{V_m} + \left[\frac{E}{V_m} - \cos\phi\cos(\theta_1 - \phi)\right]e^{-R(\omega t - \pi/3 - \theta_1)/\omega L}\right\} \tag{5.19}$$

Application of Fourier analysis to these current pulses indicates that the fifth harmonic can reach peak levels of up to three times those of the rectangular waveshape with the same fundamental component.

An approximate solution [2] on the assumption of a negligible commutation overlap is shown in Figure 5.9.

5.2.5 Effect of Transformer and System Impedance

In practice, the existence of reactance in the commutation circuit causes conduction overlap of the incoming and outgoing phases.

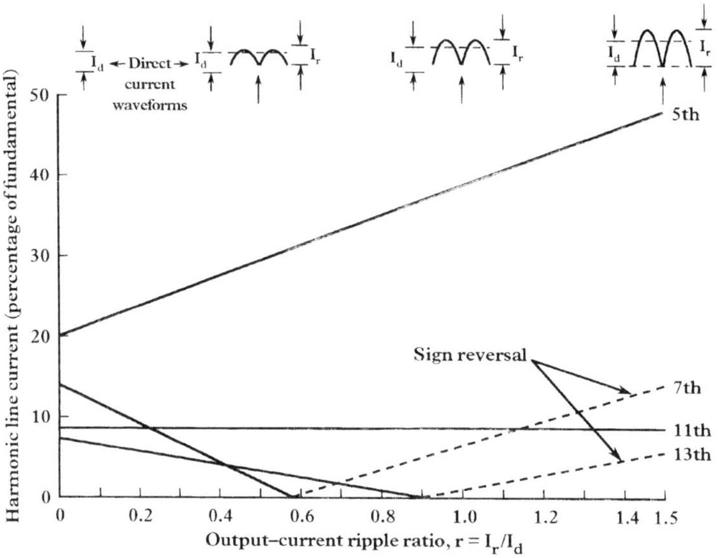

Figure 5.9 Harmonic content of supply current for six-pulse converter with finite inductive load

As we have seen in previous sections, high-pulse configurations are combinations of six-pulse groups, i.e. the commutation overlaps are those of the six-pulse group as shown by the broken lines in Figure 5.4.

The current waveform has now lost the even symmetry with respect to the centre of the idealised rectangular pulse. Using as a reference the corresponding commutation voltage (i.e. the zero voltage crossing) and assuming a purely inductive commutation circuit, the following expression defines the commutating current [1].

$$i_c = \frac{V_c}{\sqrt{2}X_c}[\cos\alpha - \cos\omega t] \tag{5.20}$$

where X_c is the reactance (per phase) of the commutation circuit, which is largely determined by the transformer leakage reactance. At the end of the commutation $i_c = I_d$ and $\omega t = \alpha + \mu$, and Equation 5.20 becomes

$$I_d = \frac{V_c}{\sqrt{2}X_c}[\cos\alpha - \cos(\alpha + \mu)] \tag{5.21}$$

Dividing Equation 5.20 by Equation 5.21

$$i_c = I_d\left(\frac{\cos\alpha - \cos\omega t}{\cos\alpha - \cos(\alpha + \mu)}\right) \tag{5.22}$$

and this expression applies for $\alpha < \omega t < \alpha + \mu$. The rest of the positive current pulse is defined by

$$i = I_d \quad \text{for } \alpha + \mu < \omega t < \alpha + \frac{2\pi}{3} \tag{5.23}$$

and

$$i = I_d - I_d \left[\frac{\cos\left(\alpha + \frac{2\pi}{3}\right) - \cos \omega t}{\cos\left(\alpha + \frac{2\pi}{3}\right) - \cos\left(\alpha + \frac{2\pi}{3} + \mu\right)} \right] \quad \text{for} \quad \alpha + \frac{2\pi}{3} < \omega t < \alpha + \frac{2\pi}{3} + \mu \tag{5.24}$$

The negative current pulse still possesses half-wave symmetry and, therefore, only odd-ordered harmonics are present. These can now be expressed in terms of the delay(firing) and overlap angles; the magnitude of the fifth harmonic related to the fundamental component is illustrated in Figure 5.10. In summary, the existence of system impedance is seen to reduce the harmonic content of the current waveform. The effect of commutation overlap on the dc voltage waveform is illustrated in Figure 5.3.

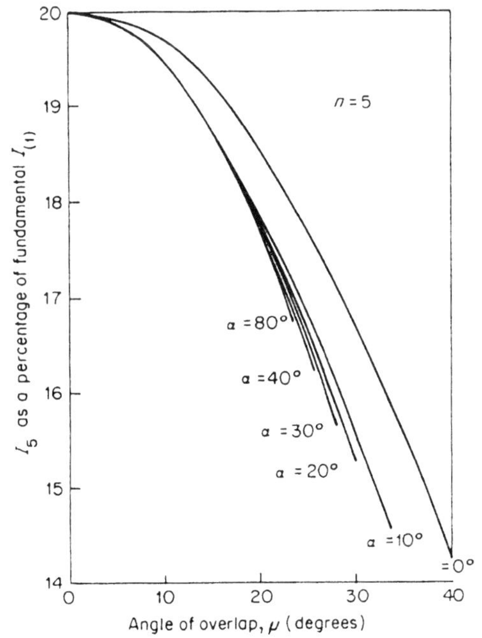

Figure 5.10 Variation of fifth harmonic current in relation to angle of delay and overlap

5.3 Frequency Domain Model

The previous analyses assume a steady firing delay angle, either no commutation period or one of unvarying duration, and undistorted ac system voltage conditions. In practice this is rarely the case. Harmonic voltages and/or unbalance will exist on the ac side, and current ripple will exist on the dc side. Through the constant current control, the firing angle will not be steady, and the commutation period duration will also be varying. Therefore not only will harmonic voltages and currents be transferred through the converter, but through the variation of thyristor switching instants they may be amplified. These interactions have particular relevance for non-characteristic harmonics, ie of different order to those discussed in the previous section.

A more accurate technique for the analysis of static converters in the frequency domain is the transfer function concept. The six-pulse converter transfer functions $Y_{\psi dc}$ and $Y_{\psi ac}$ describe the interconnection between the dc and ac sides of the converter. The dc voltage is calculated by summing each phase voltage multiplied by its associated transfer function, as shown in Equation 5.25.

$$v_d = N \sum_{\psi} Y_{\psi dc} v_{\psi} \tag{5.25}$$

where $\psi =$ 0, 120, and 240 degrees for phases a, b and c, N is the converter transformer ratio (converter to ac system side), and v_{ψ} are the three phase voltages. $Y_{\psi dc}$ has values between -1 and 1, where 1 signifies a connection of the dc side positive bus to the phase in question, -1 signifies a connection of the dc side negative bus to the phase in question, and 0 indicates no connection. By assigning the transfer function a value of 0.5 for the two commutating phases the dc voltage is correctly represented during the commutation process.

The ac current in each phase can be defined by Equation 5.26

$$i_{\psi} = N Y_{\psi ac} i_{dc} \tag{5.26}$$

where i_{dc} is the dc side current, and $Y_{\psi ac}$ is similar to $Y_{\psi dc}$, except that during the commutation period the ac current rises or falls in a continuous manner. In this analysis it is approximated by a linear transfer of the dc side current from one phase to the next.

Both transfer functions are built up by the summation of a basic function (no commutation period, steady firing angle), a firing angle variation function, and a commutation function. The process is demonstrated graphically in Figure 5.11, in which the dotted line represents the basic transfer function, the dashed line the function revised to include a firing angle variation of $\Delta\alpha$, and the solid line the function further revised to include the effect of a commutation period. Breaking up the transfer functions in this way allows the frequency spectra to be more easily written.

The firing angle variation function is characterised as a set of pulses, with fixed leading edges and variable trailing edges. For $Y_{\psi dc}$, the commutation function comprises a set of rectangular pulses, of which the leading edges match the firing angle variation, and trailing edges vary somewhat differently. For $Y_{\psi ac}$, the

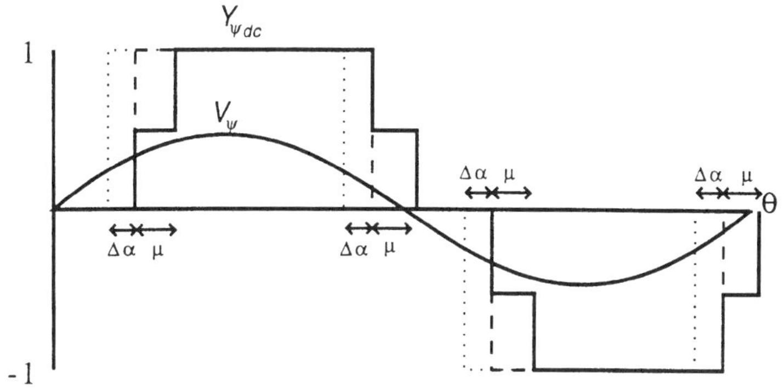

(a) Transfer function to dc voltage

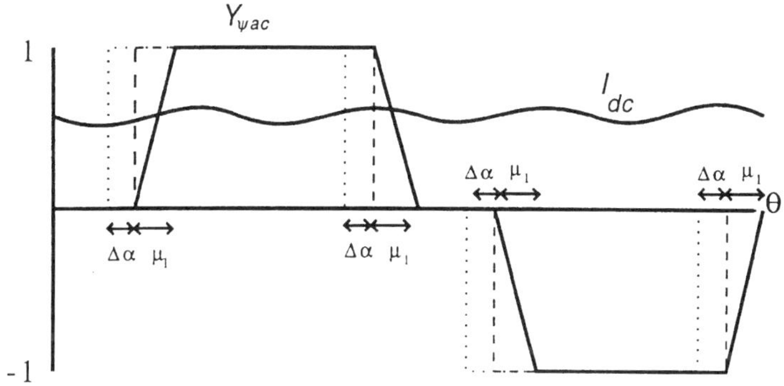

(b) Transfer function to ac current

Figure 5.11 Transfer functions (a) $Y_{\psi dc}$ and (b) $Y_{\psi ac}$

commutation function comprises a set of sawtooth pulses, of which the leading and trailing edges match the firing angle variation. When the spectrum of this waveform is written, the current-time area of the commutation function has the dominant effect. An effective commutation period duration μ_1 is defined, such that the area of the $Y\Psi_{ac}$ commutation function matches the area of the true commutation waveform. In addition, a small variable triangular pulse is added to account for the variation in this area consequential to the ac voltage, dc current, or firing angle variations.

If the fundamental frequency positive sequence ac voltage component at the converter terminals is described in the form

$$V_\psi = V_1 \cos(\omega_0 t - \psi), \tag{5.27}$$

the firing angle defined by

$$\alpha = \alpha_0 + b_\alpha \cos(k\omega_0 t + \delta_{k\alpha}) \tag{5.28}$$

at the instant of firing, and the end of the commutation period defined by the expression

$$\alpha + \mu = \alpha_0 + \mu_0 + b_e \cos(k\omega_0 t + \delta_{ke}) \tag{5.29}$$

also at the instant of firing, the frequency spectrum of the transfer function $Y_{\psi dc}$, derived from spectra developed in [5], and Appendices II and III can be written

$$\begin{aligned}
Y_{\psi dc}(t) = {} & \frac{2\sqrt{3}}{\pi}\sum_m (\pm)\left(\frac{J_0(mb_\alpha) + 1 + 2J_0(mb_\alpha)\sin\left(\frac{m\mu_0}{2}\right)\Big/ -\frac{m\mu_0}{2} - \frac{\pi}{2}}{2m}\right)\cos[m(\omega_0 t - \alpha_0 - \psi)] \\
& + \frac{\sqrt{3}}{\pi}\sum_m \sum_{n=1}^{\infty} (\pm)\frac{J_n(mb_e)}{m}\cos\left[(m+nk)\omega_0 t - m(\alpha_0 + \mu_0) + n\left(\delta_{ke} - k\mu_0 - \frac{\pi}{2}\right) - m\psi\right] \\
& + \frac{\sqrt{3}}{\pi}\sum_m \sum_{n=1}^{\infty} (\pm)\frac{J_n(mb_e)}{m}\cos\left[(m-nk)\omega_0 t - m(\alpha_0 + \mu_0) - n\left(\delta_{ke} - k\mu_0 + \frac{\pi}{2}\right) - m\psi\right] \\
& + \frac{\sqrt{3}}{\pi}\sum_m \sum_{n=1}^{\infty} (\pm)\frac{J_n(mb_\alpha)}{m}\cos\left[(m+nk)\omega_0 t - m\alpha_0 + n\left(\delta_{k\alpha} - \frac{\pi}{2}\right) - m\psi\right] \\
& + \frac{\sqrt{3}}{\pi}\sum_m \sum_{n=1}^{\infty} (\pm)\frac{J_n(mb_\alpha)}{m}\cos\left[(m-nk)\omega_0 t - m\alpha_0 - n\left(\delta_{k\alpha} + \frac{\pi}{2}\right) - m\psi\right]
\end{aligned} \tag{5.30}$$

where $m = 1, 5, 7, 11$ etc., J_n is the Bessel function of the first kind, and

$$(\pm) = \sin\left(\frac{m\pi}{2}\right). \tag{5.31}$$

For simplicity this is written in a form that assumes that both the commutation period variation and the firing angle variation are at the frequency $k\omega_0$.

The frequency spectrum of the transfer function $Y_{\psi ac}$ can be similarly written

$$\begin{aligned}
Y_\psi(t) = {} & \frac{2\sqrt{3}}{\pi}\sum_m (\pm)\left(\frac{1 - J_0(mb_\alpha)}{2m} + \frac{J_0(mb_\alpha)}{m}\frac{2}{m\mu_0}\sin\left(\frac{m\mu_0}{2}\right)\Big/ -m\mu_0/2\right) \times \\
& \cos[m(\omega_0 t - \alpha_0 - \psi)] + \left\{\frac{2\sqrt{3}}{\pi}\sum_m \sum_{n=1}^{\infty} (\pm)\frac{J_n(mb_\alpha)}{m}\,\frac{2\sin\left[\frac{(m+nk)\mu_1}{2}\right]}{(m+nk)\mu_1}\times\right. \\
& \left.\cos\left[(m+nk)\omega_0 t - m\left(\alpha_0 + \frac{\mu_1}{2}\right) + n\left(\delta_{k\alpha} - \frac{\pi}{2} - \frac{k\mu_1}{2}\right) - m\psi\right]\right\}
\end{aligned}$$

$$+\left\{\frac{2\sqrt{3}}{\pi}\sum_{m}\sum_{n=1}^{\infty}(\pm)\frac{J_n(mb_\alpha)}{m}\,\frac{2\sin\left[\dfrac{(m-nk)\mu_1}{2}\right]}{(m-nk)\mu_1}\times\right.$$

$$\left.\cos\left[(m-nk)\omega_0 t - m\left(\alpha_0+\frac{\mu_1}{2}\right)-n\left(\delta_{k\alpha}+\frac{\pi}{2}-\frac{k\mu_1}{2}\right)-m\psi\right]\right\}$$

$$+\frac{\sqrt{3}}{\pi}\frac{b_d}{2}\sum_{m}\sum_{n=-\infty}^{\infty}(\pm)(m+nk)\left\{\frac{\sin\left[\dfrac{(m+nk)\mu_1}{4}\right]}{\dfrac{(m+nk)\mu_1}{4}}\right\}^2\frac{J_n(mb_\alpha)}{m}\times$$

$$\left\{\cos\left[(m+nk+k)\omega_0 t - m\left(\alpha_0+\frac{\mu_1}{\sqrt{x}}\right)+n\left(\delta_{k\alpha}-\frac{k\mu_1}{\sqrt{x}}-\frac{\pi}{2}\right)+\delta_{kd}-\frac{k\mu_1}{\sqrt{x}}-\frac{\pi}{2}-m\psi\right]\right.$$

$$\left.+\cos\left[(m+nk-k)\omega_0 t - m\left(\alpha_0+\frac{\mu_1}{\sqrt{x}}\right)+n\left(\delta_{k\alpha}-\frac{k\mu_1}{\sqrt{x}}-\frac{\pi}{2}\right)-\delta_{kd}+\frac{k\mu_1}{\sqrt{x}}-\frac{\pi}{2}-m\psi\right]\right\}\tag{5.32}$$

where b_d and δ_{kd} define a current-time correction term such that the change in the effective commutation period duration is

$$d\mu_1 = b_d\cos(k\omega_0 t+\delta_{kd})\tag{5.33}$$

at the instant of firing, and x approximates the average angular position of the correction pulse, being 1 for an inverter and 2 for a rectifier.

These spectra contain both characteristic and non-characteristic harmonics. As the dc voltage comprises a summation of these transfer functions multiplied by the three ac voltages, the ac current comprises the dc current multiplied by the appropriate transfer function, non-characteristic frequencies in the transfer function will lead to non-characteristic frequencies in the converter currents and voltages. It is necessary to examine the transfer function more closely, via the control and commutation process, to determine these spectra.

5.3.1 Commutation Analysis

The spectrum of the transfer functions depends on the variation of the commutation period parameters, and this variation must be described. The commutation circuit itself is simple, as shown in Figure 5.12. Writing the circuit equations and integrating from the time of firing $\omega_0 t_i$ to $\omega_0 t$, gives

$$\int_{\omega_0 t_i}^{\omega_0 t} v_{com}\,\mathrm{d}\omega_0 t = 2X_c\int_0^{i_2(\omega_0 t)}\mathrm{d}i_2 - X_c\int_{i_d(\omega_0 t_i)}^{i_d(\omega_0 t)}\mathrm{d}i_d\tag{5.34}$$

This forms the basis of the commutation period variation analysis.

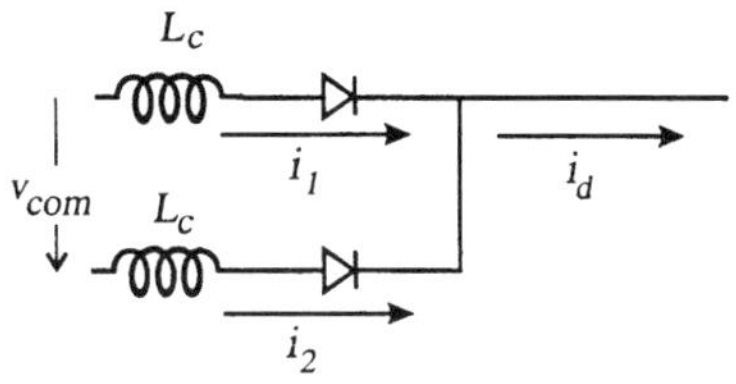

Figure 5.12 The commutation circuit

Average commutation period Assuming undistorted waveforms and solving Equation 5.34 results in

$$\mu_0 = \cos^{-1}\left[\cos\alpha_0 - \frac{2X_c I_d}{\sqrt{3}NV_1}\right] - \alpha_0 \tag{5.35}$$

where V_1 is the peak single phase ac voltage at the converter transformer primary and X_c is the converter transformer leakage reactance referred to the converter transformer secondary.

Equating the area of the equivalent sawtooth pulse of duration μ_1 to the integration of the commutation current between α_0 and $\alpha_0 + \mu_0$ results in the following term

$$\mu_1 = 2\mu_0 - \frac{\sqrt{3}NV_1}{X_c I_d}[\mu_0\cos(\alpha_0) + \sin(\alpha_0) - \sin(\alpha_0 + \mu_0)] \tag{5.36}$$

Sensitivity to firing angle From Equation 5.35 it is apparent that variation in the firing angle will cause a variation in the commutation period duration. Differentiating Equation 5.35 with respect to the firing angle, keeping the dc current and ac voltage undistorted, yields the following

$$\frac{\partial\mu}{\partial\alpha(\omega_0 t_i)} = \frac{\sin(\alpha_0)}{\sqrt{1 - \left(\cos\alpha_0 - \dfrac{2X_c I_d}{\sqrt{3}NV_1}\right)^2}} - 1 \tag{5.37}$$

Differentiating Equation 5.36 with respect to the firing angle yields, with a small signal assumption,

$$\frac{\partial\mu_1}{\partial\alpha(\omega_0 t_i)} = \frac{\sqrt{3}NV_1}{X_c I_d}\mu_0\sin(\alpha_0) - 2 \tag{5.38}$$

Sensitivity to dc current Letting the dc current distortion tend towards zero, solving Equation 5.34 for a constant firing angle and an undistorted commutating voltage, and differentiating with respect to the dc current distortion at the instant of firing results in

$$\frac{\partial\mu}{\partial I_d(\omega_0 t_i)} = \frac{2X_c \cos\left(\frac{k\mu_0}{2}\right)}{\sqrt{3}NV_1\sqrt{1-\left(\cos\alpha_0 - \frac{2X_c I_d}{\sqrt{3}NV_1}\right)^2}} \underline{/k\mu_0/2} \tag{5.39}$$

Integrating the ac current over the commutation period yields the current-time area of the commutation current waveform, and differentiating with respect to the dc current at the instant of firing yields the sensitivity of the commutation period current-time area to the dc current. However, some of this sensitivity is already accounted for by the unmodulated transfer function spectrum, which must be subtracted to yield

$$\frac{\partial\mu_1}{\partial I_d(\omega_0 t_i)} = \frac{\mu_0}{I_d}\left[1 + \frac{2}{k\mu_0}\sin\left(\frac{k\mu_0}{2}\right)\underline{/k\mu_0/2}\right] + \frac{2}{kI_d}\left[\frac{2}{k\mu_1}\sin\left(\frac{k\mu_1}{2}\right)\underline{/k\mu_1/2} - 1\right]\underline{/\pi/2} \tag{5.40}$$

This represents the sensitivity of the effective commutation period to a distorted dc current, excluding that already described by the steady commutation period converter description.

Sensitivity to ac voltage A non-sinusoidal set of commutating voltages also results in a varying commutation period. Initially, the ac side voltage distortion is assumed to consist of a single positive sequence frequency as described below

$$v_\psi = V_1\cos(\omega_0 t - \psi) + \mathrm{d}V_\psi(\omega_0 t) \tag{5.41}$$

where

$$\mathrm{d}V_\psi(\omega_0 t) = V_k\cos(k\omega_0 t + \delta_k - \psi) \tag{5.42}$$

At each valve switching the commutating voltage is comprised of a different combination of the three phases on the ac side. The contribution of a positive sequence distortion to the commutation voltage at the beginning of the commutation process can be written

$$\mathrm{d}V_{com}(\omega_0 t) = \sqrt{3}V_k \sin[(k-1)\omega_0 t + \alpha_0 + \delta_k] \tag{5.43}$$

The frequency of the voltage that interferes with the commutation process is less than the positive sequence frequency on the three phase ac voltage, by the fundamental frequency.

Taking into account the variation of the ac voltage waveform over the commutation period, assuming the distortion level tends towards zero, and keeping a constant dc current and firing angle, allows the following relationship between the commutation period duration and ac voltage distortion

$$\frac{\partial \mu}{\partial V_{\psi}(\omega_0 t_i)} = \left[\frac{-\mu_0}{V_1 \sqrt{1 - \left[\cos(\alpha_0) - \frac{2X_c I_d}{\sqrt{3} N V_1} \right]^2}} \right] \frac{2}{k\mu_0} \sin\left(\frac{k\mu_0}{2} \right) \angle \alpha_0 + k\mu_0/2 - \pi/2 \tag{5.44}$$

Implicit in this equation is the frequency transformation already discussed. The related expression for a negative sequence voltage distortion is derived in a similar way, but with an implicit frequency transformation from the distorting frequency to the distorting frequency plus the fundamental frequency

$$\frac{\partial \mu}{\partial V_{\psi}(\omega_0 t_i)} = \left[\frac{-\mu_0}{V_1 \sqrt{1 - \left[\cos(\alpha_0) - \frac{2X_c I_d}{\sqrt{3} V_1} \right]^2}} \right] \frac{2}{k\mu_0} \sin\left(\frac{k\mu_0}{2} \right) \angle -\alpha_0 + k\mu_0/2 + \pi/2 \tag{5.45}$$

Similar equations can be derived relating to the current-time area of the commutation waveform, with the same implicit frequency transformations. For a positive sequence distortion

$$\frac{\partial \mu_1}{\partial V_{\psi}(\omega_0 t_i)} = \frac{3NV_1\mu_0}{2(X_c I_d)^2} [\sin(\alpha_0 + \mu_0) - \sin(\alpha_0) - \mu_0 \cos(\alpha_0)] \times \frac{2}{k\mu_0} \sin\left(\frac{k\mu_0}{2} \right) \angle \alpha_0 + k\mu_0/2 - \pi/2 \tag{5.46}$$

and for a negative sequence distortion

$$\frac{\partial \mu_1}{\partial V_{\psi}(\omega_0 t_i)} = \frac{3NV_1\mu_0}{2(X_c I_d)^2} [\sin(\alpha_0 + \mu_0) - \sin(\alpha_0) - \mu_0 \cos(\alpha_0)] \times \frac{2}{k\mu_0} \sin\left(\frac{k\mu_0}{2} \right) \angle -\alpha_0 + k\mu_0/2 + \pi/2 \tag{5.47}$$

The terms developed describe the relationships between the commutation period characteristics and variation in the dc current, ac voltage, and firing angle. Maintaining the assumption that all these variations are very small, the commutation period characteristics can be defined by summing the multiple of each sensitivity factor and its relevant variable.

5.3.2 Control Transfer Functions

If the converter is controlled, the firing angle variation can be related to waveform distortion through the controller characteristics. As an example, consider the case of a converter under constant current control. For a controller of the proportional/integral (PI) type, the transfer function for dc current to firing angle is

$$G_1(k) = P + \frac{1}{jk\omega_0 T} \tag{5.48}$$

where P is the proportional gain, and T is the integral time constant, such that $G_1(k)$ is in radians per kiloampere. The transfer function of the current transducer will be significant at the higher frequencies, and should be included in this equation.

5.3.3 Transfer of Waveform Distortion

The transfer functions $Y_{\psi dc}$ and $Y_{\psi ac}$, in conjunction with the firing angle and commutation period variability terms allow the prediction of voltage waveform distortion on the dc side of the converter, and current waveform distortion on the ac side of the converter. The characteristic distortion levels are little changed from the simplified analysis. As becomes apparent, many frequencies are generated as a result of just one distortion source. However the most significant frequencies are limited to a set of three, being one on the dc side, at frequency $k\omega_0$ and two on the ac side, being at frequencies $(k+1)\omega_0$ in positive sequence and $(k-1)\omega_0$ in negative sequence. Approximation to these three frequencies leads to the three port model [4], and verification is limited to these terms. Extension is made to a twelve pulse converter by doubling the magnitude of the transfer function and using the terms for $m = 1, 11, 13, 23, 25$ etc.

To verify the predictions of the model, and establish the importance of the described mechanisms, comparative results are gained by another method. In this case, a time domain dynamic simulation program based on the EMTP formulation [3] is used. An HVdc rectifier is modelled, based on the CIGRE benchmark HVdc test system [6] described in Appendix VI. With infinite ac and dc side busbars, single distorting frequencies are injected on the firing angle order, the ac side voltage, and the dc side current. After a short settling period, a Fast Fourier Transform is made of the required variables, and the results compared in the frequency domain. In all the figures, the harmonic transfer obtained from dynamic simulation, the predicted transfer not allowing for commutation period variation, and the prediction allowing for commutation period variation are shown.

Sensitivity to firing angle Expanding Equation 5.25 out over ψ, and making the reasonable approximation that $J_n[(m-1)b_\alpha]/(m-1) = J_n(mb_\alpha)/m = J_n[(m+1)b_\alpha]/(m+1)$, yields the non-characteristic dc voltage distortion

$$v_d = \left\{ NV_1 \frac{6\sqrt{3}}{\pi} \sum_{n=1}^{\infty} \left[J_n(b_\alpha) \cos\left(\alpha_0 + \frac{n\pi}{2}\right) \underline{/n\delta_{k\alpha}} + J_n(b_e) \cos\left(\alpha_0 + \mu_0 + \frac{n\pi}{2}\right) \underline{/n\delta_{ke} - n\mu_0} \right] \times \right.$$
$$\left. \cos(nk\omega_0 t) \right\} + \left\{ NV_1 \frac{6\sqrt{3}}{\pi} \sum_m \sum_{n=1}^{\infty} \left[\frac{J_n(mb_\alpha)}{m} \sin(\alpha_0) \underline{/n\delta_{k\alpha}} + \right. \right.$$
$$\left. \left. \frac{J_n(mb_e)}{m} \sin(\alpha_0 + \mu_0) \underline{/n\delta_{ke} - (m + nk)\mu_0} \right] \times \sin\left[(m + nk)\omega_0 t - m\alpha_0 - \frac{n\pi}{2} \right] \right\} +$$
$$\left\{ NV_1 \frac{6\sqrt{3}}{\pi} \sum_m \sum_{n=1}^{\infty} \left[\frac{J_n(mb_\alpha)}{m} \sin(\alpha_0) \underline{/-n\delta_{k\alpha}} + \right. \right.$$
$$\left. \left. \frac{J_n(mb_e)}{m} \sin(\alpha_0 + \mu_0) \underline{/-n\delta_{ke} - (m - nk)\mu_0} \right] \times \sin[(m - nk)\omega_0 t - m\alpha_0 - \frac{n\pi}{2}] \right\} \tag{5.49}$$

for $m = 12, 24$, etc.

The ac current distortion is simply the transfer function $Y_{\psi ac}$ multiplied by the dc current. The firing angle and commutation period variation is described by Equations 5.28, 5.29, 5.33, 5.37 and 5.38.

Considering only the terms for $n = 1$ simplifies the dc spectrum to the modulated frequency and the dc side characteristic harmonics plus or minus the modulated frequency, and the ac spectrum to the modulated frequency plus or minus the fundamental and the characteristics plus or minus the modulated frequency.

For a small modulating signal, the Bessel function $J_1(mb)/m$ can be closely approximated by $b/2$, linearizing the relationship. Figures 5.13 and 5.14 show the

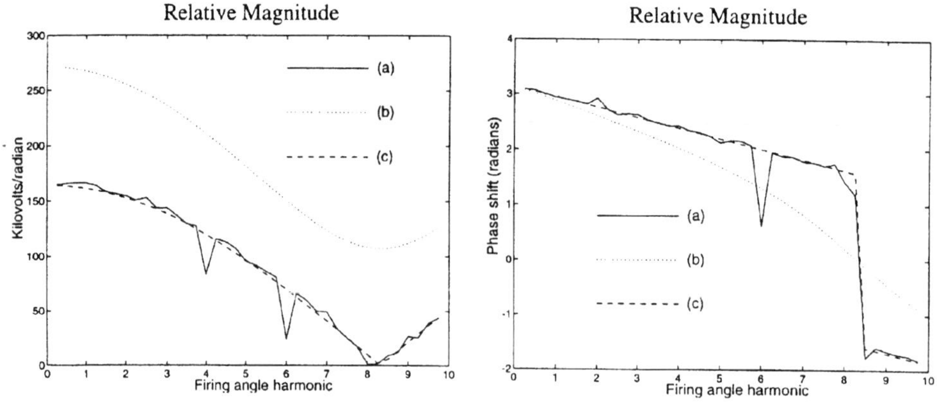

Figure 5.13 CIGRE rectifier harmonic transfer, firing angle modulation to dc voltage. (a) dynamic simulation (b) prediction without commutation period variation (c) prediction with commutation period variation

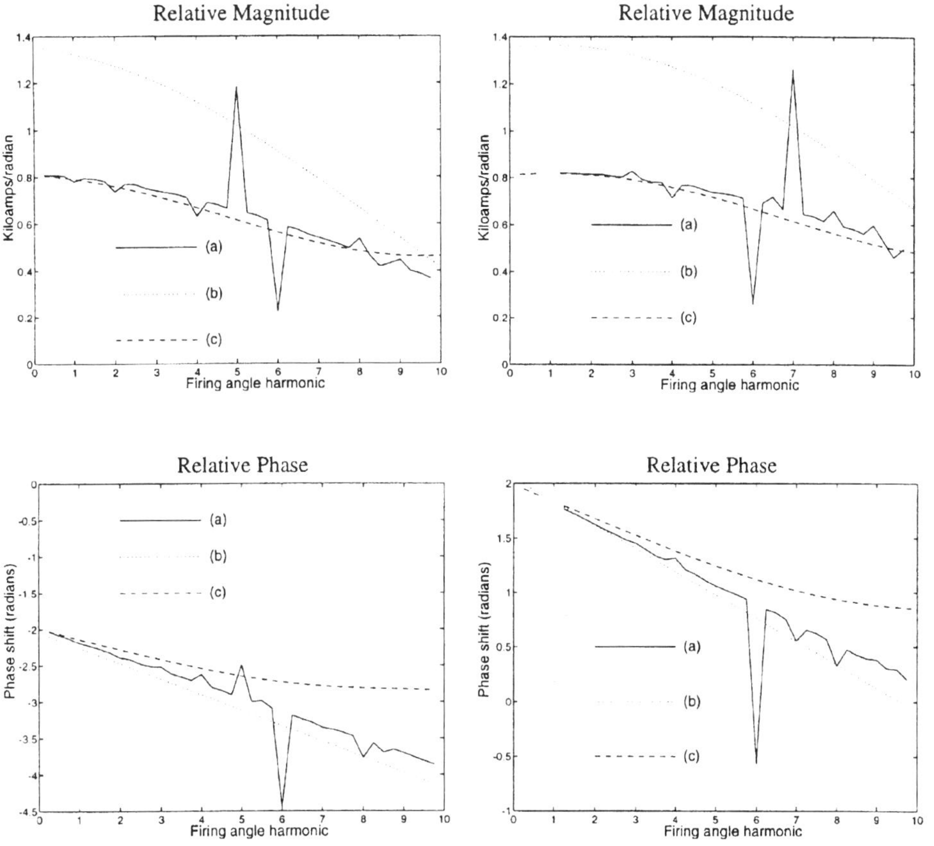

positive sequence current negative sequence current

Figure 5.14 CIGRE rectifier harmonic transfer, firing angle modulation to +ve and −ve sequence ac current. For key see Figure 5.13

predicted and measured distortions resulting from a 3° peak to peak firing angle modulation. Only the lowest order components are shown.

A particular feature of this spectrum is that the magnitude of the harmonics generated decrease relatively slowly with m. The commutation period variation has a strong effect on the resulting distortion, both in magnitude and phase. The spikes on the curves occur as the higher order components were neglected.

Sensitivity to ac voltage The contributions to the converter dc side voltage from the ac side voltage can be divided into two; the direct transfer due to the unmodulated transfer function, and the indirect transfer via the commutation period modulation.

Taking the direct transfer portion, and expanding out over ψ, for $m = 12, 24, 36$ etc. gives, for a positive sequence ac voltage

$$
\begin{aligned}
v_d = {} & N\left\{V_k \frac{6\sqrt{3}}{\pi}\cos\left(\frac{\mu_0}{2}\right)\cos\left[(k-1)\omega_0 t + \alpha_0 + \frac{\mu_0}{2} + \delta_k\right]\right\} \\
& - N\left\{\frac{6\sqrt{3}}{\pi}\sum_m V_k \frac{\cos((m-1)\mu_0/2)}{m-1}\cos\left[(m+k-1)\omega_0 t - (m-1)\left(\alpha_0 + \frac{\mu_0}{2}\right) + \delta_k\right]\right\} \\
& + N\left\{\frac{6\sqrt{3}}{\pi}\sum_m V_k \frac{\cos((m+1)\mu_0/2)}{m+1}\cos\left[(m-k+1)\omega_0 t - (m+1)\left(\alpha_0 + \frac{\mu_0}{2}\right) - \delta_k\right]\right\}
\end{aligned}
\tag{5.50}
$$

and for a negative sequence ac voltage

$$
\begin{aligned}
v_d = {} & N\left\{\frac{V_k 6\sqrt{3}}{\pi}\cos\left(\frac{\mu_0}{2}\right)\cos\left[(k+1)\omega_0 t - \left(\alpha_0 + \frac{\mu_0}{2}\right) + \delta_k\right]\right\} \\
& - N\left\{\frac{V_k 6\sqrt{3}}{\pi}\sum_m \frac{\cos((m-1)\mu_0/2)}{m-1}\cos\left[(m-k-1)\omega_0 t - (m-1)\left(\alpha_0 + \frac{\mu_0}{2}\right) - \delta_k\right]\right\} \\
& + N\left\{\frac{V_k 6\sqrt{3}}{\pi}\sum_m \frac{\cos((m+1)\mu_0/2)}{m+1}\cos\left[(m+k+1)\omega_0 t - (m+1)\left(\alpha_0 + \frac{\mu_0}{2}\right) + \delta_k\right]\right\}
\end{aligned}
\tag{5.51}
$$

The second contribution can be obtained by substitution of the commutation period variation Equation 5.29 and 5.44 or 5.45 into Equation 5.49.

The ac side current is simply obtained by substituting the effective commutation period variation terms from Equations 5.33 and 5.46 or 5.47 to $Y_{\psi ac}$, and multiplying by I_d.

Observing the two contributions to the dc side voltage distortion, and noting that the direct term is the most important, followed by the $n = 1$ term of the modulation series shows that a positive sequence frequency on the ac side is transformed to that frequency minus the fundamental on the dc side, and a negative sequence frequency on the ac side is transformed to that frequency plus the fundamental on the dc side. At the higher orders, terms appear at the characteristic harmonics plus or minus the first order frequency on the dc side. Zero sequence frequencies on the ac side do not affect the dc side.

AC side terms appear at the same frequencies as those caused by a firing angle modulation at the first order dc side frequency.

Figure 5.15 shows the predicted magnitude and phase of transfer from ac voltage distortion to dc voltage distortion, for the first order terms only, compared with the results obtained by dynamic simulation of the CIGRE model rectifier (Appendix VI).

Sensitivity to dc current There are two ways the dc current variation affects the ac side current, directly from transfer via the unmodulated transfer function, and

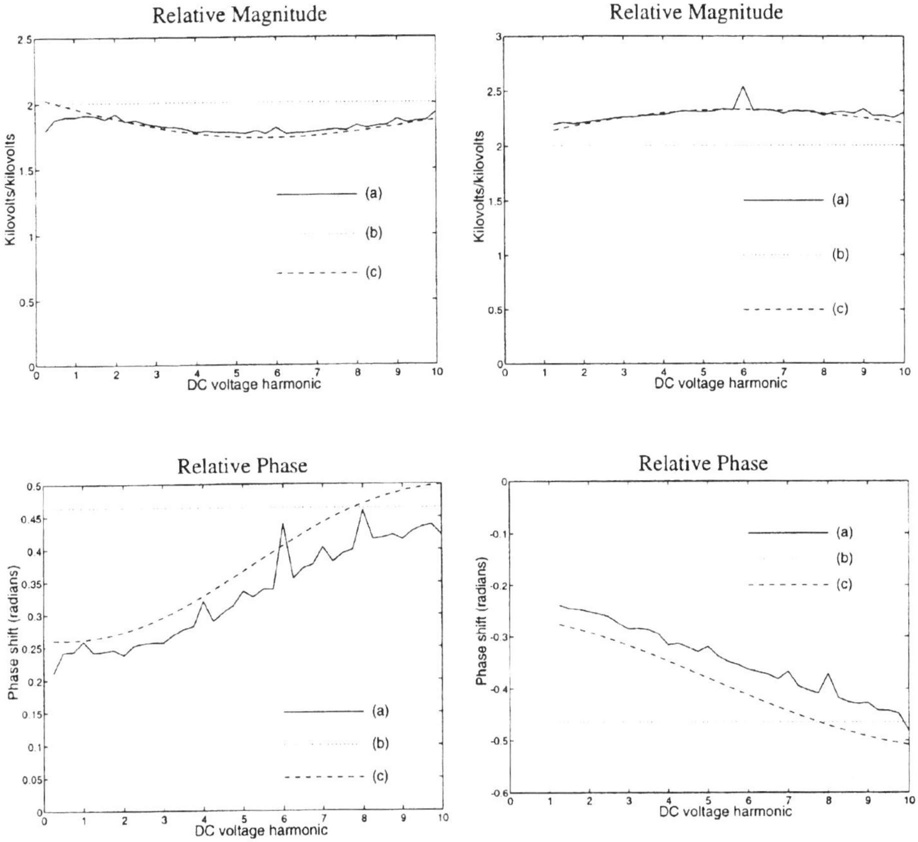

from +ve sequence voltage   from -ve sequence voltage

Figure 5.15 CIGRE rectifier harmonic transfer, -ve and +ve sequence ac voltage to dc voltage. For key see Figure 13

indirectly from the consequential commutation period modulation. The spectrum from the first mechanism is as follows

$$I_{ac_\psi} = N\left\{\frac{2\sqrt{3}}{\pi}\sum_m(\pm)\frac{I_k}{m}\frac{2}{m\mu_0}\sin\left(\frac{m\mu_0}{2}\right)\cos\left[(m+k)\omega_0 t - m\left(\alpha_0+\frac{\mu_0}{2}\right)+\delta_k - m\psi\right]\right\}$$
$$+N\left\{\frac{2\sqrt{3}}{\pi}\sum_m(\pm)\frac{I_k}{m}\frac{2}{m\mu_0}\sin\left(\frac{m\mu_0}{2}\right)\cos\left[(m-k)\omega_0 t - m\left(\alpha_0+\frac{\mu_0}{2}\right)-\delta_k - m\psi\right]\right\} \tag{5.52}$$

for $m = 1, 11, 13, 23$, etc. The spectrum from the second mechanism can be defined by substituting the commutation period variation term from Equations 5.33 and 5.40 into $Y_{\psi ac}$, and multiplying by I_d.

For the test case there are two ways that dc current distortion affects the converter dc side voltage. Firstly and most simply the commutation period modulation directly reflects a distortion onto the dc voltage. This can be defined by substituting the commutation period variation from Equations 5.29 and 5.39 into equation 5.49.

Secondly, the dc current distortion that is directly transferred through the converter passes through the converter transformer leakage reactance. This in turn generates a voltage distortion at the valve terminals, which is transferred back onto the dc side by the converter transfer function. To determine these terms Equation 5.52 is multiplied by the frequency dependent impedance of the converter transformers (for a 12-pulse converter, the leakage reactance of both transformers in parallel) to yield an ac voltage distortion, which is in turn passed back to the dc side by substitution of all the terms into Equations 5.50 and 5.51 as appropriate. This results in a proliferation of terms, most of which are not significant. The term of most interest, that at the original distorting frequency, is as follows

$$\begin{aligned} v_d = & \left\{ \frac{2\sqrt{3}}{\pi} \frac{(1+k)X_c}{2} I_k \frac{2}{\mu_0} \sin\left(\frac{\mu_0}{2}\right) \frac{6\sqrt{3}}{\pi} \cos\left(\frac{\mu_0}{2}\right) \cos\left(k\omega_0 t + \delta_k - \frac{\pi}{2}\right) \right\} \\ & + \left\{ \frac{2\sqrt{3}}{\pi} \frac{(k-1)X_c}{2} I_k \frac{2}{\mu_0} \sin\left(\frac{\mu_0}{2}\right) \frac{6\sqrt{3}}{\pi} \cos\left(\frac{\mu_0}{2}\right) \cos\left(k\omega_0 t + \delta_k - \frac{\pi}{2}\right) \right\} \end{aligned} \tag{5.53}$$

The full spectrum has been derived in [7].

Figure 5.16 shows the predicted and simulated ac current distortion resulting from a dc current distortion. Again, only the lowest order components are shown.

5.3.4 Discussion

Each of the plotted graphs demonstrate several important points concerning transfer of distortion through the ac–dc converter. One is the importance of the dynamic variation of the commutation period on transfer of waveform distortion. This is shown by the the difference between the curves showing the transfer function predictions including and excluding commutation period variation. For the case of firing angle modulation to dc voltage, shown in Figure 5.13, the commutation period duration variation was at least as important as its absolute duration, and both have a significant effect. For other transfers, the commutation period variation affects the distortion transfer by up to 20%.

Reasonable agreement is demonstrated between the transfer function predictions and the time domain simulation verification. Agreement between such different approaches to converter modelling implies that the linearizations made in the transfer function model are justifiable. The validity of the linearizing assumptions decreases with increasing distortion magnitude and frequency, which can be seen in the loss of accuracy with increasing frequency in almost all the predictions.

Some disagreement is exhibited between the frequency domain calculated and the time domain simulated results at integer multiples of the fundamental frequency. Integer harmonics have an associated spectrum that includes higher order terms that reflect back to the same frequency. These terms are most significant for the firing angle and commutation period variation effects, as the magnitudes of the higher order terms from this mechanism of transfer are far greater than the higher order terms from the more direct transfer of distortion through the converter. These terms

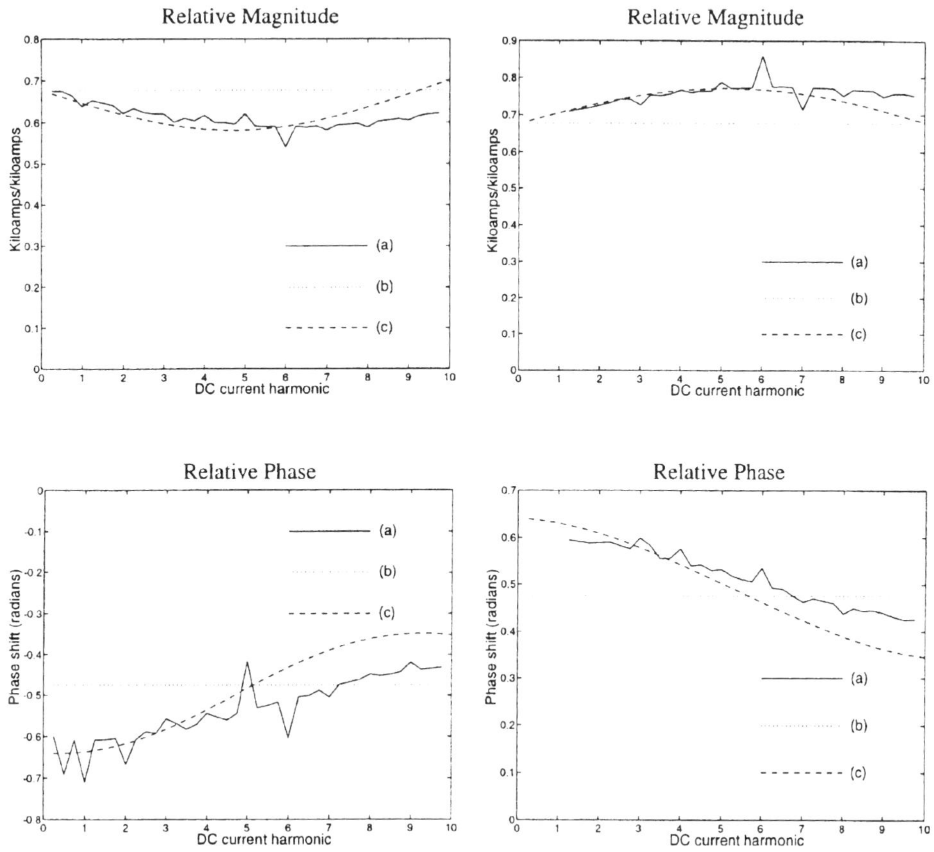

Figure 5.16 CIGRE rectifier harmonic transfer, dc current to +ve and -ve sequence ac current. For key see Figure 13

were not included in the frequency domain predictions. Interestingly, many of the higher order terms undergo a magnitude and phase transformation that makes the final harmonic transfer magnitude and phase dependent. This dependence makes the accurate prediction or simulation of integer harmonic transfer through the ac–dc converter substantially more difficult.

Although the effect of unbalance in the three phase supply is not discussed, the analysis is still directly applicable, by dividing the unbalance into positive, negative, and zero sequence components. Zero sequence components have no effect.

5.4 The Converter Frequency Dependent Equivalent

This section introduces the concept of a converter frequency dependent equivalent, and shows how through the relationships developed in Section 5.3 the converter equivalent can be constructed.

For the purpose of this section, the ac and dc systems around the converter are reduced to their equivalents, with the ac system modelled by a frequency dependent Thevenin source, and the dc system modelled by a frequency dependent Norton source. Cross-coupling between phases is not accounted for. The equivalents are illustrated in Figure 5.17(a).

A primary goal of this section is to reduce the converter and its associated ac system, or the converter and its associated dc system to a Thevenin or Norton equivalent. With the accompanying system equivalents, the harmonic interactions around the converter can be simply defined. This is illustrated in Figure 5.17(b) and (c).

The equivalents developed, due to the non-linear nature of the converter, will have cross-coupling between frequencies, and can be written in matrix notation.

The harmonic relationships for the converter, ac and dc systems shown are written in the following form

$$\overline{V}_{dc} = [A].\overline{V}_{ac} + [B].\overline{\beta} + [C].\overline{I_{dc}} \tag{5.54}$$

$$\overline{I}_{ac} = [D].\overline{I}_{dc} + [E].\overline{\beta} + [F].\overline{V_{ac}} \tag{5.55}$$

$$\overline{V}_{ac} = [Z_{ac}].\overline{I}_{ac} + \overline{V}_{ac_0} \tag{5.56}$$

$$\overline{I}_{dc} = [Y_{dc}].\overline{V}_{dc} + \overline{I}_{dc_0} \tag{5.57}$$

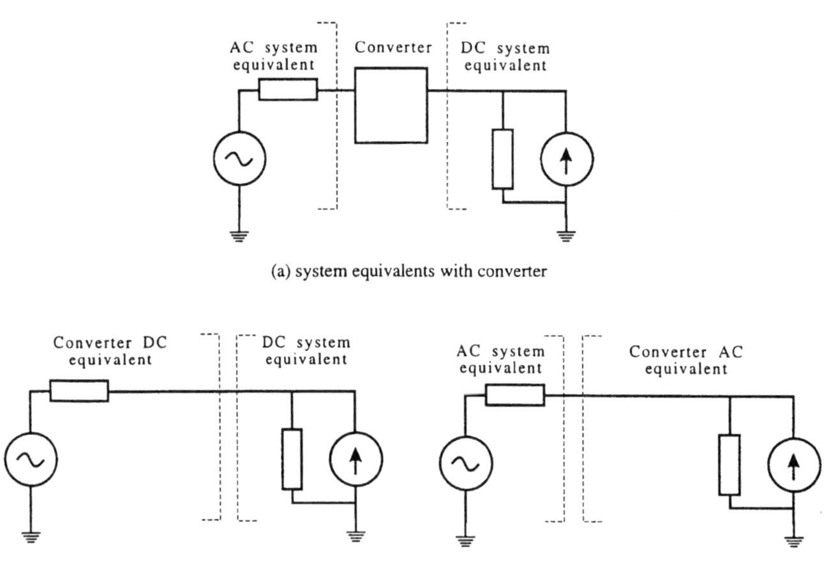

(a) system equivalents with converter

(b) DC side equivalents

(c) AC side equivalents

Figure 5.17 System equivalents with converter equivalents

where

$\overline{V}_{dc}$	is the vector of dc side harmonic voltages,
$\overline{I}_{dc}$	is the vector of dc side harmonic currents,
$\overline{V}_{ac}$	is the vector of ac side positive, negative and zero sequence harmonic voltages,
$\overline{I}_{ac}$	is the vector of ac side positive, negative and zero sequence harmonic currents,
$\overline{I}_{dc_0}$	is the vector of harmonic current sources on the dc side,
$\overline{V}_{ac_0}$	is the vector of harmonic voltage sources on the ac side,
$\overline{\beta}$	is the vector of firing angle harmonics,
Y_{dc}	is the diagonal matrix of dc side harmonic admittances, and
Z_{ac}	is the diagonal matrix of ac side harmonic impedances.

Matrices A,B,C,D,E, and F represent the converter transfer functions from ac voltage to dc voltage, firing angle to dc voltage, dc current to dc voltage, dc current to ac current, firing angle to ac current, and ac voltage to ac current respectively. As the frequency and phase sequence transformations are implicit in the structure of the matrices, each element need only describe an amplitude change and a phase shift, and can be written as a complex number. For some terms the phase angle of the harmonic is reversed as well as phase shifted, and the element will include a complex conjugation operator.

If the firing angle modulation can be related by a linear or linearised function to the dc voltage or current, Equations 5.54 to 5.57 can be manipulated into the form

$$\overline{V}_{dc} = [matrix_1].\overline{I}_{dc} + [matrix_2].\overline{V}_{ac_0} \tag{5.58}$$

and

$$\overline{I}_{ac} = [matrix_3].\overline{V}_{ac} + [matrix_4].\overline{I}_{dc_0} \tag{5.59}$$

The converter from the dc side can be viewed as an impedance with a voltage source, and from the ac side can be viewed as an admittance with a current source. If the harmonic sources on the dc side are zero, then from the ac side the converter will appear as an admittance matrix, and if the harmonic sources on the ac side are zero, from the dc side the converter will appear as an impedance matrix. This needs to be interpreted very carefully. Given only one external harmonic source, many harmonics will be generated. However, all these harmonics are approximately linearly related to the originating harmonic, and the response of the converter to the original harmonic, at the original harmonic, remains approximately linear. This is why the relationships can be represented as impedances or admittances.

The elements of these matrices can be derived by numerical methods [4], or by the relationships developed in Section 5.3, as is done here.

5.4.1 Frequency Dependent Impedance

Even a single non-characteristic frequency, when applied to a converter in the steady state, results in the generation of a multitude of non-characteristic frequencies. For a complete analysis, many frequencies need to be defined and many interactions considered.

However, if the converter is the main non-linear element in the network, it is of particular interest to consider the response of the converter to a non-characteristic or characteristic frequency, at that same frequency. This is very useful for simplified harmonic analyses, and particularly filter design.

A useful measure is the frequency dependent impedance of the converter. It is defined here as the returned current or voltage distortion as a result of an applied voltage or current distortion, at the same frequency. The response can be expressed as an admittance or an impedance. This impedance may have positive or negative real and imaginary values, which when taken in conjunction with the connected harmonic impedances may indicate light or even negative damping at some frequencies. This is basically a describing function technique.

The impedance discussed is the impedance of the converter Thevenin equivalent, or the admittance of the converter Norton equivalent, at the frequency of interest. Also to be defined is the voltage or current harmonic source that may exist behind the converter. To define this, all other externally generated harmonics need to be considered, along with their interactions, to see if an independent source exists at the frequency of interest. However, this source's very independence means that it does not affect the damping of the frequency in question, only the final steady state value.

From the transfer functions derived earlier, let us take only the most significant terms and write them as a set of simultaneous equations. As in the matrices, implicit in these equations is that the positive sequence V_{acp} and I_{acp} are at one harmonic higher than V_{dch} and I_{dch}, which are in turn one harmonic higher than the negative sequence V_{acn} and I_{acn}. This enables the relationships to be written as vector multiplications.

$$V_{dch} = a_1N(V_{acp} - a_5Z_{xp}NI_{dch}) + a_2N(V_{acn} - a_8Z_{xn}NI_{dch}) + a_3tfm_\alpha + a_4tfm_e \quad (5.60)$$

$$I_{acp} = a_5NI_{dch} + a_6Ntfm_\alpha + a_7Ntfm_d \quad (5.61)$$

$$I_{acn} = a_8NI_{dch} + a_9Ntfm_\alpha + a_{10}Ntfm_d \quad (5.62)$$

$$tfm_e = a_{11}NV_{acp} + a_{12}NV_{acn} + a_{13}I_{dch} + a_{14}tfm_\alpha \quad (5.63)$$

$$tfm_d = a_{15}NV_{acp} + a_{16}NV_{acn} + a_{17}I_{dch} + a_{18}tfm_\alpha \quad (5.64)$$

$$tfm_\alpha = a_{19}NV_{acp} + a_{20}NV_{acn} + a_{21}I_{dch} \quad (5.65)$$

$$V_{acp} = -I_{acp}Z_{acp} + V_{acp_0} \quad (5.66)$$

$$V_{acn} = -I_{acn}Z_{acn} + V_{acn_0} \quad (5.67)$$

$$I_{dch} = V_{dch}/Z_{dch} + I_{dch_0} \quad (5.68)$$

tfm is the transfer function modulation, subscript α indicating firing angle modulation, e indicating end of commutation period modulation, and d indicating duration of effective commutation period μ_1 modulation. Letting k be the frequency on the dc side of a 12-pulse converter, the terms a_1 to a_{18} have the following values

$$a_1 = \frac{6\sqrt{3}}{\pi}\cos\left(\frac{\mu_0}{2}\right)\angle \alpha_0 + \mu_0/2 \tag{5.69}$$

$$a_2 = \frac{6\sqrt{3}}{\pi}\cos\left(\frac{\mu_0}{2}\right)\angle -\alpha_0 - \mu_0/2 \tag{5.70}$$

$$a_3 = -\frac{NV_1 3\sqrt{3}}{\pi}\sin(\alpha_0) \tag{5.71}$$

$$a_4 = -\frac{NV_1 3\sqrt{3}}{\pi}\sin(\alpha_0 + \mu_0)\angle -k\mu_0 \tag{5.72}$$

$$a_5 = \frac{2\sqrt{3}}{\pi}\frac{2}{\mu_0}\sin\left(\frac{\mu_0}{2}\right)\angle -\alpha_0 - \mu_0/2 \tag{5.73}$$

$$a_6 = I_d\frac{2\sqrt{3}}{\pi}\frac{2\sin\left(\dfrac{(k+1)\mu_1}{2}\right)}{(k+1)\mu_1}\angle -\alpha_0 - \mu_1/2 - k\mu_1/2 - \pi/2 \tag{5.74}$$

$$a_7 = I_d\frac{\sqrt{3}}{\pi}\left(\frac{\sin\left(\dfrac{\mu_1}{4}\right)}{\dfrac{\mu_1}{4}}\right)^2\angle -\alpha_0 - \mu_1/\sqrt{x} - k\mu_1/\sqrt{x} - \pi/2 \tag{5.75}$$

$$a_8 = \frac{2\sqrt{3}}{\pi}\frac{2}{\mu_0}\sin\left(\frac{\mu_0}{2}\right)\angle \alpha_0 + \mu_0/2 \tag{5.76}$$

$$a_9 = I_d\frac{2\sqrt{3}}{\pi}\frac{2\sin\left(\dfrac{(k-1)\mu_1}{2}\right)}{(k-1)\mu_1}\angle \alpha_0 + \mu_1/2 - k\mu_1/2 + \pi/2 \tag{5.77}$$

$$a_{10} = I_d\frac{\sqrt{3}}{\pi}\left(\frac{\sin\left(\dfrac{\mu_1}{4}\right)}{\dfrac{\mu_1}{4}}\right)^2\angle \alpha_0 + \mu_1/\sqrt{x} - k\mu_1/\sqrt{x} + \pi/2 \tag{5.78}$$

$$a_{11} = \left[\frac{-\mu_0}{NV_1\sqrt{1-\left[\cos(\alpha_0) - \dfrac{2X_c I_d}{\sqrt{3}NV_1}\right]^2}}\right]\times$$
$$\left[\frac{2}{(k+1)\mu_0}\sin\left(\frac{(k+1)\mu_0}{2}\right)\right]\angle \alpha_0 + (k+1)\mu_0/2 - \pi/2 \tag{5.79}$$

$$a_{12} = \left\{ \left[\frac{-\mu_0}{NV_1\sqrt{1-\left[\cos(\alpha_0) - \dfrac{2X_cI_d}{\sqrt{3}NV_1}\right]^2}} \right] \times \left[\frac{2}{(k-1)\mu_0} \sin\left(\frac{(k-1)\mu_0}{2}\right) \right] \right\}$$

$$\underline{/-\alpha_0 + (k-1)\mu_0/2 + \pi/2} \tag{5.81}$$

$$a_{13} = \frac{2X_c \cos\left(\dfrac{k\mu_0}{2}\right)}{\sqrt{3}NV_1\sqrt{1-\left(\cos\alpha_0 - \dfrac{2X_cI_d}{\sqrt{3}NV_1}\right)^2}} \underline{/k\mu_0/2} \tag{5.81}$$

$$a_{14} = \frac{\sin(\alpha_0)}{\sqrt{1-\left(\cos\alpha_0 - \dfrac{2X_cI_d}{\sqrt{3}NV_1}\right)^2}} \tag{5.82}$$

$$a_{15} = \left\{ \frac{3NV_1\mu_0}{2(X_cI_d)^2}[\sin(\alpha_0+\mu_0) - \sin(\alpha_0) - \mu_0\cos(\alpha_0)] \times \frac{2}{(k+1)\mu_0}\sin\left(\frac{(k+1)\mu_0}{2}\right) \right\} \underline{/\alpha_0 + (k+1)\mu_0/2 - \pi/2} \tag{5.83}$$

$$a_{16} = \left\{ \frac{3NV_1\mu_0}{2(X_cI_d)^2}[\sin(\alpha_0+\mu_0) - \sin(\alpha_0) - \mu_0\cos(\alpha_0)] \times \left[\frac{2}{(k-1)\mu_0}\sin\left(\frac{(k-1)\mu_0}{2}\right)\right] \right\} \underline{/-\alpha_0 + (k-1)\mu_0/2 + \pi/2} \tag{5.84}$$

$$a_{17} = \frac{\mu_0}{I_d}\left[1 + \frac{2}{k\mu_0}\sin\left(\frac{k\mu_0}{2}\right)\underline{/k\mu_0/2}\right] + \frac{2}{kI_d}\left[\frac{2}{k\mu_1}\sin\left(\frac{k\mu_1}{2}\right)\underline{/k\mu_1/2} - 1\right]\underline{/\pi/2} \tag{5.85}$$

$$a_{18} = \frac{\sqrt{3}NV_1}{X_cI_d}\mu_0\sin(\alpha_0) - 2 \tag{5.86}$$

These terms depend on the actual and effective commutation periods, defined as follows

$$\mu_0 = \cos^{-1}\left[\cos\alpha_0 - \frac{2X_cI_d}{\sqrt{3}NV_1}\right] - \alpha_0 \tag{5.87}$$

$$\mu_1 = 2\mu_0 - \frac{\sqrt{3}NV_1}{X_cI_d}[\mu_0\cos(\alpha_0) + \sin(\alpha_0) - \sin(\alpha_0+\mu_0)] \tag{5.88}$$

where V_1 is the peak single phase ac voltage at the converter transformer primary, X_c is the converter transformer leakage reactance referred to the converter transformer secondary, and I_d is the average dc current.

a_{19}, a_{20} and a_{21} depend on the converter firing angle controller, and must be user specified. They are specified here for the type of control described in section 5.3.3. For the constant current control

$$a_{19} = 0 \tag{5.89}$$

$$a_{20} = 0 \tag{5.90}$$

$$a_{21} = G_1 \tag{5.91}$$

These equations describe the interaction as shown in Figure 5.18. This is similar to the three port network model proposed by [4]. Linear superposition of the different effects is assumed, which means for example that commutation period modulation due to ac voltage does not affect the commutation period enough to significantly modify the effect of dc current modulation of the commutation period. This assumption is reasonable for small distortion levels. Interactions at higher order frequencies also exist, but at insignificant levels when their reflections back onto the original frequency are considered.

The frequencies depicted in Figure 5.18 are of primary importance. In a similar way to the commutation period variation, firing angle modulation will be mainly confined to the frequency one harmonic less than the positive sequence ac side distortion, one harmonic more than the ac side negative sequence distortion, and the same frequency as the dc side distortion.

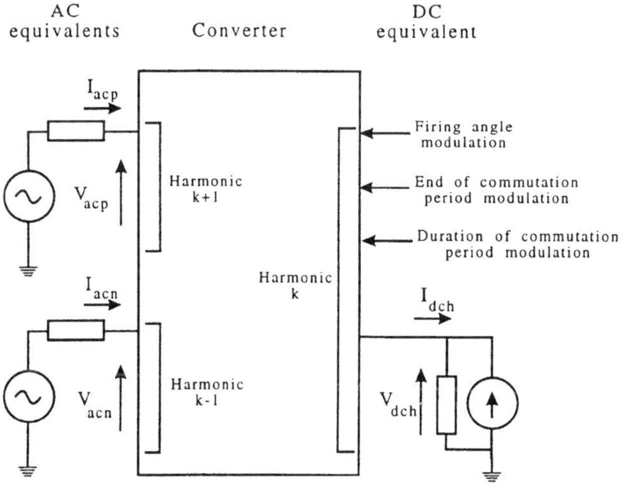

Figure 5.18 First order non-characteristic frequency interactions around a converter

The interrelationship between the three harmonic sequences around the converter expressed by Equations 5.60–5.68 can be reduced and rewritten by a 3 × 3 matrix as follows [8]

$$\begin{bmatrix} V_{dch} \\ I_{acp} \\ I_{acn} \end{bmatrix} = \begin{bmatrix} a & b & c \\ d & e & f \\ g & h & i \end{bmatrix} \begin{bmatrix} V_{acp} \\ V_{acn} \\ I_{dch} \end{bmatrix} \tag{5.92}$$

where the following assignments are made

$$\begin{aligned}
a &= a_1N + a_3a_{19}N + a_4a_{11}N + a_4a_{14}a_{19}N \\
b &= a_2N + a_3a_{20}N + a_4a_{12}N + a_4a_{14}a_{20}N \\
c &= -a_1a_5N^2Z_{xp} - a_2a_8N^2Z_{xn} + a_3a_{21} + a_4a_{13} + a_4a_{14}a_{21} \\
d &= a_6a_{19}N^2 + a_7a_{15}N^2 + a_7a_{18}a_{19}N^2 \\
e &= a_6a_{20}N^2 + a_7a_{16}N^2 + a_7a_{18}a_{20}N^2 \\
f &= a_5N + a_6a_{21}N + a_7a_{17}N + a_7a_{18}a_{21}N \\
g &= a_9a_{19}N^2 + a_{10}a_{15}N^2 + a_{10}a_{18}a_{19}N^2 \\
h &= a_9a_{20}N^2 + a_{10}a_{16}N^2 + a_{10}a_{18}a_{20}N^2 \\
i &= a_8N + a_9a_{21}N + a_{10}a_{17}N + a_{10}a_{18}a_{21}N
\end{aligned} \tag{5.93}$$

5.4.2 Converter DC Side Impedances

If no ac side distorting sources are allowed, Equations 5.66 and 5.67 can be rewritten

$$V_{acp} = -I_{acp}Z_{acp} \tag{5.94}$$

$$V_{acn} = -I_{acn}Z_{acn} \tag{5.95}$$

The solution of these combined with the Equation Set 5.92 is straightforward, and if the following assignments are made

$$\begin{aligned}
A &= -aZ_{acp} + \frac{bgZ_{acn}Z_{acp}}{1 + hZ_{acn}} \\
B &= c - \frac{biZ_{acn}}{1 + hZ_{acn}} \\
C &= \frac{egZ_{acn}Z_{acp}}{(1 + dZ_{acp})(1 + dZ_{acn})} \\
D &= \frac{f}{1 + dZ_{acp}} - \frac{eiZ_{acn}}{(1 + dZ_{acp})(1 + hZ_{acn})}
\end{aligned} \tag{5.96}$$

then the impedance of the dc side of the converter can be written

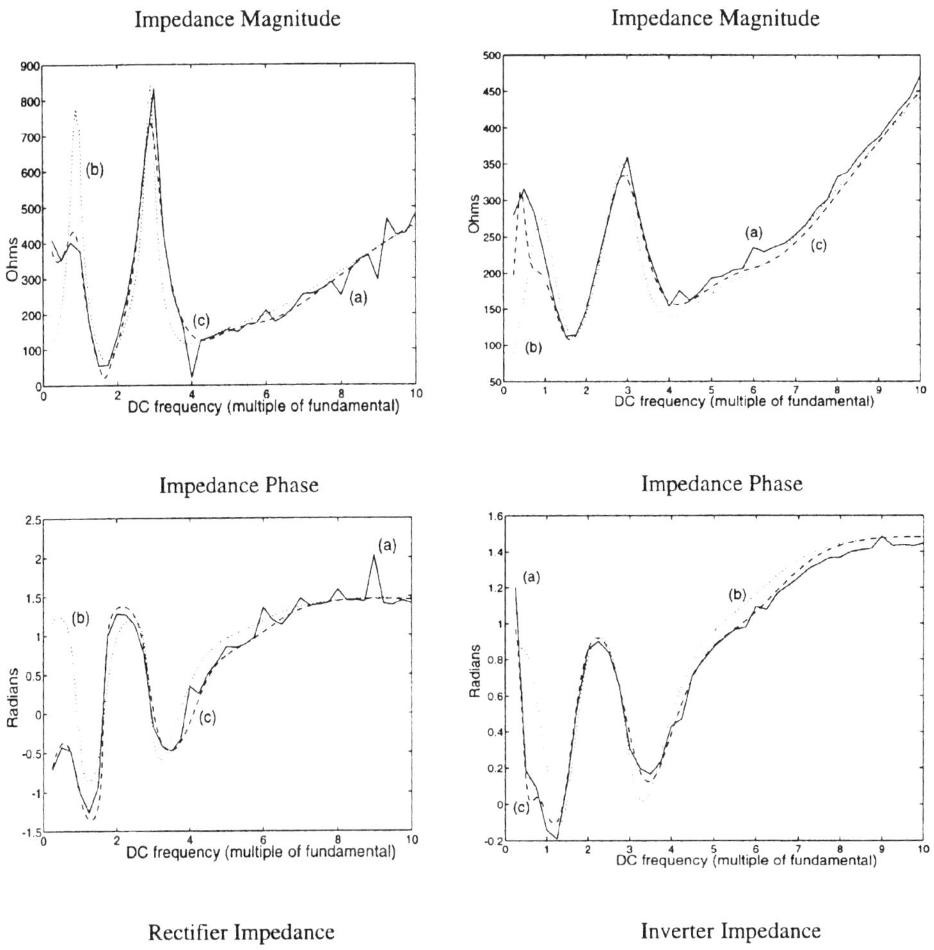

Figure 5.19 CIGRE rectifier and inverter dc side impedances

$$\frac{V_{dch}}{-I_{dch}} = -\left(\frac{AD}{1-C} + B\right) \tag{5.97}$$

In Figure 5.19 the measured and predicted impedances are shown for the CIGRE model rectifier and inverter. The rectifier is operating in constant current control, and the inverter is in minimum gamma control.

In each figure curve (a) represents the converter impedance measured from a dynamic simulation, curve (b) the impedance prediction without firing angle variation, and curve (c) the full impedance prediction. Useful agreement is obtained for both the current controlled rectifier and the minimum gamma controlled inverter. Numerical noise, and higher order terms in the simulated results give the jagged appearance of the plotted curve.

5.4.3 Converter AC Side Positive Sequence Impedances

If no dc side distorting source and no ac side negative sequence distorting source are allowed, Equations 5.67 and 5.68 can be rewritten

$$V_{dch} = I_{dch} Z_{dch} \tag{5.98}$$

and

$$V_{acn} = -I_{acn} Z_{acn} \tag{5.99}$$

The solution of these equations combined with Equation Set 5.92 is again straightforward, and if the following assignments are made

$$\begin{aligned} A &= a - \frac{bgZ_{acn}}{1 + hZ_{acn}} \\ B &= c - \frac{biZ_{acn}}{1 + hZ_{acn}} \\ C &= d - \frac{egZ_{acn}}{1 + hZ_{acn}} \\ D &= f - \frac{eiZ_{acn}}{1 + hZ_{acn}} \end{aligned} \tag{5.100}$$

then the positive sequence admittance of the ac side of the converter can be written

$$\frac{I_{acp}}{V_{acp}} = C + \frac{DA}{Z_{dch} - B} \tag{5.101}$$

The converter impedance can be obtained merely by inverting this.

5.4.4 Converter AC Side Negative Sequence Impedances

If no dc side harmonic source and no ac side positive sequence harmonic source is allowed, Equations 5.66 and 5.68 can be rewritten

$$V_{dch} = I_{dch} Z_{dch} \tag{5.102}$$

and

$$V_{acp} = -I_{acp} Z_{acp} \tag{5.103}$$

The solution of these equations combined with Equation Set 5.92 is again straightforward, and if the following assignments are made

$$A = b - \frac{aeZ_{acp}}{1 + dZ_{acp}}$$

$$B = c - \frac{afZ_{acp}}{1 + dZ_{acp}}$$

$$C = h - \frac{geZ_{acp}}{1 + dZ_{acp}}$$

$$D = i - \frac{gfZ_{acp}}{1 + dZ_{acp}} \tag{5.104}$$

then the negative sequence admittance of the ac side of the converter can be written

$$\frac{I_{acn}}{V_{acn}} = C + \frac{DA}{Z_{dch} - B} \tag{5.105}$$

The converter impedance can be obtained merely by inverting this.

In Figure 5.20 the measured and predicted impedances are shown for the CIGRE model rectifier and inverter. The rectifier is operating in constant current control, and the inverter is in minimum gamma control.

In the figures curve (a) represents the converter impedance derived from a dynamic simulation, curve (b) the simple impedance prediction with no firing angle or commutation period variation, and curve (c) the full predicted impedance. Agreement is reasonable, and the converter commutation period and control interactions are shown to be significant.

5.4.5 Simplified Converter Impedances

The converter dc side impedance comprises three components, being the ac side plus transformer impedance at the higher positive sequence frequency, the ac side plus transformer impedance at the lower negative sequence frequency, scaled approximately by $(6/\pi)^2 N^2$, and a final more complicated contribution from the transfer function variation terms. Just the scaled sum of the ac side impedances gives a good approximation at all but the lowest frequencies.

Whenever there is significant smoothing reactance on the dc side of the converter, this component dominates the converter ac side impedance. The converter impedance at harmonic order k can then be simplified to a multiple of the dc side impedance, i.e.

$$Z_{+ve}(k) = (k - 1) \times Z_X \times C \tag{5.106}$$

$$Z_{-ve}(k) = (k + 1) \times Z_X \times C \tag{5.107}$$

$$\tag{5.108}$$

where Z_X is the dc system impedance (including the smoothing reactor) at fundamental frequency, and the constant C is determined by the converter characteristics. If the transfer function modulation could be ignored, the constant

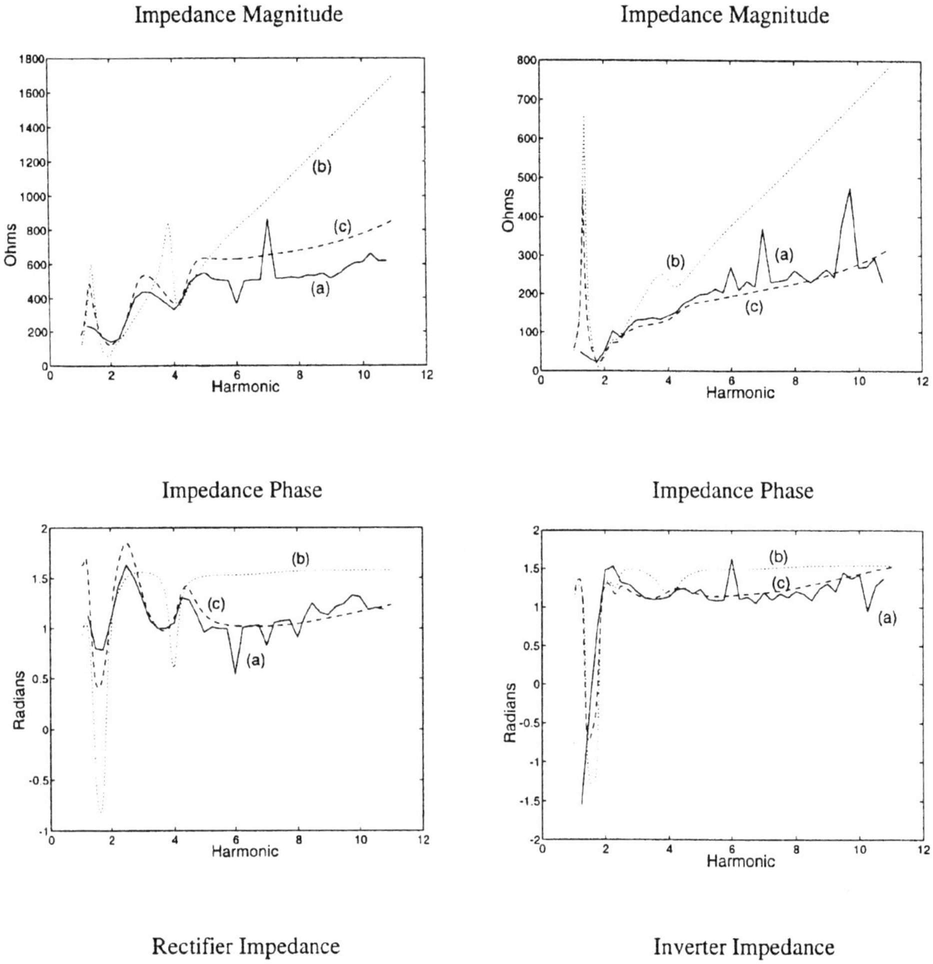

Figure 5.20 CIGRE rectifier ac side harmonic impedance, with firing angle control

C would be approximately $(\pi/6)^2.(1/N)^2$, but as is visible in Figure 5.20, transfer function modulation can have a substantial effect. For a 500 MW HVdc model [6], $Z_X \times C$ has a value of $44.0 \angle 68.75° \Omega$. The zero sequence impedance is dependent only on the transformer connections, and the presence of a delta winding.

5.4.6 Example of Application of the Impedance Models

To demonstrate the effect of converter representation on the harmonic impedances, the test system of Figure 5.21 is used. It is a simplified version of the lower South Island of the New Zealand system. A fictitious ac–dc converter is placed at the Tiwai bus with a rating of 500 MW and a set of twelve pulse related filters (11th, 13th and

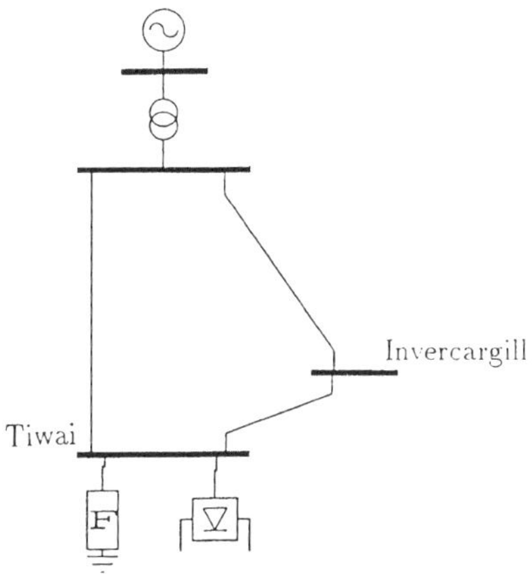

Figure 5.21 Lower South Island Test System

high pass) are also connected at this bus. It is assumed that a new non-linear load is proposed for connection at Invercargill, requiring local filtering. It is, therefore, necessary to calculate harmonic impedance information viewing the system from Invercargill, as an aid to ac filter design.

Figures 5.22 and 5.23 show the system impedance and reactance, respectively, calculated with models described in Chapter 4 (graphs (i) and (iv)) and those derived in this section (graphs (ii) and (iii)). The detailed and simplified cross-coupling model give very similar results. This can be attributed to a high dc side impedance due to the presence of a large dc smoothing reactor. Greater differences between these two models would be expected for a small dc side impedance.

As expected, near the harmonic filters tuned frequencies, the results are the same regardless of the converter model due to the dominance of the harmonic filters. However, outside these frequencies there are large differences between the models, particularly in the frequency range 100 Hz to 500 Hz. This is very significant since it is in this frequency range where important uncharacteristic harmonic resonances may occur.

5.5 Summary

Algebraic predictions of distortion transfer around a controlled 12-pulse ac–dc converter have been developed using frequency-domain defined transfer functions, and verified for low-order terms by comparison with results obtained by a time-domain technique. It has been established that to model waveform distortion transfer through an ac–dc converter with reasonable accuracy, it is necessary to include the influence of waveform distortion on the dynamics of the commutation period.

The transfer function analysis has also been used to calculate the harmonic impedances of ac–dc converters and the results have been used to assess the validity

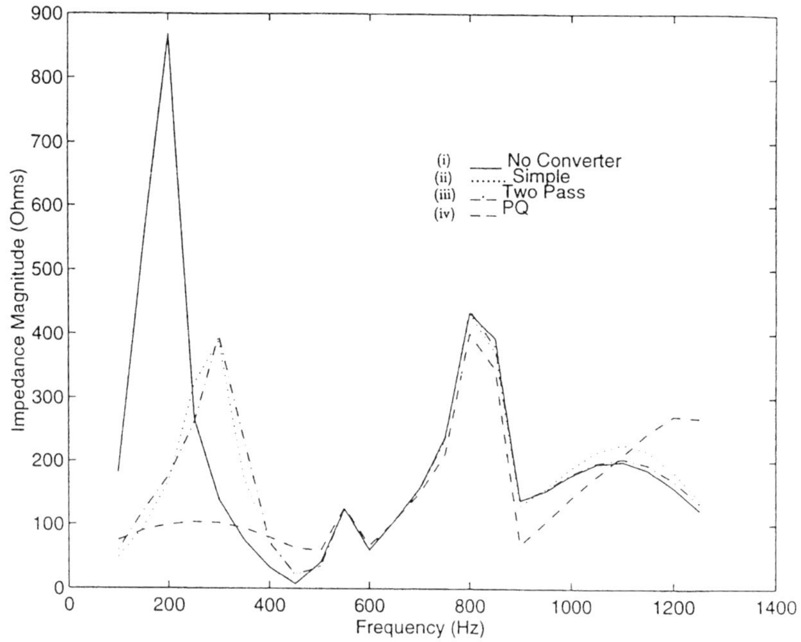

Figure 5.22 System impedance as seen from Invercargill

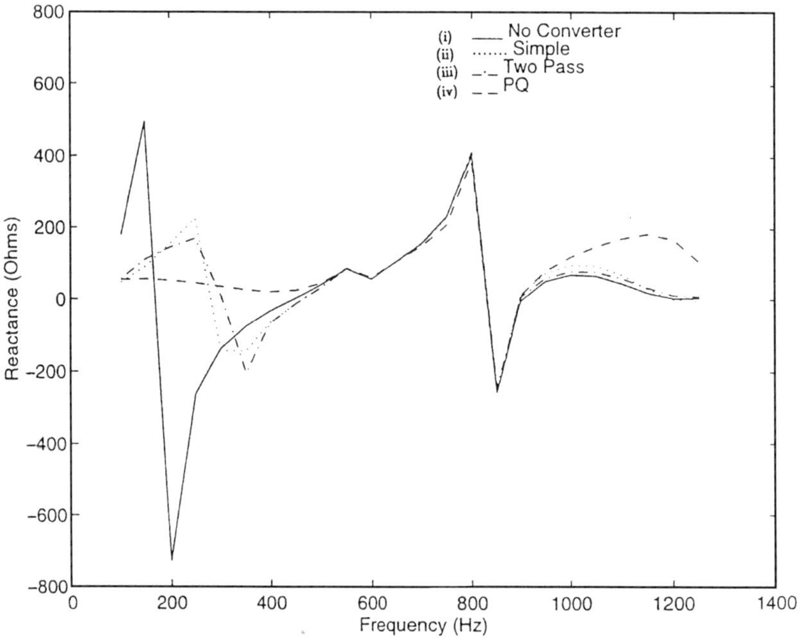

Figure 5.23 System reactance as seen from Invercargill

of other, less rigorous models. The current practice of either ignoring the presence of ac–dc converters, or using load-flow information, when calculating network harmonic impedances has been shown to be highly inaccurate at all frequencies in the absence of local filters. A simplified method derived from the rigorous analysis, on the other hand, has been shown to provide reasonable results for the frequency range of interest to harmonic filtering.

5.6 References

1. Arrillaga, J, Bradley, DA and Bodger, PS, (1985). *Power System Harmonics*. John Wiley & Sons, Chichester, UK.
2. Dobinson, LG, (1975). Closer accord on harmonics, *Electronics and Power*, **21**, 567–572.
3. Dommel, HW, (1969). Digital computer simulation of electromagnetic transients in single and multiphase networks, *IEEE Transactions on Power Apparatus and Systems*, **PAS–88**(4), 388–399.
4. Larson, EV, Baker, DH and McIver, JC, (1989). Low order harmonic interaction on ac/dc systems, *IEEE Transactions on Power Delivery*, **4**(1), 493–501.
5. Schwarz, M, Bennett, WR and Stein, S, (1966). *Communication Systems and Techniques*, McGraw-Hill, New York.
6. Szechtman, M, Weiss, T and Thio, CV, (1991). First benchmark model for hvdc control studies, *Electra*, **135**, 55–75.
7. Wood, AR, (1993). An analysis of non-ideal HVdc convertor behaviour in the frequency domain, and a new control proposal. PhD thesis, University of Canterbury, NZ.
8. Wood, AR and Arrillaga, J, (1994). The frequency dependent impedance of an hvdc converter, *ICHPs Conference, 21–24 Sept. 1994, Bologna, Italy*, September.

6

HARMONIC INSTABILITIES

6.1 Introduction

Ac–dc systems with low short circuit ratios (SCR) often experience problems of instability in the form of waveform distortion. The low SCR indicates a high ac system impedance, whose inductance may resonate with the reactive compensation capacitors and the harmonic filters installed at converter terminals. These resonant frequencies can be low, possibly as low as the second harmonic. The resonances can be excited under certain operating conditions or in the event of fault, and the small initial distortion may develop to an instability. Instability related to the interaction of harmonics (or any frequencies) has been customarily referred to as harmonic instability.

The problem of harmonic instability in ac–dc systems was first identified in relation to the individual valve firing control [1] and an alternative control principle, the phase locked oscillator [2], was adopted for new installations. Since then, other forms of harmonic instability have been identified, involving complementary resonances, composite resonances, cross-modulation and transformer core saturation.

The problem has in the past been discovered during or after the system had been commissioned and usually occurs only under certain unfavourable conditions. Ideally, the problem should be detected at the early planning stage to introduce counter-measures and to avoid costly modifications later.

The term complementary resonance has been used to describe the situation where a parallel resonance at a harmonic on the ac side is closely coupled to a series resonance on the dc side, via the three port frequency transforming characteristics of a converter. Although, strictly speaking, this does not imply an instability, a small remote injection can excite harmonic levels sufficient to compromise the operation of the converter.

Cross modulation is used to describe the frequencies generated when two ends of a dc link are operating at slightly different frequencies. The low frequencies generated can excite generator or motor shaft mechanical oscillation modes, which can similarly lead to shut-down of the plant.

Composite resonance takes an overall view of the converter embedded in its ac and dc systems, and looks for a true instability that requires only a very small excitation for the instability to grow in a self-sustaining manner. This may or may not include a contribution from converter transformer core magnetic saturation.

This chapter examines composite resonance without a core saturation contribution from a simple circuit analysis approach, and also directly analyses the instability that incorporates a converter transformer core saturation contribution.

6.2 Composite Resonance — A Circuit Approach

The term resonance is most often used with reference to isolated parts of an overall system, usually being either the ac or the dc system. This sort of electrical resonance is well defined, being the frequency at which the capacitive and inductive reactances of the circuit impedance are equal. At the resonant frequency, a parallel resonance has a high impedance and a series resonance has a low impedance. This approach has led to the concept of a complementary resonance; a high impedance parallel resonance on the ac side coupled with a low impedance series resonance at an associated frequency on the dc side.

A resonant circuit has another feature related to the circuit energy loss per period of the resonant frequency. In general terms, a quality factor Q can be defined as the total energy stored in the circuit divided by the total energy lost per period, multiplied by 2π. The quality factor gives a measure of damping at the resonant frequency.

When the dc and ac systems are interconnected by a static converter, the system impedances interact via the converter characteristics to create entirely different resonant frequencies. The term 'composite resonance' is used here to describe this sort of resonance, emphasising its dependence on all the components of the system. A special case of composite resonance, involving a converter transformer core saturation contribution is discussed in Section 6.4.

The composite resonant frequencies are relatively easy to determine. A point near the converter is selected (in this instance the dc terminals of the converter), and the equivalent admittances looking each way (i.e. looking into the dc system, and looking into the converter) are added. The simplified circuit is shown in Figure 6.1. The resulting admittance indicates a composite resonance when it is purely resistive.

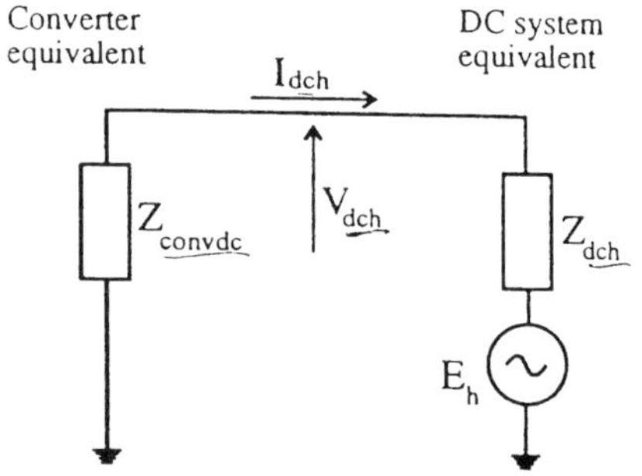

Figure 6.1 Converter and dc system electrical equivalent

A composite resonance may be excited by a relatively small distortion source in the system, or by an imbalance in the converter components or control. The amplification of a small source by the resonant characteristics of the system can present problems, and should be taken into account if steady state harmonic sources are expected.

Further to this, the converter impedance viewed from the dc side is comprised of several contributions. Firstly there is the ac side and converter transformer impedance, which usually sums to be largely inductive. Secondly, there is the end of commutation period dynamics, which is such that if the dc current out of the converter increases, the dc voltage reduces. This impedance looks mainly resistive. Finally, the constant current control modifies the converter dc terminal voltage according to the dc current. This can also be described as an impedance, although over a range of frequencies, the resistive component of this impedance will be negative.

A true instability results when, at the composite resonant frequency, the resistance of the overall circuit is negative. This can occur at non-integer frequencies, and is driven by conversion from the fundamental frequency and dc components to the composite resonance frequency via the converter control. Light damping, or ringing, during fault recovery, indicates that the negative resistance offered by the current controller is close to the natural resistance of the circuit.

The actual damping (positive or negative) of the network is easily determined, by approximation to a second order system. If the impedance of a series RLC circuit is plotted against angular frequency ω, the slope of reactance versus frequency at the resonant frequency is just 2L. If a series RLC circuit is fabricated that has the same resonant frequency, the same resistance at that frequency, and the same reactance versus frequency slope, then this circuit is easily solved to determine the damping in the more complex system. This is how the composite resonance damping factor of Table 6.1 was derived, for the test cases.

6.2.1 The Effect of Firing Angle Control on Converter Impedance

The approach outlined above requires a converter equivalent impedance to be defined, in this case being the converter impedance viewed from its dc terminals. This section quantifies how the control response affects the converter dc side impedance, for the case of constant current control

The expression for the converter dc side impedance, from equation (5.100) can be rewritten, extracting the terms associated with the type of control. For example, for a constant current controller, only the control term a_{21} has to be defined, and the dc side impedance of the converter can be written

$$Z_{\mathrm{convdc}} = -\left(\frac{AD}{1-C} + B\right) + fz.a_{21} \tag{6.1}$$

where

$$fz = \frac{A}{(1-C)(1+dZ_{acp})}\left(- a_6N - a_7a_{18}N + \frac{eZ_{acn}(a_9N + a_{10}a_{18}N)}{1+hZ_{acn}}\right)$$
$$- a_3 - a_4a_{14} + \frac{bZ_{acn}(a_9N + a_{10}a_{18}N)}{1+hZ_{acn}}. \qquad (6.2)$$

Using this relationship, any dc side converter impedance and composite resonance damping factor can be specified, within controller limitations.

The limitation that a controller cannot see into the future, coupled with the phase delay inherent in the added impedance function fz, limits the range of achievable impedances. In Figure 6.2, the frequency dependent transfer function of fz is plotted for the CIGRE benchmark test system rectifier [5] which shows clearly how a significant phase delay is introduced.

A feature of an uncontrolled converter dc side impedance is that while a low resistance is possible, a negative resistance is extremely unlikely. Simple constant current control will degrade the damping of distortion by adding a negative resistance at some frequencies, and a design goal should be to minimise the degradation near the composite resonant frequencies.

6.2.2 Test Case

The example used is the CIGRE model HVdc link [3]. The model was designed to present a difficult case for control, having a parallel resonance at the second harmonic on the ac side, and a series resonance at fundamental frequency on the dc side. The rectifier dc terminals are chosen as the point where the system impedances are added. Two alternative constant current control gains are used at the rectifier, selected to have slightly negative and slightly positive damping factors. A third case

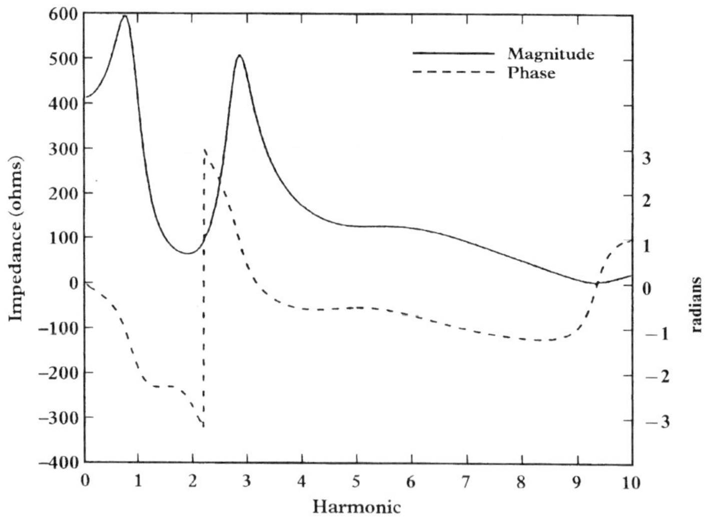

Figure 6.2 Transfer function for impedance contribution of constant current control

Table 6.1 Constant current control gains for test case examples

Ex. No	Prop. gain rad/kA	Int. time const.	HP filter	Composite res. damping factor
1	1.0714	0.0093	no	−5.1
2	0.9341	0.0107	no	+2.2
3	1.0714	0.0093	yes	+43

is presented where a high pass filter is placed in parallel with the PI control path, having a characteristic such that its gain can be written

$$HP = \frac{Gs^2/\omega_1^2}{1 + 2\zeta/\omega_1 + s^2/\omega_1^2} \tag{6.3}$$

where $G = 0.9$ rad/kA, $\zeta = 0.3$ rad/s, and $\omega_1 = 200\pi$ rad/sec. This was chosen to increase the damping factor at the composite resonance, without affecting the transient response at low frequencies. The dc current transducers have a time constant of 1 ms, and the inverter is operating in minimum gamma control.

Dynamic simulations are run for each case, in two cases involving a three phase fault on the inverter ac busbar for 70 ms to excite the composite resonance. The control gains for all three cases are given in Table 6.1.

Figure 6.3 shows the calculated real and reactive components of the series composite resonant circuit for examples 1 and 3. The underdamped resonant point, indicated by a zero reactive component and a low resistance, occurs at about 1.5 times the fundamental frequency, or 75 Hz. This frequency is mainly defined by the ac and dc networks, with small changes in constant current control gain having little effect.

The constant current control gain has a strong effect on the composite resonance damping factor through its effect on the circuit resistance. Table 6.1 gives the selected constant current control gains and the calculated composite resonance damping factors. Rather higher gains than would normally be employed are chosen to show light positive and negative damping. The effect of the high pass filter in example 3 is to increase the circuit resistance at the composite resonance, at the cost of introducing a significant negative resistance at a higher frequency and an additional, but well damped, composite resonance at about 180 Hz. However, both these features are less critical than the original lightly damped composite resonance.

The results of dynamic simulation runs are shown in Figure 6.4. Example 1 demonstrates the composite instability predicted by the analysis, and example 2 shows the lightly damped resonance also predicted by the analysis. Example 3 represents the response for the example 1 system including the additional high pass filter control block. Because the high pass filter block allows a higher PI gain to be used, transient overcurrents can be substantially reduced.

Table 6.2 presents both the calculated and simulated time for the waveform distortion to halve (for positive damping) or double (for negative damping).

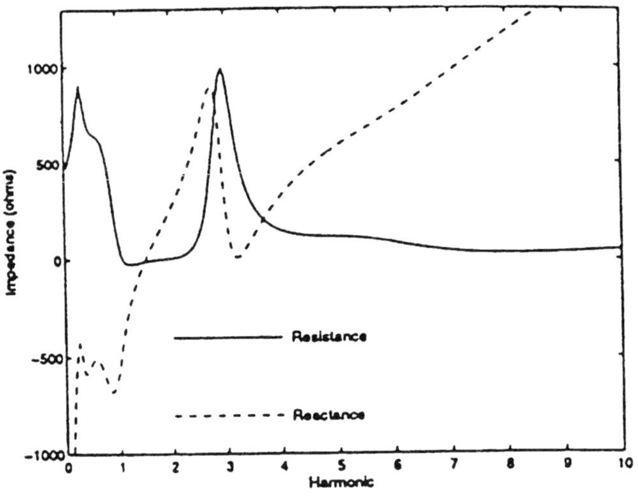

(a) Example 1

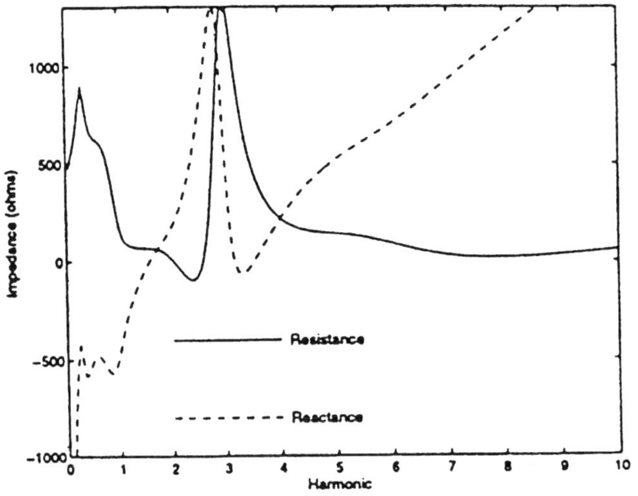

(b) Example 2

Figure 6.3 Composite circuit series impedance

Agreement is shown for all the examples. The resonant frequency for the simulation is at 67.5 Hz, some 7.5 Hz lower than predicted. The explanation for this is unclear, although it could arise from approximations in the transfer function coefficients or the estimated impedance of the inverter, or in the dynamic simulation.

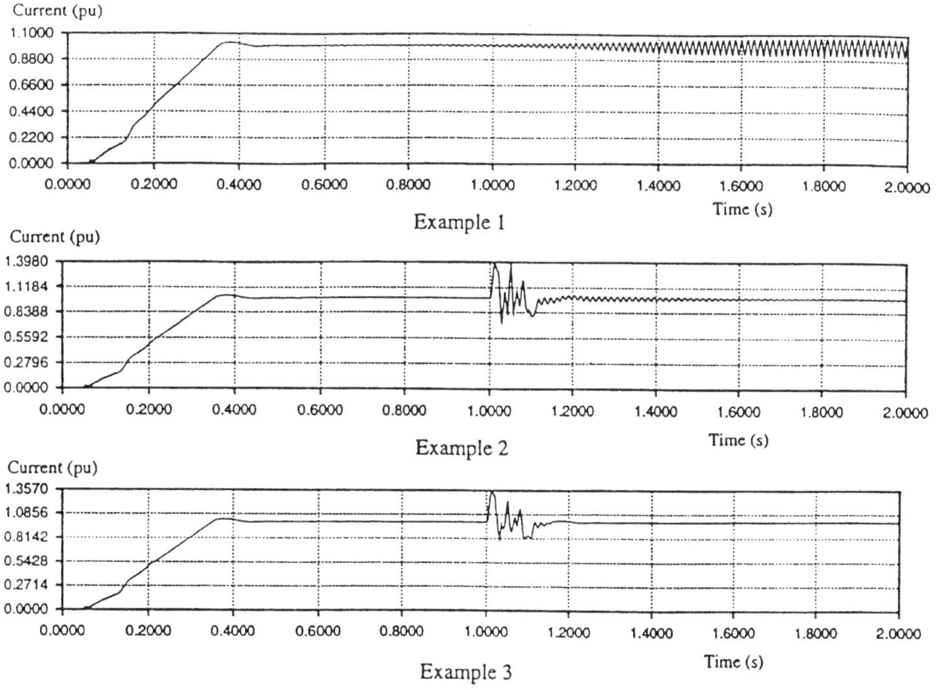

Figure 6.4 Rectifier dc current transient responses

6.2.3 Discussion

The dynamic simulation indicates that a predicted instability rises only to a certain level and then stabilizes. As the firing angle modulation amplitude rises, its effect does not rise proportionally, due to the non-linearity of the effect of firing angle modulation on the converter. The exception to this is when there is another source of waveform distortion, such as valve commutation failure or transformer core saturation, both of which render this analysis inadequate.

The algebraic analysis indicated that an additional control block that reduces the control gain around the composite resonance would improve the system response. A

Table 6.2 Comparison of calculated and simulated waveform distortion damping

Example No	Calculated time for distortion to halve/double	Simulated time for distortion to halve/double
1	0.14 sec	0.15 sec
2	0.32 sec	0.3 sec
3	0.02 sec	0.02 sec

control block was incorporated to do this, which damped the instability whilst keeping the high gain and, therefore, high speed PI control. This control solution was not optimised.

The frequency domain analysis presented takes into account the inverter dc side impedance, but with a very crude model of the minimum gamma control response, which is of importance below the fundamental frequency. The control system of the converter at the other end of the dc link must be accounted for if it is effective at the frequency in question. Typically, a minimum gamma controller generates a significant negative resistance which may contribute to a lightly damped low frequency sub-harmonic in the dc link transient response.

It should be noted that the equations used in this analysis can be converted into a controlled system block diagram format, so that more traditional control design tools can be used.

6.3 Transformer Core Related Harmonic Instability in AC–DC Systems

Transformer core saturation often has an additional amplifying effect for harmonic instability. There have been several reported incidences of core saturation instability, at the Kingsnorth [4], Nelson River [5] Blackwater back-to-back intertie [6] and Chateauguay [7] schemes. Despite these incidences, there is little information on the nature of the phenomenon and this may have led to some incidences being misinterpreted as another type of harmonic instability or resonance. The control solutions are very similar, typically involving some sort of firing angle modulation and in some cases the installation of additional harmonic filters. However, the above contributions provide no information on how to predict the likelihood of a HVdc scheme experiencing such an instability. To carry out such predictions requires a thorough understanding of the mechanisms involved.

The objective of this section is to provide an overview and facilitate the understanding of transformer core related instability, as well as suggesting the criteria to predict the system susceptibility to this type of instability

6.3.1 AC–DC Frequency Interactions

The mechanism of ac-dc harmonic interaction has already been discussed in Chapter 5 and is summarised in Figure 6.5. The presence of a harmonic distortion at k times fundamental frequency on the dc side of a 12-pulse HVdc converter will produce on the ac side, positive sequence harmonics of orders $k+1$, $13 \pm k$, $25 \pm k$, $12n + 1 \pm k$; $n = 3, 4, \ldots$, and negative sequence harmonics of orders $k - 1$, $11 \pm k$, $23 \pm k$, $12n - 1 \pm k$; $n = 3, 4 \ldots$ These harmonic sequences are reflected back to the dc side as the kth harmonic and various high order harmonics of $12 \pm k$, $24 \pm k$, $12n \pm k$; $n = 3, 4 \ldots$ Among these harmonics, the most significant terms are the first order kth harmonic on the dc side, and the positive sequence $k + 1$ and negative sequence $k - 1$ harmonics on the ac side. The higher harmonics are an order of magnitude smaller than the lower order harmonics. Therefore, for most analyses, particularly those

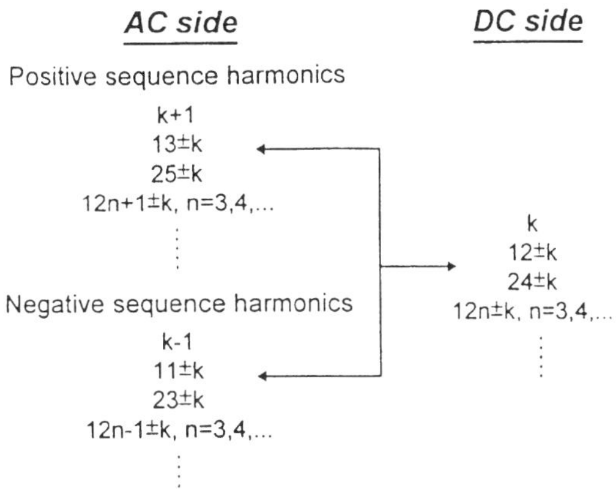

Figure. 6.5 AC-dc harmonic interactions

with small distortion levels, it is reasonable to ignore the contribution from high order harmonics. Figure 6.5 gives a representation of the harmonic orders that can be expected on each side of an HVdc converter.

With the high order harmonics ignored, the presence of a 2nd harmonic distortion on the dc side will result in a positive sequence 3rd harmonic and a negative sequence fundamental frequency component on the ac side.

The interaction mechanism can be extended to non-harmonic frequencies. For instance, if there is a distortion near the fundamental frequency such as at 51 Hz on the dc side, the distortions on the ac side would be near the second harmonic (101 Hz) for the positive sequence component and near dc (1 Hz) for the negative sequence component. As the frequency of the dc side distortion approaches fundamental frequency, the lower corresponding frequency component on the ac side, which is in the negative sequence format, will be approaching 0 Hz, i.e. approaching dc.

If the dc side distortion is exactly at fundamental frequency, the negative sequence component on the ac side is true dc, but with different levels in the three phases, which is customarily regarded as 'unbalanced dc' generated by the converter. However, the sum of the dc distortions in the three phases will be zero, and these distortions can be written mathematically in a negative sequence format as follows:

$$
\begin{aligned}
I_a &= |I|\cos(0.t + \delta + 0^\circ) \\
I_b &= |I|\cos(0.t + \delta + 120^\circ) \\
I_c &= |I|\cos(0.t + \delta + 240^\circ).
\end{aligned} \tag{6.4}
$$

This form of distortion can be represented by three stationary vectors of similar length, and oriented in the negative sequence format as shown in Figure 6.6. This form of dc generated by the converter is arbitrarily called 'negative sequence dc'.

The negative sequence dc concept is important in the analysis of converter transformer core saturation instability because the distortion on the converter dc side, related to this instability, is close to the fundamental frequency. Therefore, the

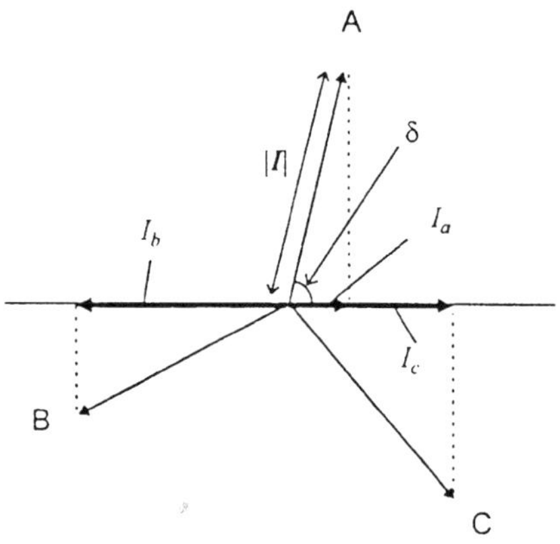

Figure 6.6 The form of negative sequence dc produced by HVdc converter

significant harmonic distortions on the converter ac side concerning this instability are the positive sequence second harmonic and the negative sequence dc. The dc will tend to saturate the transformer core which may ultimately lead to an instability.

6.3.2 Instability Mechanism

The mechanism of the converter transformer core saturation instability phenomenon can be demonstrated using the block diagram of Figure 6.7. If a small level of positive sequence second harmonic voltage distortion exists on the ac side of the converter, a fundamental frequency distortion will appear on the dc side. Through the dc side impedance, a fundamental frequency current will flow, resulting in a positive sequence second harmonic current and a negative sequence dc flowing on the ac side. The negative sequence dc will begin to saturate the converter transformer, resulting in a multitude of harmonic currents being generated, including a positive sequence second harmonic current. Associated with this current will be an additional contribution to the positive sequence second harmonic voltage distortion and in this way the feedback loop is completed. The stability of the system is determined by the characteristics of this feedback loop.

In Figure 6.7 the aforementioned instability feedback loop does not involve the entirety of the negative sequence dc produced by the converter. This is because, due to the dynamics of the instability, the dc side distortion is never exactly at the fundamental frequency, and therefore, the negative sequence dc is not a truc dc but is varying slowly. This variation can be visualized by the slow rotation, in clockwise or anti-clockwise direction, of the vectors in Figure 6.6. The level of dc component in the transformer valve side current will be changing and is in fact transferring itself between phases in a cyclic manner. However, since the variation is near dc, the phrase 'negative sequence dc' is used to refer to this extremely slow varying distortion which is oriented in a negative sequence format as explained in the preceding section. This variation, although sufficiently slow to cause transformer

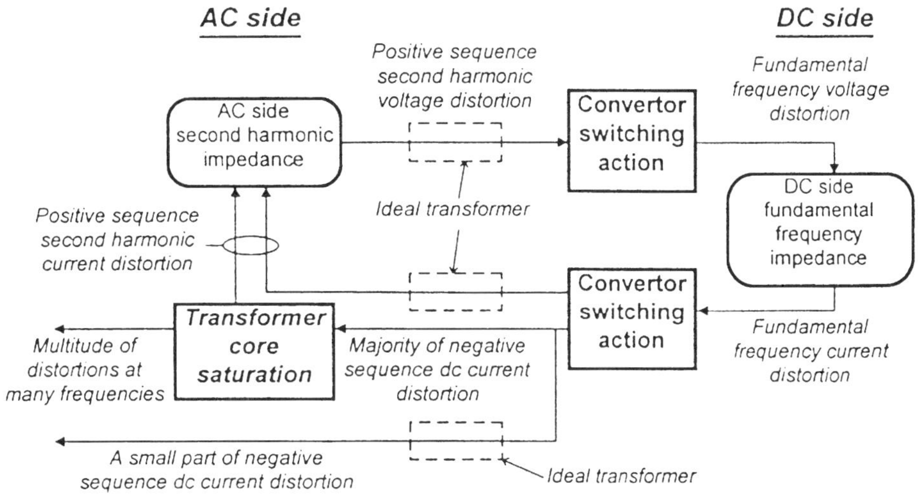

Figure 6.7 Mechanism of core saturation instability

saturation, is also sufficiently fast for a percentage of it to pass through the transformer and into the ac system. The faster the variation of this negative sequence dc, the more of it will pass through the transformer and the less of it will saturate the transformer, and vice versa. The portion that passes through the converter transformer is distributed into the ac network and may tend to dc-bias other transformers in the system, but it is unlikely to cause significant saturation to further contribute to the build-up of the instability. For accurate prediction of an instability, the division of current into magnetizing and transferred components is important, but if only an indication of the risk of instability is required, it may be neglected.

The onset of core saturation instability is closely related to the saturation level of the converter transformer. In this analysis, the instability is broadly divided into two categories, distinguished by their starting conditions. The first type has a spontaneous nature as it develops under normal operating condition without any external stimulus. System imbalances or asymmetry in the converter firings will result in low levels of transformer saturation which can ultimately develop into an instability. Study of this type of instability requires the evaluation of the transformer response at low saturation levels. The second type is referred as kick-started instability which may see substantial transformer saturation as the starting condition. Some disturbances may impress a high level of saturation on the converter transformer and consequently result in the development of core saturation instability after the disturbance. For this latter category of instability, the transformer response at high levels of saturation has to be determined.

6.3.3 Instability Analysis

The techniques used to analyse this instability can be grouped into the three categories of direct frequency domain [6,8], iterative frequency domain [7,9], and time domain simulation [10]. Due to the complicated nature of the interaction of the non-characteristic harmonics around the converter, the two latter numerical methods have been more popular than the linear approximation approach. This preference is further enhanced by advances in computer technology, enabling complicated and numerically intensive algorithms to be implemented in a short time. However, the direct frequency domain technique provides greater insight into the mechanism of the interaction and should lead to more effective control solutions. The following sections introduce a direct frequency domain method [8], which has been used to reveal various characteristics of the ac–dc systems susceptible to this instability [11].

To analyse the instability mechanism as shown in Figure 6.7, the ac–dc system is simplified to the equivalent circuit of Figure 6.8 consisting of a converter block interconnecting the ac and dc side impedances at the relevant frequencies. On the ac side, the positive sequence second harmonic current I_{acp} generated by the converter flows into the ac system second harmonic impedance Z_{acp}, while the negative sequence dc I_{acn} flows into a parallel circuit of transformer magnetizing inductance L_m and ac side impedance at the low frequency of I_{acn} variation. Assuming that the ac side impedance remains fairly constant around 0 Hz, the impedance can be simplified to the ac system resistance at 0 Hz of R_{acn}. On the dc side, the presence of fundamental frequency voltage distortion causes an equivalent current distortion to flow through the dc side impedance.

The converter block accounts for the transfer of ac side voltage distortion to the dc side and the dc side current distortion to the ac side. In reference [8], this operation of the converter in the frequency domain is represented with a linearised transfer function describing the conduction and non-conduction periods of the thyristors. Considering only the most significant low order harmonics, these interactions have been analysed in Section 5.4.1 with the help of a 3 × 3 matrix, which is reproduced in

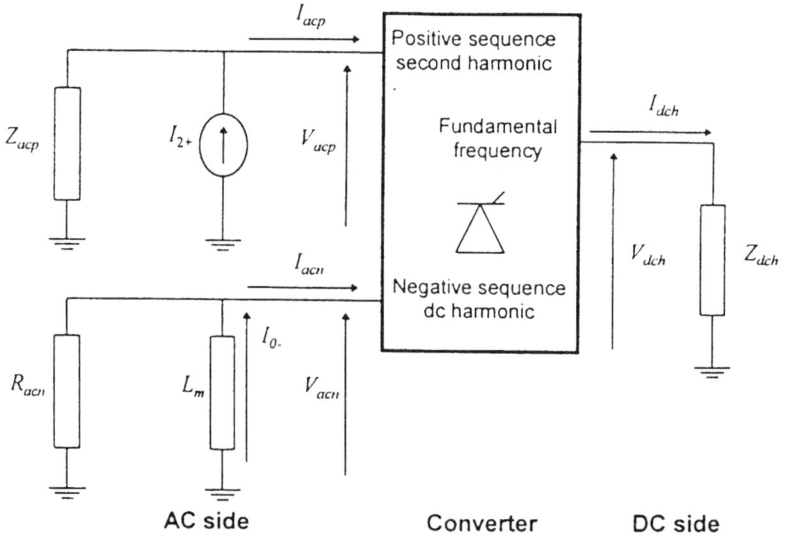

Figure 6.8 Equivalent circuit for analysis of core saturation instability

Equation (6.5). The elements within the matrix describe the amplitude change and phase shift introduced during the transfer of distortions from either sides of a converter. The response of the converter controller and the signal transducers is included in the matrix.

$$\begin{bmatrix} V_{dch} \\ I_{acp} \\ I_{acn} \end{bmatrix} = \begin{bmatrix} a & b & c \\ d & e & f \\ g & h & I \end{bmatrix} \cdot \begin{bmatrix} V_{acp} \\ V_{acn} \\ I_{dch} \end{bmatrix}. \tag{6.5}$$

From Figure 6.7, the contribution from the transformer saturation to the instability feedback loop comes in the form of an additional positive sequence second harmonic current resulting from the saturation. This effect is modelled as a positive sequence second harmonic current injection I_{2+} into the second harmonic part of the equivalent circuit. The depth of the saturation is calculated according to the amount of negative sequence dc from the converter that is flowing into the transformer magnetizing inductance L_m (i.e. I_{0-}).

The transformer saturation related to this instability can be regarded as asymmetrical saturation since the transformer is only saturated in one-half of a fundamental cycle, as shown in Figure 6.9. As the transformer is coming in and out of saturation, the magnetising inductance L_m is non-linear, but it is only in saturation for a short period of each cycle as indicated by the width of the distorted magnetising current pulse. Therefore, it is reasonable to assume L_m as the unsaturated magnetising reactance, as indeed it has this value most of the time.

Under the worst case scenario, the transformer magnetisation flux is taken as reaching the limits of the non-saturated part of its magnetisation characteristic, and the magnetisation characteristic of I_{mag}/flux is approaching infinity in the saturated region, as shown in Figure 6.9. Algebraic analysis has shown that under this condition, there is a one-to-one linear relationship between the resulting positive sequence second harmonic current I_{2+} and the level of saturating negative sequence

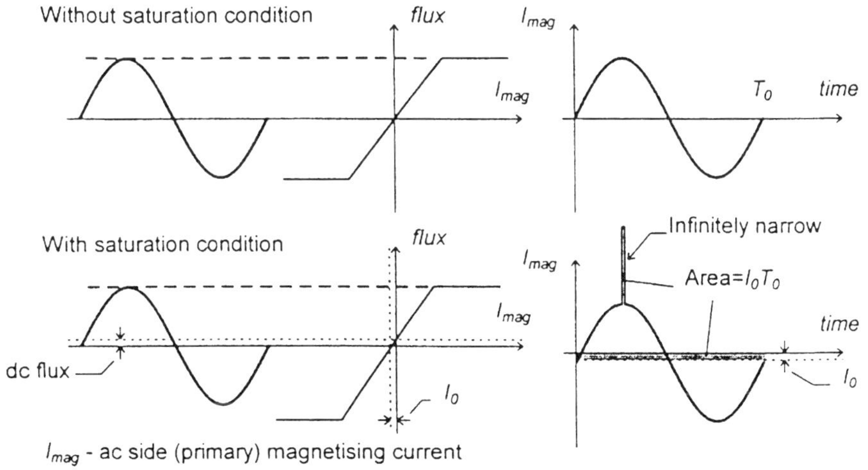

Figure 6.9 Transformer operation under worst case scenario

dc I_{0-}[8]. However, converter transformers are usually over-designed and there is always a considerable margin before reaching the saturation region. Moreover, the actual I_{mag}/flux ratio in the saturation region is far from infinite, and therefore, the relationship between I_{2+} and I_{0-} will realistically be less than one, depending on the magnetization characteristic of the transformer, i.e.

$$I_{2+} = -X.I_0 -, \qquad 0 < X < 1. \tag{6.6}$$

The reference point for phase is, as per the converter analysis, at the peak of the red phase fundamental voltage component.

Dynamic simulations are utilised to approximate this ratio X. Using a three-phase star-star transformer, a series of negative sequence dc currents are injected into the transformer secondary while measuring the amount of positive sequence second harmonic flowing into the primary. The amplitude of the negative sequence dc current is ramped up in stages and the corresponding amplitude of the positive sequence second harmonic current is plotted against it as shown in Figure 6.10. The dc injections are determined as percentages of the transformer rated magnetizing current, making the extent of the transformer saturation which is determined by the dc flux levels correspond directly to the level of injected currents. The measured values show a nearly linear relationship and by applying linear approximation, the slope of the curve is calculated as the ratio X for that particular transformer.

Figure 6.10 shows that at saturation levels below 40–50%, the measured values are less than the linear approximated values. The non-linear effect at low saturation levels is determined by calculating the value of X from the slope between the measured I_{2+} and the injected I_{0-} at each saturation level. The various values of X at different low saturation levels are summarised in Table 6.3 alongside the linear approximated values.

Table 6.3 also depicts the influence of the transformer knee point voltage level on the ratio X. Transformers with higher knee point voltage will be less prone to saturation and, hence, result in lower harmonic contribution. The value of the ratio

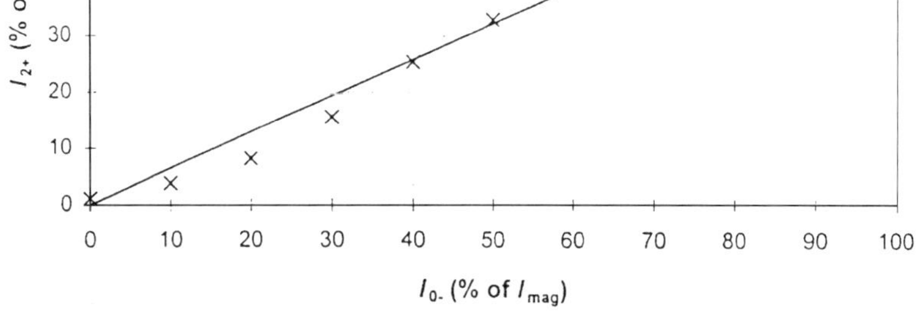

Figure 6.10 Measured and linearised I_{2+} versus injected I_{0-}

Table 6.3 X values at different saturation levels

Knee point voltage level (p.u.)	Injected I_{0-} (% of I_{mag})					Linear Approx.
	10	20	30	40	50	
1.05	0.51	0.60	0.80	0.92	0.96	0.85
1.15	0.38	0.47	0.63	0.74	0.80	0.74
1.25	0.33	0.42	0.53	0.60	0.65	0.65
1.35	0.32	0.38	0.49	0.51	0.55	0.56
1.45	0.29	0.37	0.43	0.46	0.47	0.49

X will also vary according to other characteristics of the transformer, including its levels of magnetising current and saturated reactance.

Therefore, depending on the category of core saturation instability (i.e. spontaneous or kick-started) under study, a series of simulations need to be undertaken on the particular transformer under analysis, and the calculated relationship linearized and applied to the model to predict the system stability.

With the 3×3 matrix describing the converter operation, and the current injection modelling the transformer saturation effect, the stability of the system can be determined from the roots of the system characteristic equations. By expressing one of the concerned distortions, such as I_{acn}, in the exponential vector form $I_{acn}^{t=0}.e^{-(\alpha+j\beta)t}$ with $I_{acn}^{t=0}$ as the initial condition, the distortion will decay away resulting in stable systems if the real part of the root $-\alpha$ is negative (i.e. α is positive). This term α has been defined as the *Saturation Stability Factor* and used to indicate the susceptibility of an ac–dc system to the development of core saturation instability [8].

6.3.4 Dynamic Verification

This *Saturation Stability Factor* technique has been verified against dynamic simulations [8], but for completeness, a further validation is presented here. The parameters of the CIGRE benchmark model [3] were modified to create an unstable test case [12]. The resultant *Saturation Stability Factor* was lowered to become negative at −0.152 and the presence of instability was verified by the EMTDC dynamic simulation results of Figure 6.11. The system was run for one second to achieve a steady state, and then a fundamental frequency modulation was added to the rectifier firing angle order. This modulation caused harmonic distortions on both sides of the converter, including the negative sequence dc, which began to saturate the transformer. The system was subjected to this modulation for 0.5 second, and its stability assessed by the growth or decay of the harmonic distortions after the external stimulus is being removed.

The gradual rise of the distortion in the dc current indicates the onset of the instability. As the instability develops, the level of transformer saturation increases which tends to accelerate the development of the instability as indicated by the large increase in the amplitude of the saturating negative sequence dc after 3.5 seconds. The waveforms of the transformer magnetising current show that the three phases

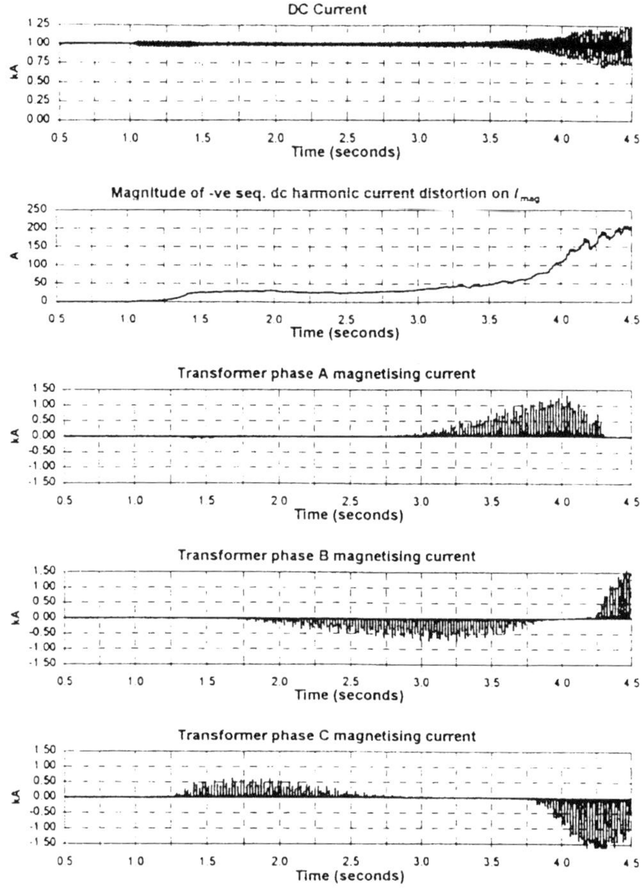

Figure 6.11 EMTDC simulation results of the modified CIGRE HVdc benchmark model

are not dc-biased to the same offset level, and there is a pseudo-sinusoidal variation alongside the increase in the saturating dc. This observation further validates the vector form of solution ($e^{-\alpha(+j\beta)t}$) proposed by the *Saturation Stability Factor* approach that the distorting harmonic sequences not only vary in their magnitude but also rotate over time. In this test system, the vector rotation term of β was found to be positive at 0.188, which implies that the negative sequence dc vector of Figure 6.6 will rotate in the clockwise direction, forcing the dc component in the magnetizing current to vary in the phase order of A, B, C, as shown by the simulation results.

6.3.5 Characteristics of the Instability

The direct frequency domain approach described above possesses great potential for the evaluation of the properties exhibited by systems prone to this type of instability. Its minimal computational burden and hence quick solution provides an effective way of investigating the system characteristics under such unstable conditions. With the use of linear approximation and a direct solution, each individual factor can be easily altered to unravel its particular effect on the instability mechanism. Furthermore, by considering several factors simultaneously, it is possible to find out the dominant factor contributing to the build up of the instability. Analysis using the *Saturation Stability Factor*, described in [11] has revealed that a vulnerable HVdc rectifier system is likely to have the following impedance profile:

- A low and predominantly capacitive dc side impedance at the fundamental frequency with the presence of a series resonance near to but higher than the fundamental frequency,
- A high and predominantly inductive ac side second harmonic impedance with the presence of a parallel resonance near to but higher than the second harmonic frequency,
- A high ac side resistance near 0 Hz.

On the other hand, a susceptible invertor system will possess opposite reactive characteristic with inductive dc side impedance at fundamental frequency, and capacitive ac side second harmonic impedance. A high 0 Hz resistance is also observed at the unstable invertor station but the two reactive components have the dominant role in determining the system stability.

The common use of HVdc back to back interties to interconnect large and weak ac networks has resulted in low order resonances at the converter terminals, making them prone to core saturation instability. However, with comparable network sizes at both the rectifier and invertor ends, this harmonic instability is most likely to occur only at one end of the scheme. This is due to the opposite reactive requirements of the impedances for the instability to occur at either ends. Moreover, the high resistance at the unstable end will be reflected onto the dc side as additional damping which tends to stabilise the opposite end system. Therefore, for the back-to-back scheme, it is necessary to consider the consequential impact on the stability of the remote end system when undertaking any modification at the local end.

Besides the system impedances, the stability of the ac–dc system is strongly dependent on the response of the converter controller. Considering the stringent reactive requirements for the instability to develop, the onset of this harmonic instability almost certainly involves a destabilising contribution from the converter controller. This suggests the possibility of preventing the onset of the instability through proper tuning of the converter controller.

6.3.6 Control of the Instability

The converter transformer core saturation instability can be prevented by operating the system away from the unstable conditions or in other words, providing sufficient damping at the relevant frequencies. This type of action may involve modification to the system impedances by filter retuning, tuning of the converter controller parameters or the converter steady state operating parameters. Although the purpose of the changes to these parameters is to modify the system response at the frequencies related to this instability, it usually affects the system response at other frequencies as well. The design of such preventative measures has to ensure that other system requirements or constraints are still met after the modifications. These actions can be broadly regarded as passive measures.

On the other hand, active measures can be applied to stabilise the system when the development of the instability is detected. This type of solution has been used to prevent core saturation instability in existing schemes, with some sensing instruments estimating the level of transformer saturation and appropriate action taken in accordance with the extent of the saturation. Active measures should be designed to function at certain limited range of frequencies without altering the system response significantly under normal operating conditions. This will allow the system to be operated as usual, but with the added security of some stabilizing action when the instability is suspected.

Due to the great differences in the characteristics of the various HVdc systems, it is difficult to pinpoint which is the best solution to counter this instability. The high dependency of the system stability on the properties of the HVdc scheme suggests that the most appropriate solution for one system may not suit the others. Moreover, each HVdc scheme usually has its own unique requirements and restrictions that have to be fulfilled. It is therefore necessary to undertake independent analysis for different systems or for a similar system under different operating conditions.

The addition of a high pass filter to the converter controller has been found to be effective for this particular test case, but had to be properly tuned to avoid exciting a composite resonance at about 70 Hz [12] With the high pass filter, the system Saturation Stability Factor was evaluated to be positive at 0.168 indicating stability, which is confirmed by the dynamic simulation results of Figure 6.12.

6.4 Summary

The linearized direct frequency domain model, described in Chapter 5, has been used to simulate the converter and a composite resonance concept has been introduced,

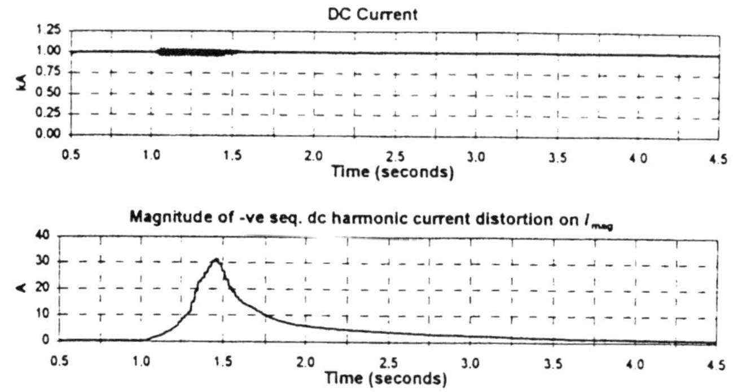

Figure 6.12 EMTDC simulation results with the addition of an auxiliary high pass filter to the converter controller

involving the converter, converter control, and ac and dc side harmonic impedances as an integrated whole. The ac–dc system has been represented as an electrical network.

It has been shown that the converter control system is a critical factor in determining system waveform stability, and that the dominant frequency is not limited to integer multiples of the fundamental frequency.

A description of the interactions of three frequencies around the converter has helped in predicting the system susceptibility to converter transformer core saturation instability.

Due to the sensitive dependence of the system stability on numerous parameters, it is necessary to undertake detailed analysis on the particular installation in order to reach the most appropriate solution.

Although in the simple example a constant current control is used, the set of equations can be manipulated in a similar way to analyse the effect of any control strategy. Further to this, the design of special control blocks to manage composite resonances at higher frequencies may be undertaken.

The technique developed here should prove a useful tool in optimizing converter control systems, with particular regard to recovery from system transients and avoidance of harmonic instabilities.

6.5 References

1. Ainsworth, JD, (1967). Harmonic Instability Between Controlled Static Converters and a.c. Networks, *IEE Proceedings*, **114**(7), 949–957.
2. Ainsworth, JD, (1968). The Phase Locked Oscillator—A New Control System for Controlled Static Converters, *IEEE Transaction on Power Apparatus and Systems*, **PAS-87**(3), 859–865.
3. Szechtman, M, Wess T and Thio, CV, (1991). First benchmark model for HVdc control studies, *Electra*, **135**, 55–75.
4. Ainsworth, JD, (1977). Core Saturation Instability in the Kingsnorth HV-d.c. Link, Paper presented to CIGRE study committee No.14, Winnipeg, Canada.
5. Chand, J and Tang, D, (1987). Experience with Resonances and Oscillations in the Nelson River HVdc System, HVdc System Operating Conference, Winnipeg, Canada.
6. Stemmler, H, (1987). HVdc Back-to-back Interties on Weak a.c. Systems, Second Harmonic Problems, Analyses and Solutions, CIGRE Symposium, **09-87**, Paper No. 300-08, 1–5, Boston.
7. Hammad, AE, (1992). Analysis of Second Harmonic Instability for the Chateauguay HVdc/SVC Scheme, *IEEE Transactions on Power Delivery*,. **7**(1), 410–415.
8. Chen, S Wood, AR and Arrillaga, J, (1996). HVdc Converter Transformer Core Saturation Instability: A Frequency Domain Analysis, *IEE Proceedings–Generation; Transmission; Distribution*, **143**(1), 75–81.
9. Yacamini, R and de Oliveira, JC, (1980). Instability in HVdc Schemes at Low Integer Harmonics, *IEE Proceedings Pt. C*, **127**(3), 179–188.
10. Burton, RS, (1994). Report on Harmonic Effects on HVdc Control and Performance, *CEA No. 337 T 750*, Prepared by Manitoba HVdc Research Centre.
11. Chen, S, Wood, AR and Arrillaga, J, (1996). A Direct Frequency Domain Investigation of the Properties of Converter Transformer Core Saturation Instability, *Accepted for*

presentation at the IEE Sixth International Conference on 'a.c. and d.c. Transmission', London, U.K., 173–178.
12. Chen, S. (1996). Analysis of HVdc Converter Transformer Core Saturation Instability, and Design of a Data Acquisition System for its Assessment. PhD Thesis, University of Canterbury, Christchurch, New Zealand.

7

MACHINE NON-LINEARITIES—HARMONIC DOMAIN

7.1 Introduction

In the models described in Chapter 4 the generators and transformers have been represented by frequency dependent impedances at harmonic frequencies.

In practice, design restrictions and cost effectiveness will result in some distortion of the machine's internal emf's. Moreover, the use of iron cores makes the electromagnetic coupling between windings of rotating machines and transformers non-linear due to saturation. Also, the frequency conversion process of the non-ideal rotating machine will react to the presence of system asymmetry or distortion by returning other frequency components.

Under certain operating conditions some of these effects may influence the network harmonic content and need to be represented in the analysis.

As in the case of the ac/dc converter, discussed in Chapters 5 and 6, machine non-linearities interact with the network parameters and operating conditions and, in general, iterative rather than direct solutions are required. The iterative methods are discussed in Chapter 9 but the machine models involved in the solution are described here, with particular regard to the synchronous generator and transformer. These are described in a Harmonic, rather than Frequency, Domain where the interaction between all the relevant harmonics is solved in a common unified algorithm.

7.2 Synchronous Machine

The windings of synchronous machines are distributed along the stator surface and, therefore, the mmf's of the coils are displaced from each other in space. Moreover, the displacement angle is different for the various harmonics, since their pole pitches are different. This causes odd non-triplen mmf harmonics in the three-phase machine, the fifth travelling in the negative direction, the seventh in the positive direction, etc.

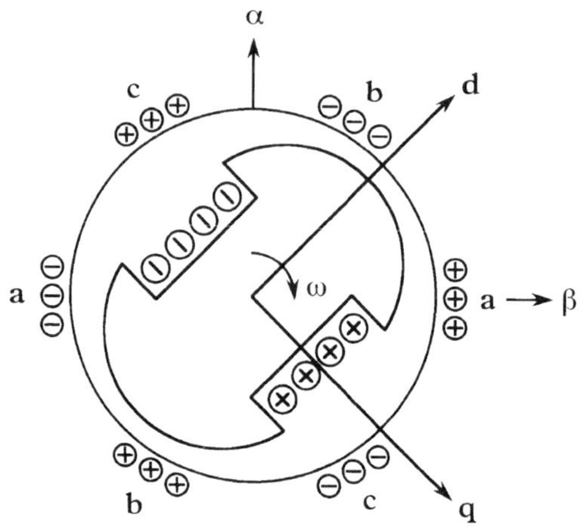

Figure 7.1 Basic machine representation

Armature slotting is another source of waveform distortion. If the machine has mg slots per pole, the corresponding emf components will have frequencies of orders 2 mg ± 1.

However, with the increased size of modern generating plant and greatly improved synchronous machine designs, the voltage distortion due to slotting and winding distribution have become less significant, and will rarely need to be taken into account in system studies.

On the other hand, the rotor saliency effect interacting with system asymmetry and/or distortion may produce non-negligible harmonic content, and needs to be taken into account.

Following Park's two-reaction theory [1,2], the modelling of a salient pole synchronous machine was based on the transformation from phase (*a,b,c*) to direct and quadrature axes (d,q) components as depicted in Figure 7.1.

The use of $\alpha\beta 0$ components was later proposed (Clarke [3]) as a natural transition between *abc* quantities and *dq*0 components and finally $\alpha\beta\gamma$ components [4] were used to modify the zero sequence quantities and obtain a power invariant transformation.

The harmonic representation of the synchronous machine based on empirical models or as an emf in series with hx''_d, or the average of hx''_d and hx''_q, as proposed in Chapter 4, do not take into account the process of harmonic conversion; a more rigorous model is required. Moreover, the process of frequency conversion is also affected by saturation and by the characteristics of the external network.

7.2.1 The Frequency Conversion Process

The asymmetry of the winding distribution and structure of the rotor, coupled to some unbalance or distortion in the stator currents, create harmonic mmf's in the rotor which in turn induce harmonic voltages in the stator.

In a cylindrical rotor machine, a negative sequence current of order h injected to the stator will create a flux of order $h+1$ in the rotor, which in turn will induce a negative sequence voltage of order h. For the case of a positive sequence current of order h injected to the stator a flux of order $h-1$ is produced, which in turn produces a positive sequence voltage of order h.

For the case of a salient pole machine, a negative sequence current of order h produces two counter-rotating fluxes of order $h+1$ in the rotor; one inducing a negative sequence voltage of order h and the other a positive sequence voltage of order $h+2$ in the stator. Similarly, a positive sequence current of order h in the stator produces counter rotating fluxes of order $h-1$, which in turn induce a positive sequence voltage of order h and a negative sequence voltage of order $h-2$. The same process is repeated for a current injection of order $h-2$ and of higher harmonic orders in the stator. This is illustrated in Figure 7.2 for cases where the stator current contains either positive or negative seventh harmonic content.

In general terms, it can be noted that in phase co-ordinates any harmonic current of order h injected in the stator will produce harmonic voltages of orders $h-2$, h and $h+2$.

7.2.2 Harmonic Model in *dq* Axes

The synchronous machine periodic behaviour is represented by a cross-coupled, multi-harmonic three-phase model derived from quantities in dq axes using accepted variable notations.

The differential equations describing the machine behaviour in dq axes [5] are

$$v_d = Ri_d + p(L_d i_d + M_{df} i_f + M_{ds} i_s) - \omega(L_q i_q + M_q i_t) \tag{7.1a}$$

$$v_q = Ri_q + p(L_q i_q + M_{qt} i_t) + \omega(L_d i_d + M_{df} i_f + M_{ds} i_s) \tag{7.1b}$$

$$v_f = 0 = R_f i_f + p(L_f i_f + M_{df} i_d + M_{fs} i_s) \tag{7.1c}$$

$$v_s = 0 = R_s i_s + p(L_s i_s + M_{ds} i_d + M_{fs} i_f) \tag{7.1d}$$

$$v_t = 0 = R_t i_t + p(L_t i_t + M_{qt} i_q) \tag{7.1e}$$

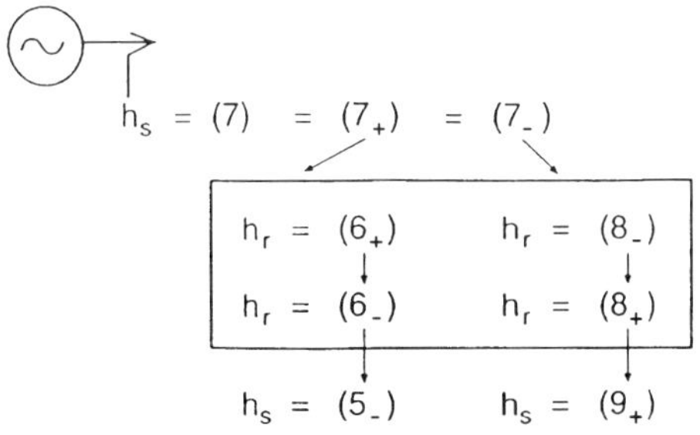

Figure 7.2 Response of a salient pole generator to the presence of a harmonic current

The variables s and t in Equation (7.1) represent closed damper windings in axes d and q, respectively. In the analysis to follow, it is assumed that the applied field voltage v_f contains no harmonics and, therefore, this variable has been set to zero in Equation (7.1c). The harmonic domain linearization process requires a phasor representation of Equations (7.1). At a particular frequency h, the required variables of Equations (7.1) are described in phasor form as follows [6]:

$$w_h = Re\{w_h e^{jh\omega t}\} \tag{7.2}$$

where phasor w_h is

$$w_h = w'_h + jw''_h \tag{7.3}$$

In the harmonic domain the operator p is defined as

$$p = jh\omega \tag{7.4}$$

Writing Equations (7.1) in phasor form and solving for dq quantities yields

$$\bar{V}_{dq} = \{[Z_{dq}] - [Z_{dq-fst}]^{-1}[Z_{fst-dq}]\}\bar{I}_{dq} \tag{7.5}$$

For harmonic h Equation (7.5) is written in compact form as

$$\bar{V}_{dqh} = [Z_{dqh}]\bar{I}_{dqh} \tag{7.6}$$

7.2.3 Two-phase Transformation *dq* to $\alpha\beta$

The relationship between the space-stationary orthogonal dq axes and a three-phase (abc) winding is determined depending on whether this winding is stationary or rotating. To account for both conditions, a winding transformation has been defined [3] between the three-phase (abc) and hypothetical two-phase orthogonal ($\alpha\beta$) windings, such that the phase separation between the winding axes remains constant irrespective of winding rotation. The α axis is rigidly fixed to phase a and the β axis is 90 degrees ahead, in the direction of rotation. If the three-phase winding is stationary, the $\alpha\beta$ and dq axes are presumed coincident. However, if winding rotation is to be considered, then a second transformation between the rotating two-phase $\alpha\beta$ and stationary dq windings is required. This case, associated with a practical synchronous machine in operation, is modelled in the analysis to follow.

The transformation between $\alpha\beta$ and dq components of voltage can be determined with the relationship given by Clarke [3]. Using the convention $h' = h \pm 1$ for the resultant harmonics, the matrix equation is written as

$$\begin{bmatrix} V'_{\alpha h} \\ V'_{\beta h} \end{bmatrix} = \begin{bmatrix} \cos\omega t & -\sin\omega t \\ \sin\omega t & \cos\omega t \end{bmatrix} \begin{bmatrix} V_{dh} \\ V_{qh} \end{bmatrix} \tag{7.7}$$

Phasors V_{dh} and V_{qh} can be expressed in the trigonometric form of Equation (7.7) as

$$V_{dh} = V'_{dh} \cos h\omega t - V''_{dh} \sin h\omega t \tag{7.8a}$$

$$V_{qh} = V'_{qh} \cos h\omega t - V''_{qh} \sin h\omega t \tag{7.8b}$$

where

$$\cos h\omega t = \tfrac{1}{2}(\mathrm{e}^{\mathrm{j}h\omega t} + \mathrm{e}^{-\mathrm{j}h\omega t}) \tag{7.9a}$$

$$\sin h\omega t = \tfrac{1}{2}\mathrm{j}(\mathrm{e}^{\mathrm{j}h\omega t} - \mathrm{e}^{-\mathrm{j}h\omega t}) \tag{7.9b}$$

$$\mathrm{e}^{\mathrm{j}h\omega t} = \cos h\omega t + \mathrm{j} \sin h\omega t \tag{7.9c}$$

$$\mathrm{e}^{-\mathrm{j}h\omega t} = \cos h\omega t - \mathrm{j} \sin h\omega t \tag{7.9d}$$

Similar expressions apply for cos ωt and sin ωt. Substitution of Equations (7.8) and (7.9) into (7.7) yields, when combined into phasors of the form given by Equation (7.3), the following equations

$$V^{h}_{\alpha h-1} = \tfrac{1}{2}(V_{dh} - \mathrm{j}V_{qh}) \tag{7.10a}$$

$$V^{h}_{\alpha h+1} = \tfrac{1}{2}(V_{dh} + \mathrm{j}V_{qh}) \tag{7.10b}$$

$$V^{h}_{\beta h-1} = \tfrac{1}{2}(\mathrm{j}V_{dh} + V_{qh}) \tag{7.10c}$$

$$V^{h}_{\beta h+1} = \tfrac{1}{2}(-\mathrm{j}V_{dh} + V_{qh}) \tag{7.10d}$$

A particular solution of Equation (7.7) for $h = 0$ will result in Equations (7.10) multiplied by a factor of 2. The transformation between $\alpha\beta$ and dq quantities is obtained assembling together Equations (7.10), which in matrix form can be expressed as

$$\bar{V}_{\alpha\beta} = [C]\bar{V}_{dq} \tag{7.11}$$

An identical transformation applies to the currents. For a practical harmonic spectrum $-n$ to $+n$, where n is the number of harmonics to be analysed, matrices $\bar{V}_{\alpha\beta}$, $\bar{V}_{dq}$ and $[C]$ have the form [7]

$$\bar{V}_{\alpha\beta} = [V_{\alpha-n}, V_{\beta-n}, \ldots, V_{\alpha-1}, V_{\beta-1}, V_{\alpha o}, V_{\beta o}, V_{\alpha 1}, V_{\beta 1}, \ldots, V_{\alpha n}, V_{\beta n}]^T \tag{7.12a}$$

$$\bar{V}_{dq} = [V_{d-n}, V_{q-n}, \ldots, V_{d-1}, V_{q-1}, V_{do}, V_{qo}, V_{d1}, V_{q1}, \ldots, V_{dn}, V_{qn}]^T \tag{7.12b}$$

$$[C] = \tfrac{1}{2}\begin{bmatrix} & M^{*} & & & & & \\ N^{*} & & M^{*} & & & & \\ & N^{*} & & 2M^{*} & & & \\ & & N^{*} & & N & & \\ & & & 2M & & N & \\ & & & & M & & N \\ & & & & & M & \end{bmatrix} \tag{7.12c}$$

where

$$M = N^* = \begin{bmatrix} 1 & \mathrm{j} \\ -\mathrm{j} & 1 \end{bmatrix} \tag{7.13}$$

7.2.4 Admittance Matrix $[Y_{\alpha\beta}]$

To derive $[Y_{\alpha\beta}]$ the general matrix equation is first written for all harmonics as

$$\bar{I}_{dq} = [V_{dq}]\bar{V}_{dq} \tag{7.14}$$

Variables $\bar{V}_{dq}$ are determined by inverse transformation of Equation (7.11). An identical procedure is followed to calculate variables $\bar{I}_{dq}$ and, thus, the following expressions are obtained

$$\bar{V}_{dq} = [C]^* \bar{V}_{\alpha\beta} \tag{7.15a}$$

$$\bar{I}_{dq} = [C]^* \bar{I}_{\alpha\beta} \tag{7.15b}$$

Substitution of Equations (7.15) into Equation (7.14) gives

$$\bar{I}_{\alpha\beta} = [Y_{\alpha\beta}]\bar{V}_{\alpha\beta} \tag{7.16}$$

where

$$[Y_{\alpha\beta}] = [C][Y_{dq}][C]^* \tag{7.17}$$

Matrix $[Y_{\alpha\beta}]$ has the form

$$[Y_{\alpha\beta}] =$$

$$\begin{array}{c|ccccccccccccccc|}
 & -n & -n+1 & \cdots & -h & \cdots & -2 & -1 & 0 & 1 & 2 & \cdots & h & \cdots & n-1 & n \\
\hline
-n & A_n^* & & B_{n-1}^* & & & & & & & & & & & & \\
-n+1 & & A_{n-1}^* & & & & & & & & & & & & & \\
 & D_{n-1}^* & & & & B_h^* & & & & & & & & & & \\
-h & & & & A_h^* & & & & & & & & & & & \\
 & & & D_h^* & & & & B_2^* & & & & & & & & \\
-2 & & & & & & A_2^* & & B_1^* & & & & & & & \\
-1 & & & & & D_2^* & & A_1^* & & D_o & & & & & & \\
0 & & & & & & D_1^* & & A_o & & D_1 & & & & & \\
1 & & & & & & & B_o & & A_1 & & D_2 & & & & \\
2 & & & & & & & & B_1 & & A_2 & & & & & \\
 & & & & & & & & & B_2 & & & & D_h & & \\
h & & & & & & & & & & & & A_h & & & \\
 & & & & & & & & & & & B_h & & & & D_{n-1} \\
n-1 & & & & & & & & & & & & & & A_{n-1} & \\
n & & & & & & & & & & & & & B_{n-1} & & A_n \\
\hline
\end{array} \tag{7.18}$$

where

$$A_o = (MY_{dq-1}M + NY_{dq1}N)/2 \quad (7.19a)$$

$$A_1 = (2MY_{dqo}M + NY_{dq2}N)/4 \quad (7.19b)$$

$$A_h = (MY_{dq(h-1)}M + NY_{dq(h+1)}N)/4 \qquad h = 2, \ldots, n-1 \quad (7.19c)$$

$$A_n = MY_{dqn-1}M/4 \quad (7.19d)$$

$$B_o = MY_{dqo}N/2 \quad (7.19e)$$

$$B_1 = MY_{dq1}N/2 \quad (7.19f)$$

$$B_h = MY_{dqh}N/4 \qquad h = 2, \ldots, n-1 \quad (7.19g)$$

$$D_o = NY_{dqo}M/2 \quad (7.19h)$$

$$D_h = NY_{dqh}M/4 \qquad h = 2, \ldots, n-1 \quad (7.19i)$$

and M and N are defined in Equation (7.13)

The expanded representation of $\bar{I}_{\alpha\beta}$ has the same form as $\bar{V}_{\alpha\beta}$ of Equation (7.12a). Note from Equation (7.19a) that the dc component A_o depends on negative and positive components of fundamental frequency. Terms B_o and $D_o = B^*_o$ play an important role in the mechanism of frequency conversion and their effect should be included. Terms A_h depend on negative and positive harmonics $h-1$ and $h+1$, respectively. On the other hand, B_h and D_h only depend on harmonic order h. Also note, from Equation (7.18), that no cross-coupling exists between even and odd harmonics.

It can be observed from Equation (7.16), taking the expanded matrices $\bar{I}_{\alpha\beta}$, $[Y_{\alpha\beta}]$ and $\bar{V}_{\alpha\beta}$, that a particular current $I_{\alpha\beta h}$ depends on voltages at three harmonics, i.e. $V_{\alpha\beta(h-2)}$, $V_{\alpha\beta h}$ and $V_{\alpha\beta(h+2)}$. Or, in other words, the voltage $V_{\alpha\beta h}$ produces harmonic currents of order $h-2$ and $h+2$.

7.2.5 Admittance Matrix $[Y_{abc}]$

A direct admittance transformation from the two-phase $\alpha\beta$ reference to the three-phase *abc* cannot be achieved because the conversion matrix has non-square form, thus precluding inversion. This problem is solved by adding to $[Y_{\alpha\beta}]$ a zero sequence diagonal matrix of order equal to the harmonic spectrum analysed and also adding to the transformation matrix, defined as T_h for a particular harmonic h, a third row of equal numerical constants. The augmented matrix $[Y_{\alpha\beta}]$ has the form

$$[Y_{\alpha\beta\gamma}] = \begin{bmatrix} Y_{\alpha\beta} & \\ & Y_\gamma \end{bmatrix} \quad (7.20)$$

The zero sequence component satisfies the condition of having zero contribution or being uncoupled from the machine $\alpha\beta$ and dq components [5]. With the incorporation of the zero sequence component the following relationships can be defined for any harmonic h

$$\bar{V}_{\alpha\beta\gamma h} = [T_h]\bar{V}_{abch} \tag{7.21a}$$

$$\bar{I}_{\alpha\beta\gamma h} = [T_h]\bar{I}_{abch} \tag{7.21b}$$

where

$$[T_h] = \begin{bmatrix} 1 & -\frac{1}{2} & -\frac{1}{2} \\ 0 & \frac{\sqrt{3}}{2} & -\frac{\sqrt{3}}{2} \\ \frac{1}{3} & \frac{1}{3} & \frac{1}{3} \end{bmatrix} \tag{7.22}$$

Equation (7.16) can be re-written in $\alpha\beta\gamma$ components as

$$\bar{I}_{\alpha\beta\gamma} = [Y_{\alpha\beta\gamma}]\bar{V}_{\alpha\beta\gamma} \tag{7.23}$$

Dropping the subscript h and substituting Equations (7.22) into Equation (7.23) leads to

$$\bar{I}_{abc} = [Y_{abc}]\bar{V}_{abc} \tag{7.24}$$

where

$$[Y_{abc}] = [T]^{-1}[Y_{\alpha\beta\gamma}][T] \tag{7.25}$$

In Equation (7.24) the terms abc are sequentially accommodated for each harmonic h in blocks of order 3. A different representation is required for the harmonic domain solution where the complete harmonic spectrum (positive and negative harmonics) is used for each phase. As an example, considering three harmonics and a dc term, the current and voltage vectors of equation (7.24) have the form

$$\bar{I}_{abc} = [I_{a-3}, \ldots, I_{ao}, \ldots, I_{a3}, I_{b-3}, \ldots, I_{bo}, \ldots, I_{b3}, I_{c-3}, \ldots\ I_{co}, \ldots, I_{c3}]^T \tag{7.26a}$$

$$\bar{V}_{abc} = [V_{a-3}, \ldots, V_{ao}, \ldots, V_{a3}, V_{b-3}, \ldots, V_{bo}, \ldots\ V_{b3}, V_{c-3}, \ldots, V_{co}, \ldots, V_{c3}]^T \tag{7.26b}$$

and the structure of matrix $[Y_{abc}]$, with dots indicating non-zero elements, is illustrated in Figure 7.3.

7.2.6 Illustration of Harmonic Impedances

An indication of the relative importance of the self and cross-coupling harmonic impedances, derived from the above formulation, is given in Figure 7.4 for the machine data on Table 7.1.

The subscripts of the impedances refer to the harmonic order of the current injection, the superscript to the harmonic order of the voltage produced and the sign of subscripts and superscripts refer to the sequence (i.e. + for positive sequence).

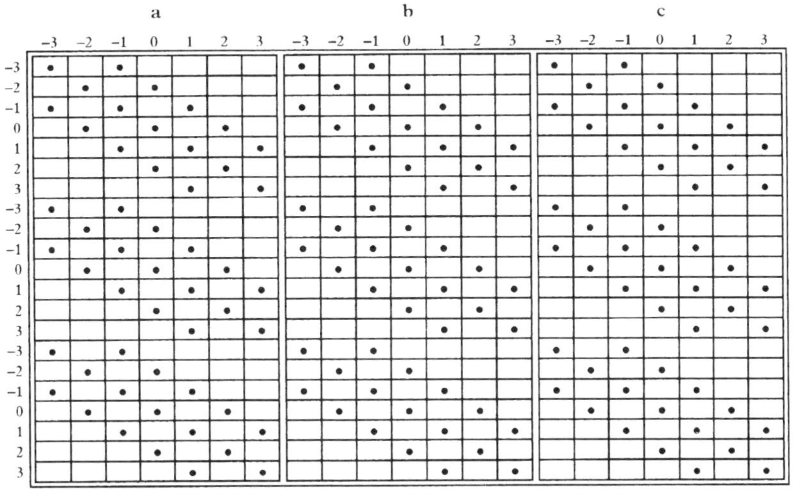

Figure 7.3 Structure of matrix $[Y_{abc}]$

Z_{+h}^{+h} and Z_{-h}^{-h} are defined as the open circuit harmonic impedances while Z_h^{h-2} and Z_{-h}^{h+2} are defined as the open circuit conversion impedances.

Figure 7.4 shows that the magnitude of the (self) harmonic inductance is very close to that of the commonly used approximation.

$$\frac{h}{2}(X_d'' + X_q'').$$

However, the magnitudes of the conversion impedances, neglected by the established models can be significant. The theoretical harmonic impedances are

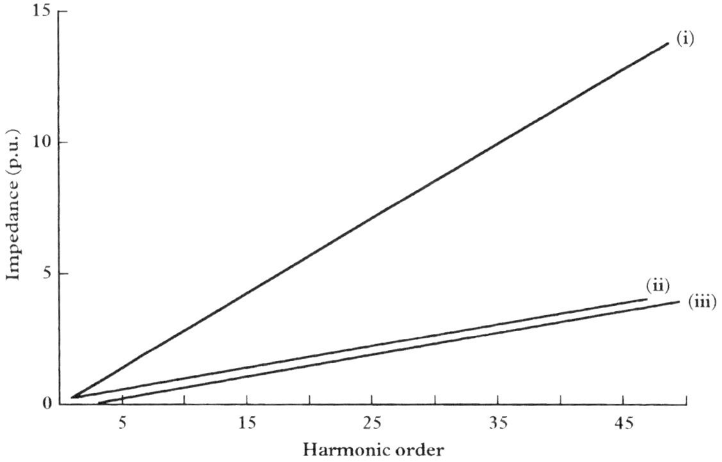

Figure 7.4 Magnitude of the hydrogenerator harmonic impedances. (i) Z_{+h}^{+h} and Z_{-h}^{-h}. (ii) $Z_{-h}^{+(h+2)}$. (iii) $Z_{+h}^{-(h-2)}$

Table 7.1 Generator Data [2]

Rating	100	MVA
Voltage	14	kV
x_d	1.2	pu
x'_d	0.3667	pu
x''_d	0.2	pu
x_q	0.8	pu
x''_q	0.367	pu
T_{do}	7.64	seconds
T''_{do}	0.0633	seconds
T''_{qo}	0.132	seconds
x_1	0.2	pu
x_o	0.05	pu
R_a	0.005	pu

very inductive, i.e. their phase angles are higher than 87° at frequencies higher than the fundamental. However, experimental information obtained from negative sequence injection tests indicates that in practice the power factors are higher. The difference is expected to be largely due to the presence of eddy currents.

7.2.7 Model Validation

The proposed formulation for the generator harmonic impedances is validated against a time domain simulation for the test generator using the PSCAD2-EMTDC program [15]. 0.15p.u. quantities of negative-sequence currents of fundamental 5th and 9th harmonics were injected simultaneously into the generator bus. Due to the generator saliency, positive-sequence 3rd, 7th and 11th voltages are expected.

Having run the simulation to steady state, the voltages and currents were measured and an FFT was performed over one fundamental cycle.

Figure 7.5 shows that the spectra derived from Time Domain and Harmonic Domain simulation match perfectly, both in magnitude and phase.

For accurate results, the dynamic simulation required a 5 μs time step and it was necessary to carry out the simulation for 700 fundamental frequency cycles to allow transients to decay. In comparison, the harmonic domain solution is very efficient computationally.

7.2.8 Accounting for Saturation

Generally, the effect of saturation is taken into account by iteratively modifying the self- and mutual inductances on the *dq* axes [8].

Saturation depends on the total mmf rather than on separate contributions of the *d* and *q* axes [9].

Using the original *dq* differential equations a phase co-ordinate model is developed here to obtain a Norton harmonic equivalent to represent magnetic

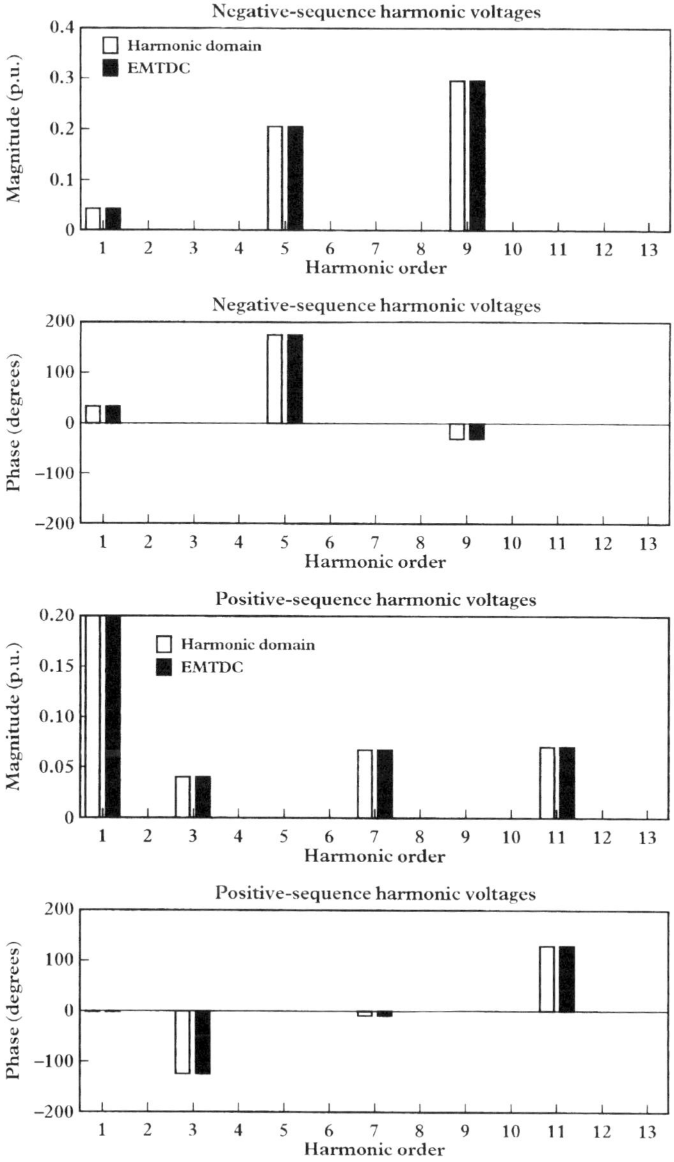

Figure 7.5 Comparison of time and harmonic domain solutions for generator harmonic voltages with harmonic current injection

saturation. The phase co-ordinate nature of the mathematical model makes unnecessary in a unified solution of the generator and power system the need for an iterative procedure to correct the *dq* inductances. Each stator phase has associated with it the same magnetizing characteristic, but saturates differently, according to the phase unbalance at generator terminals, which depends on the rest of the network contribution.

The stator magnetic and electrical parts are represented by the state equations describing the non-linear behaviour of an ideal inductor, i.e.

$$v = p\phi, \tag{7.27}$$

$$i = f(\phi). \tag{7.28}$$

Equations (7.27) and (7.28) are linearized as

$$\Delta V = \mathrm{j}h\omega\Delta\phi \tag{7.29}$$

$$\Delta I = F_g \Delta\phi \tag{7.30}$$

where

$$\Delta I = I - I_{bg}; \qquad \Delta V = V - V_{bg}, \tag{7.31}$$

I_{bg} : base current, obtained from generator magnetizing characteristic.
V_{bg} : base voltage at generator terminals.

Substitution of Equations (7.31) into Equation (7.30) allows a solution for I to be obtained, which in general form can be expressed with the matrix equation

$$\bar{I} = [H_g]\bar{V} + \bar{I}_N \tag{7.32}$$

where

$$\bar{I}_N = \bar{I}_{bg} - [H_g]\bar{V}_{bg} \tag{7.33a}$$

$$[H_g] = [F_g]\{\mathrm{j}h\omega\}^{-1}. \tag{7.33b}$$

Equation (7.32) represents the Norton harmonic equivalent for the generator stator combining together its magnetic and electrical parts. $\bar{I}_N$ is a vector of Norton harmonic currents and $[H_g]$ is a square harmonic matrix of Norton magnetic admittances.

The Norton harmonic equivalent representing the effect of saturation is combined with the machine *abc* quantities according to the winding connections. As an example, the structure of the resulting matrix $[Y_{abc}] + [H_g]$ is shown in Figure 7.6 for the case when the machine windings use a grounded Star configuration.

Figure 7.6 illustrates that explicit cross-coupling exists between even and odd harmonics in phases a, b and c. However, even harmonics are only present if hysteresis effects are taken into account, if they are externally excited or if the magnetising characteristic is asymmetrical on its positive and negative regions. For single-valued symmetrical magnetizing curves, only odd harmonics are produced and only cross-coupling between these harmonics takes place.

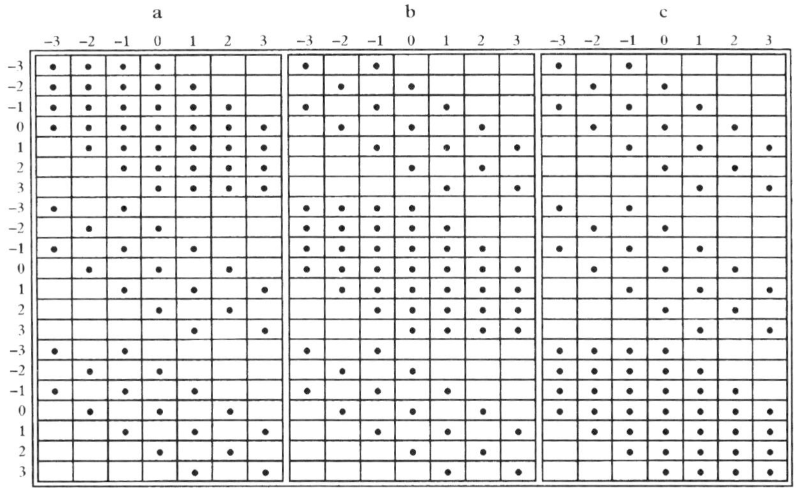

Figure 7.6 Structure of matrix $[Y_{abc}] + [H_g]$

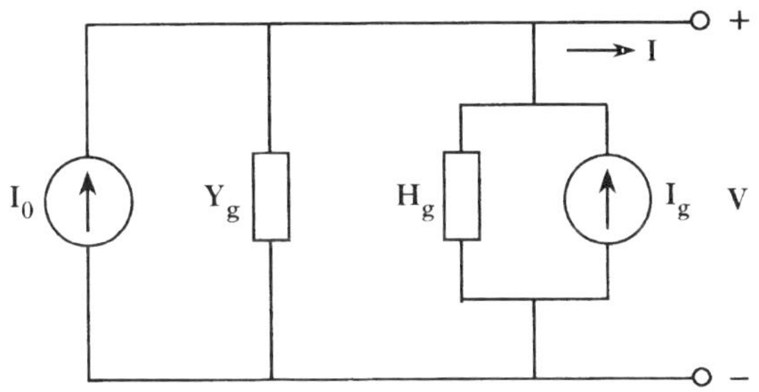

Figure 7.7 Generator Norton harmonic equivalent

7.2.9 Norton Equivalent

The synchronous machine harmonic model with frequency conversion and saturation effects included can be represented as a Norton equivalent [7]. In the analysis to follow, a generator notation is assumed for simplicity of variable manipulations $Y_g = Y_{abc}$, and brackets are dropped at intermediate algebraic steps.

Figure 7.7 illustrates the Norton equivalent representation of a synchronous generator in open-circuit.

The generator current I is calculated as

$$I = -(Y_g + H_g)V + I_0 + I_g. \tag{7.34}$$

Assuming a perfectly smooth excitation, I_0 is only affected by positive sequence fundamental frequency current. Thus, I_0 is determined as

$$I_0 = (Y_{g(+)} + H_{g(+)})V_{(+)} + I_{(+)} - I_{g(+)}. \tag{7.35}$$

Substitution of Equation (7.35) into Equation (7.34) gives

$$I = -(Y_g + H_g)V + (Y_{g(+)} + H_{g(+)})V_{(+)} + I_{(+)} - I_{g(+)} + I_g. \tag{7.36}$$

The currents $I_{(+)}$ and $I_{g(+)}$ can be written in terms of currents I and I_g as

$$I_{(+)} = TI \tag{7.37a}$$

$$I_{g(+)} = TI_g \tag{7.37b}$$

where

$$[T] = \frac{1}{3}\left[\begin{array}{ccc|c} & \textit{fundamental} & & h \\ 1 & a & a^2 & \\ a^2 & 1 & a & 0 \\ a & a^2 & 1 & \\ \hline & 0 & & 0 \end{array}\right]. \tag{7.38}$$

Substitution of Equations (7.37) into Equation (7.36) and re-arranging terms leads to the following matrix expression

$$\{[Y_g] + [H_g]\}\bar{V} = \{[T] - [U]\}(\bar{I} - \bar{I}_g) + \{[Y_{g(+)}] + [H_{g(+)}]\}\bar{V}_{(+)}. \tag{7.39}$$

where $[U]$ is the unit matrix.

If a balanced or unbalanced linear load Y_1 is connected at the terminals of the circuit of Figure 7.7, the current circulating through this component will be given by

$$I = Y_1 V. \tag{7.40}$$

Substitution of Equation (7.40) into Equation (7.34) finally leads to the following matrix equation

$$\{([U] - [T])[Y_l] + [Y_g] + H_g]\}\bar{V} = \{[Y_{g(+)}] + [H_{g(+)}]\}\bar{V}_{(+)} + \{[U] - [T]\}\bar{I}_g. \tag{7.41}$$

7.2.10 Case Studies

To illustrate the effect of the synchronous machine model on harmonic distortion, the synchronous generator is connected to an unbalanced load consisting of a resistance (phase A), an inductive reactance (phase B) and a capacitive reactance (phase C); all of them having a value of 1.0 p.u. The studies are restricted to the first nine harmonics.

Figure 7.8 illustrates the voltage waveforms obtained with the generator represented by the direct axis subtransient reactance x''_d. In this model, the saliency effect is absent and the results consist purely of sinusoidal unbalanced waveforms.

Figure 7.9 illustrates the voltage waveforms obtained when the process of frequency conversion in the generator is explicitly represented. The harmonics produced as a result of the frequency conversion process have dramatically changed the voltage waveforms with respect of those of Figures 7.8.

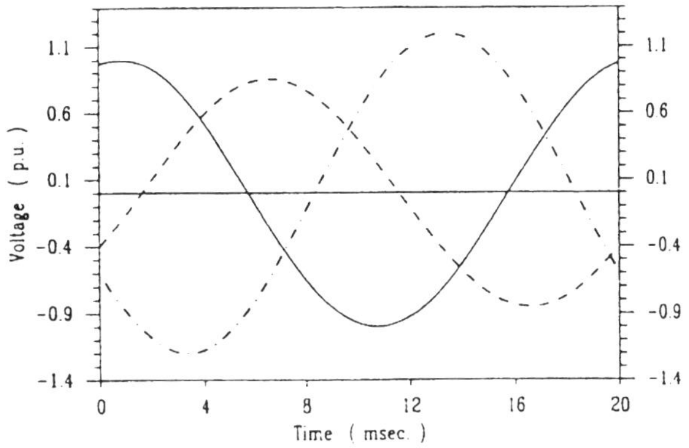

Figure 7.8 Voltage waveforms across unbalanced load; generator represented with x_d''

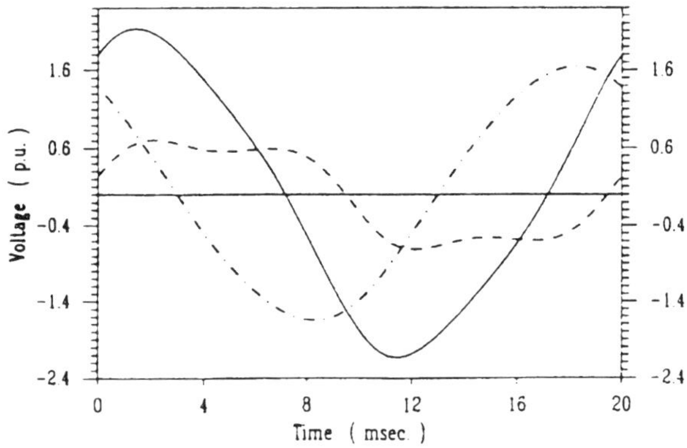

Figure 7.9 Voltage waveforms across unbalanced load; detailed generator model without saturation

To quantify the stator-rotor harmonic interaction under more realistic conditions, the test generator is connected via a linear transformer to a 220 kV line of flat configuration and varying distance.

The asymmetry of the line geometry can, even in the absence of an harmonic source, excite considerable voltage distortion at the generator terminals. This is shown in Figure 7.10, for an open circuited line positive and negative 3rd harmonic levels of 7 and 4% are seen to occur when the line distance is close to 200 km.

7.3 Transformers

When determining the harmonic impedances of a network or performing harmonic penetration studies (Chapter 4) from given harmonic current injections, the transformers were represented by linear impedances, i.e. their magnetization non-linearities were ignored on the assumption that the transformers normally operate within the linear region of the magnetizing characteristics. However, transformers

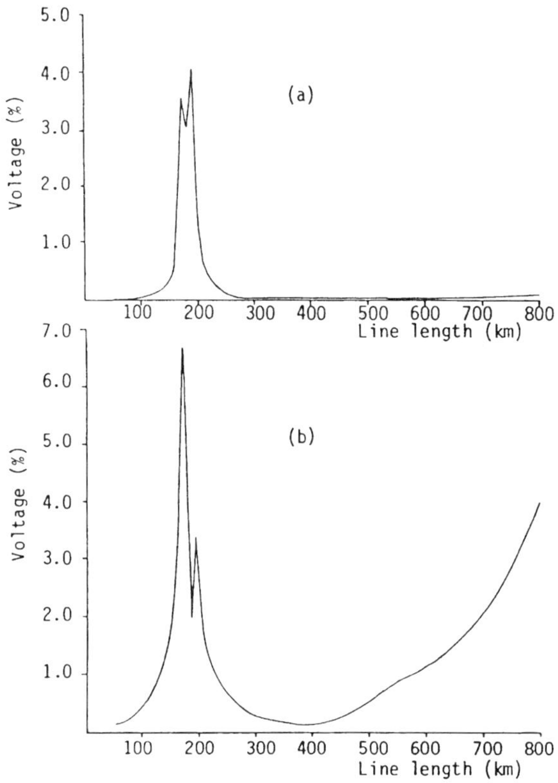

Figure 7.10 Variation of third harmonic voltage with transmission line length. (a) Negative sequence, (b) Positive sequence

are designed to operate very close to the limit of the linear characteristics and, even under small over-excitation, their contribution to the harmonic content is often important.

A more accurate model of the transformer for harmonic studies is developed in this section, describing first the case of a single phase transformer and then extending the model to multi-limb multi-winding units.

7.3.1 Representation of the Magnetisation Characteristics

Each magnetizing curve can be stored in the computer as a set of points ($\phi - i$), such that each flux value impressed in the magnetizing characteristic numerically provides the corresponding current value. However, a significant number of points is required for an accurate solution, although at the expense of increased computation time.

Alternatively, the experimental magnetizing curves can be analytically approximated.

A hyperbolic formulation [10] can satisfactorily meet these requirements. It is based in the following function

$$F(i, \phi) = (m_1 i + b_1 - \phi)(m_2 i + b_2 - \phi) - b_1 b_2 = \xi\phi \tag{7.42}$$

where

m_1, m_2 = slopes (inductances) of the unsaturated and saturated regions
b_1, b_2 = ordinates to the origin of the asymptotes to m_1 and m_2
$\xi\phi$ = correction term to modulate the knee regions.

The term $\xi\phi$ allows the modulation of the knee region. When this is not required $\xi\phi = 0$. The solution of the hyperbolic function in the first quadrant leads to the following expression for the magnetizing current

$$i = \frac{-B - \sqrt{(B^2 - 4AC)}}{2A} \tag{7.43}$$

where

$$A = m_1 m_2$$
$$B = m_1(b_2 - \phi) + m_2(b_1 - \phi)$$
$$C = \phi^2 - \phi(b_1 + b_2 + \xi)$$

Four significant digits in the specified variables m_1 and m_2 have been found sufficient to ensure an accurate determination of the magnetizing current values.

7.3.2 Norton Equivalent of the Magnetic Non-linearity [11]

A process of linearisation is used to model the effect of the magnetisation characteristic around an operating point, which in the case illustrated by Figure 7.11 is indicated by $\psi^o i^o$.

For a small increment of flux $\Delta\psi$ the corresponding increment in magnetizing current can be expressed as:

$$\Delta_i = c\Delta\psi \tag{7.44}$$

where

$$\Delta\psi = \sum_{k=-\infty}^{\infty} \Delta I\psi_k \mathrm{e}^{\mathrm{j}k\omega t}$$

$$\Delta i = \sum_{h=-\infty}^{\infty} \Delta i_h \mathrm{e}^{\mathrm{j}h\omega t}$$

$$c = f'(\psi(t)) = \sum_{i=-\infty}^{\infty} c_i \,\mathrm{e}^{\mathrm{j}k\omega t}$$

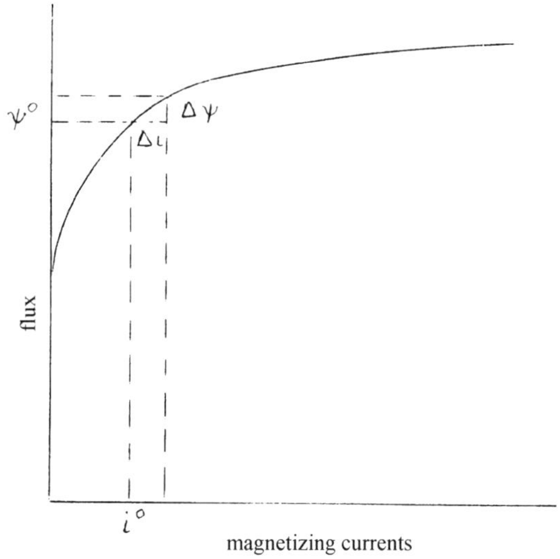

Figure 7.11 Typical transformer magnetization characteristic

$$\begin{bmatrix} \Delta i_{-h} \\ \\ \Delta i_{-2} \\ \Delta i_{-1} \\ \Delta i_o \\ \Delta i_1 \\ \Delta i_2 \\ \\ \Delta i_h \end{bmatrix} = \begin{bmatrix} c_1 & c_0 & c_{-1} & c_{-2} & c_{-3} & & & & & & & \\ & & & & & & & & & & & \\ c_3 & c_2 & c_1 & c_0 & c_{-1} & c_{-2} & c_{-3} & & & & & \\ & c_3 & c_2 & c_1 & c_0 & c_{-1} & c_{-2} & c_{-3} & & & & \\ & & c_3 & c_2 & c_1 & c_0 & c_{-1} & c_{-2} & c_{-3} & & & \\ & & & c_3 & c_2 & c_1 & c_0 & c_{-1} & c_{-2} & c_{-3} & & \\ & & & & c_3 & c_2 & c_1 & c_0 & c_{-1} & c_{-2} & c_{-3} & \\ & & & & & & & & & & & \\ & & & & & & c_3 & c_2 & c_1 & c_0 & c_{-1} & \end{bmatrix} \begin{bmatrix} \Delta\psi_{-k} \\ \\ \Delta\psi_{-2} \\ \Delta\psi_{-1} \\ \Delta\psi_0 \\ \Delta\psi_1 \\ \Delta\psi_2 \\ \\ \Delta\psi_k \end{bmatrix}. \tag{7.45}$$

$$\Delta i = [F]^{\circ}\Delta\psi. \tag{7.46}$$

The linearization is about a point $(\psi^{\circ}, i^{\circ})$ in the space of complex conjugate harmonics, where

$$\begin{aligned} \Delta\psi &= \psi - \psi^{\circ}, \\ \Delta i &= i - i^{\circ}. \end{aligned} \tag{7.47}$$

Substituting Equations (7.47) in (7.46) yields

$$\bar{I} = \left[\bar{F}^{\circ}\right]\psi + \bar{I}_N \tag{7.48}$$

$$\bar{I}_N = i^{\circ} - [F^{\circ}]\psi^{\circ} \tag{7.49}$$

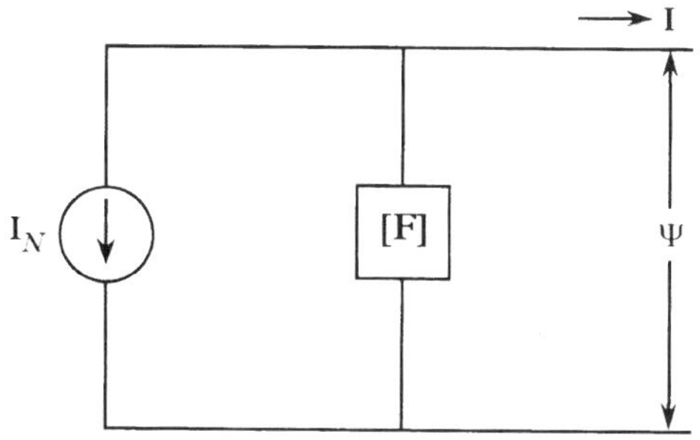

Figure 7.12 Norton equivalent circuit

which can be represented by the Norton equivalent shown in Figure 7.12.

The derivation of the Norton equivalent involves the following steps:

1 For each phase the voltage waveform is used to derive the corresponding flux wave and the latter is impressed, point by point, upon the experimental characteristic $\psi - I$, and the associated magnetizing current is then determined in the time domain.
2 By means of a Fast Fourier Transform (FFT) the magnetizing current is solved in the frequency domain and the Fourier coefficients i_h and i^*_h are assembled into a base current vector I_b.
3 Using the magnetizing current and flux as determined in step 1, the time derivative of the function $I = f(\psi)$ is evaluated.
4 The FFT is applied to the slope shape of step 3, and the Fourier coefficients c_h and c^*_h obtained from this exercise are used to assemble the Toeplitz matrices $[F]$ and $[G]$.
5 The harmonic admittance matrix $[H]$ and the Norton equivalent current source I_N, i.e. $I_N = I_b - [H]V_b$ are calculated.
6 The above linearized model can be combined with the linear network as part of an iterative solution as described in Figure 7.13.

7.3.3 Generalization of the Norton Equivalent

An accurate determination of the magnetizing currents of the multi-limb transformer should involve the complete magnetic circuit [12]. The solution is not straightforward due to the non-linearities of the magnetic reluctances, which are functions of the magnetic material and branch geometries. The yokes, zero-sequence flux paths and for a five limb transformer, the outer branches, have magnetizing characteristics considerably different from that of the winding branches.

The magnetic equivalent circuit must be described as a function of the branch fluxes and reluctances, and the magnetic circuit equations obtained by applying the magnetic circuit laws

$$\Sigma Ni = \Sigma R\phi \tag{7.50}$$

for any closed path (N being the number of turns), and

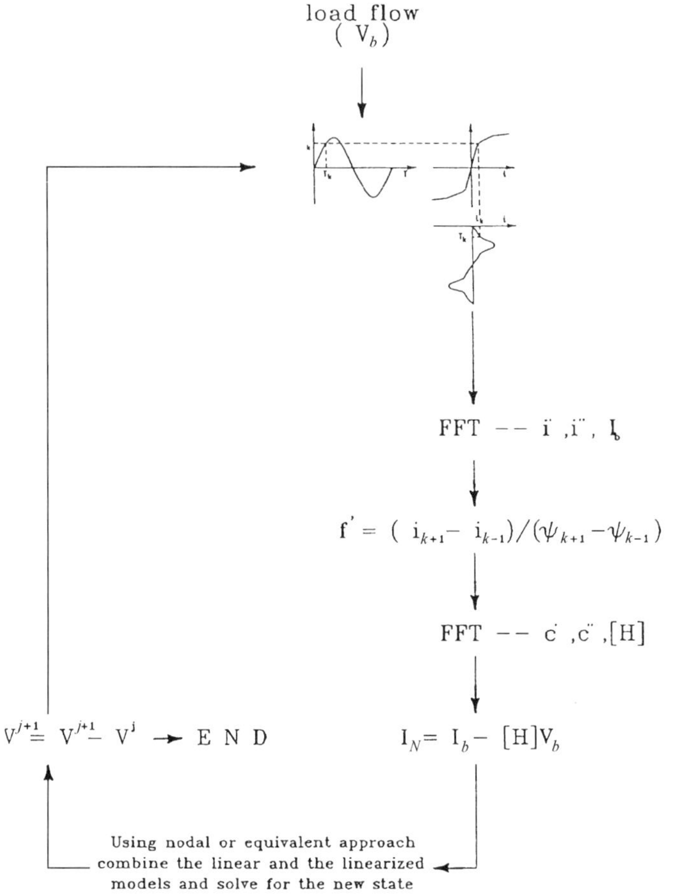

Figure 7.13 Description of the iterative algorithm

$$\Sigma\phi = 0 \tag{7.51}$$

at any junction.

For the case of the three-limb transformer shown in Figures 7.14 and 7.15, the winding flux values ϕ_1 to ϕ_3 can be calculated from Faraday's law $\phi_s = 1/N \int v_s.dt$). Therefore, the magnetic equivalent circuit will have eight unknown variables, i.e. three magnetizing currents i_1, i_2 and i_3 and five fluxes, ϕ_4 to ϕ_8. Five basic equations can be derived from the first magnetic circuit law,

$$i_1 = f_1(\phi_1) + f_6(\phi_6) \tag{7.52}$$

$$i_2 = f_2(\phi_2) + f_7(\phi_7) \tag{7.53}$$

$$i_3 = f_3(\phi_3) + f_8(\phi_8) \tag{7.54}$$

$$i_1 - i_2 = f_1(\phi_1) + f_4(\phi_4) - f_2(\phi_2) \tag{7.55}$$

$$i_2 - i_3 = f_2(\phi_2) + f_5(\phi_5) - f_3(\phi_3) \tag{7.56}$$

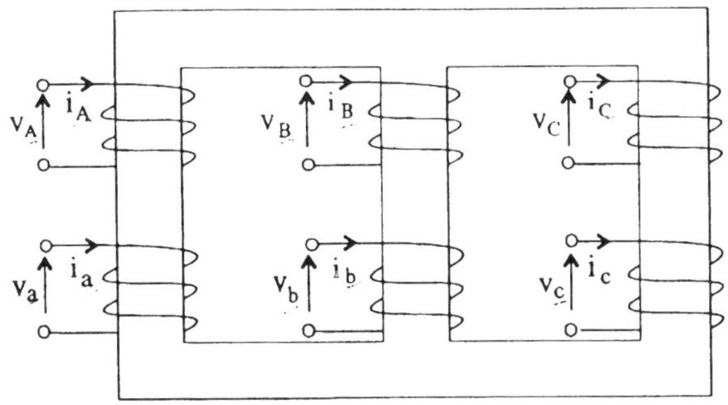

Figure 7.14 Three-limb transformer

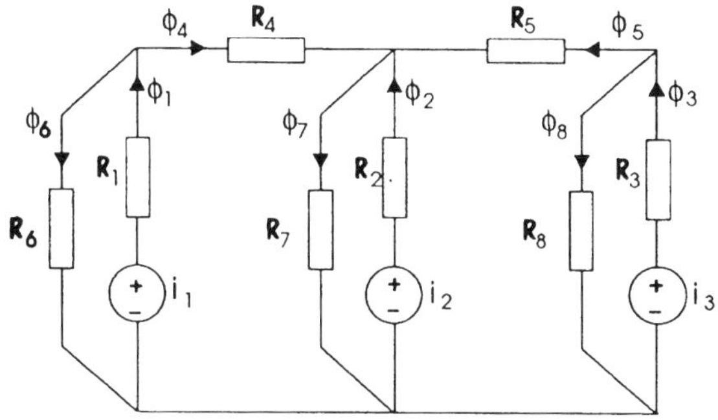

Figure 7.15 Magnetic equivalent circuit

where the branch mmf drop across each branch reluctance has been represented by the corresponding magnetizing characteristic $i = f(\phi)$. The other three necessary equations are obtained from the second magnetic circuit law

$$\phi_1 - \phi_4 - \phi_6 = 0 \tag{7.57}$$

$$\phi_2 + \phi_4 + \phi_5 - \phi_7 = 0 \tag{7.58}$$

$$\phi_3 - \phi_5 - \phi_8 = 0 \tag{7.59}$$

Applying a Taylor's expansion to Equations (7.52) to (7.56) about a base operating point, and truncating after the first order yields

$$i_1 = f_1(\phi_{1b}) + f_6(\phi_{6b}) + R_{1b}(\phi_1 - \phi_{1b}) + R_{6b}(\phi_6 - \phi_{6b}) \tag{7.60}$$

$$i_2 = f_2(\phi_{2b}) + f_7(\phi_{7b}) + R_{2b}(\phi_2 - \phi_{2b}) + R_{7b}(\phi_7 - \phi_{7b}) \tag{7.61}$$

$$i_3 = f_3(\phi_{3b}) + f_8(\phi_{8b}) + R_{3b}(\phi_3 - \phi_{3b}) + R_{8b}(\phi_8 - \phi_{8b}) \tag{7.62}$$

$$i_1 - i_2 = f_1(\phi_{1b}) + f_4(\phi_{4b}) - f_2(\phi_{2b}) + R_{1b}(\phi_1 - \phi_{1b}) + R_{4b}(\phi_4 - \phi_{4b}) - R_{2b}(\phi_2 - \phi_{2b}) \tag{7.63}$$

$$i_2 - i_3 = f_2(\phi_{2b}) + f_5(\phi_{5b}) - f_3(\phi_{3b}) + R_{2b}(\phi_2 - \phi_{2b}) - R_{5b}(\phi_5 - \phi_{5b}) - R_{3b}(\phi_3 - \phi_{3b}) \tag{7.64}$$

where the subscript b denotes the base values and the magnetic reluctance R_{kb} is the derivative of the magnetizing characteristic $i_k = f_k(\phi_k)$ with respect to the base flux $\phi_{kb}(R_{kb} = f'(\phi_{kb}), k = 1,2, \ldots, 8)$. Adding Equations (7.57) to (7.59) to this set of linearized equations, and expressing them in matrix form, with the magnetizing currents and ϕ_4 to ϕ_8 as functions of the winding fluxes, leads to

$$[H]\begin{bmatrix} i_1 \\ i_2 \\ i_3 \\ \phi_4 \\ \phi_5 \\ \phi_6 \\ \phi_7 \\ \phi_8 \end{bmatrix} = \begin{bmatrix} R_{1b}\phi_1 - a_1 \\ R_{2b}\phi_2 - a_2 \\ R_{3b}\phi_3 - a_3 \\ R_{1b}\phi_1 - R_{2b\phi 2} - a_4 \\ R_{2b}\phi_2 - R_{3b\phi 3} - a_5 \\ \phi_1 \\ \phi_2 \\ \phi_3 \end{bmatrix} \tag{7.65}$$

where

$$[H] = \begin{bmatrix} 1 & & & & & -R_{6b} & & \\ & 1 & & & & & -R_{7b} & \\ & & 1 & & & & & -R_{8b} \\ 1 & -1 & & -R_{4b} & & & & \\ & 1 & -1 & & -R_{5b} & & & \\ & & & 1 & & 1 & & \\ & & & -1 & -1 & & 1 & \\ & & & & 1 & & & 1 \end{bmatrix}$$

and

$$\begin{aligned} a_1 &= R_{1b}\phi_{1b} + R_{6b}\phi_{6b} - f_1(\phi_{1b}) - f_6(\phi_{6b}) \\ a_2 &= R_{2b}\phi_{2b} + R_{7b}\phi_{7b} - f_2(\phi_{2b}) - f_7(\phi_{7b}) \\ a_3 &= R_{3b}\phi_{3b} + R_{8b}\phi_{8b} - f_3(\phi_{3b}) - f_8(\phi_{8b}) \\ a_4 &= R_{1b}\phi_{1b} + R_{4b}\phi_{4b} - R_{2b}\phi_{2b} - f_1(\phi_{1b}) - f_4(\phi_{4b}) + f_2(\phi_{2b}) \\ a_5 &= R_{2b}\phi_{2b} + R_{5b}\phi_{5b} - R_{3b}\phi_{3b} - f_2(\phi_{2b}) - f_5(\phi_{5b}) + f_3(\phi_{3b}). \end{aligned}$$

Equation (7.65) produces the following expressions for the magnetizing currents

$$\begin{bmatrix} i_1 \\ i_2 \\ i_3 \end{bmatrix} = [d] \begin{bmatrix} R_1\phi_1 - a_1 \\ R_2\phi_2 - a_2 \\ R_3\phi_3 - a_3 \\ R_1\phi_1 - R_2\phi_2 - a_4 \\ R_2\phi_2 - R_3\phi_3 - a_5 \\ \phi_1 \\ \phi_2 \\ \phi_3 \end{bmatrix} \tag{7.66}$$

where the matrix $[d]$ comprises the first three rows of $[H]^{-1}(3 \times 8)$. These matrix equations can be re-arranged in the form

$$\begin{bmatrix} i_1 \\ i_2 \\ i_3 \end{bmatrix} = \begin{bmatrix} k_{11} & k_{12} & k_{13} \\ k_{21} & k_{22} & k_{23} \\ k_{31} & k_{32} & k_{33} \end{bmatrix} \begin{bmatrix} \phi_1 \\ \phi_1 \\ \phi_3 \end{bmatrix} + \begin{bmatrix} i_{n1} \\ i_{n2} \\ i_{n3} \end{bmatrix} \tag{7.67}$$

or

$$i_s = [K]\phi_s + i_{ns} \tag{7.68}$$

where $s = 1, 2, 3$.

The interface with the external electrical system is made by relating the magnetizing currents to the winding voltages. Using harmonic phasors and the per unit form, Faraday's law is expressed by

$$v_s = \mathrm{j}\, h\omega\phi_s. \tag{7.69}$$

Finally, substitution of Equation (7.69) into Equation (7.68) yields

$$i_s = [Y]v_s + i_{ns} \tag{7.70}$$

where

$$[Y] = [K]Diag\{\mathrm{j}h\omega\}^{-1}$$

and

$$i_{ns} = -a_1 d_{s1} - a_2 d_{s2} - a_3 d_{s3} - a_4 d_{s4} - a_5 d_{s5}$$

where $s = 1, 2, 3$.

The elements of matrix $[K]$ are defined below

$$k_{s1} = d_{s1}R_{1b} + d_{s4}R_{1b} + d_{s6}$$

$$k_{s2} = d_{s2}R_{2b} - d_{s4}R_{2b} + d_{s5}R_{2b} + d_{s7}$$

$$k_{s3} = d_{s3}R_{3b} - d_{s5}R_{3b} + d_{s8}$$

where $s = 1, 2, 3$.

Although the formulation has been derived for a three-phase three-limb transformer, it can be easily extended to multi-limb transformers with any core configuration.

7.3.4 Full Harmonic Electromagnetic Representation

The full harmonic model is obtained with the incorporation of the magnetically coupled core contribution, represented by $[Y]$ and i_{ns} in Equation (7.70). The per unit magnetizing admittance is often distributed equally between all terminals [13], i.e. halved and placed at both sides of the leakage admittance. This approach has been used here for the treatment of the self and mutual terms of the Norton admittance matrix $[Y]$ and the Norton current injections, described by i_{ns}.

However, a more rigorous distribution between primary and secondary windings at each frequency can be derived from the Steinmetz 'exact' equivalent circuit, i.e.

$$K^{stmz} = \frac{|I^*{}_{m_{sec}}|}{|I_{m_{prim}}|} = \frac{|Z^{exact}_{prim}|}{|Z^{exact*}_{sec}|} \tag{7.71}$$

where

$$Z^{exact}_{prim} = Z_{source} + R_{prim} + \mathrm{j}Xl_{prim}$$
$$Z^{exact*}_{sec} = a^2(\mathrm{j}Xl_{sec} + R_{sec} + Z_{load})$$

and superscript * is used to indicate parameters referred to the primary side. The subdivision of magnetizing current components needs to take into account the impedances of the source (primary system) and load (secondary system); thus the K^{stmz} ratio will, in general, be different for each harmonic and the magnetizing harmonic current injections of the primary and secondary will be very different under near-resonance conditions.

The application of the linear transformations given in Section 3.7 allows the determination of full harmonic electromagnetic models for multilimb power transformers. Generalized models for transformers in star-star and grounded star-delta configurations are represented by the harmonic nodal matrix Equations (7.72) and (7.73). Models for other transformer connections can be directly obtained from these basic equations. Figures 7.16 and 7.17 illustrate their associated harmonic lattice equivalents.

7.3.5 Case Study

The proposed formulation requires multi-limb three-phase transformer data which is rarely available. The test system uses a hyperbolic approximation of the magnetization characteristics obtained experimentally in reference [14].

As the main purpose in this section is the transformer model, the rest of the test system, shown in Figure 7.18, has been kept very simple. It consists of a grounded-star/grounded-star transformer connected to an infinite source specified as 1.1 p.u. via a 0.1 p.u. series inductance.

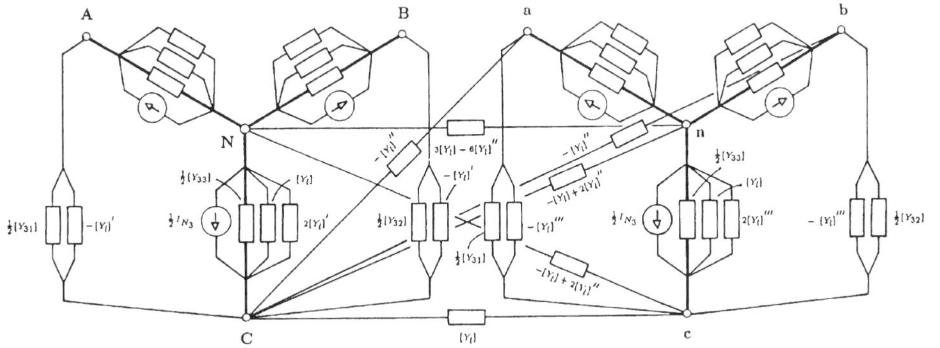

Figure 7.16 Lattice equivalent circuit for a star-star transformer

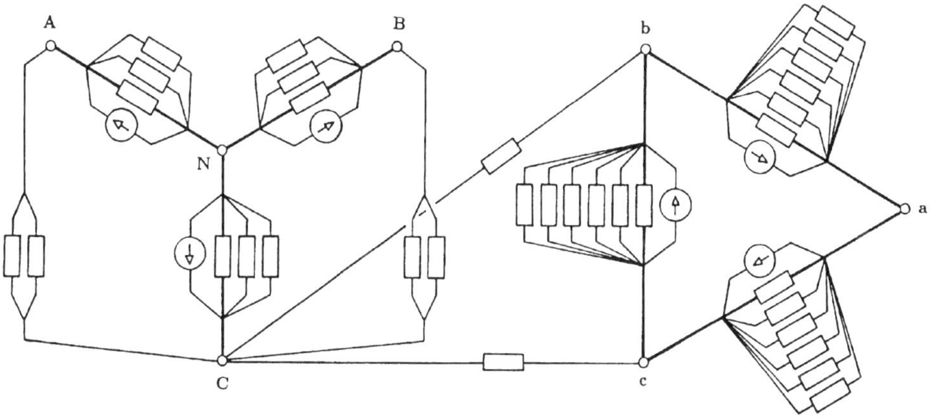

Figure 7.17 Lattice equivalent circuit for a star-delta transformer

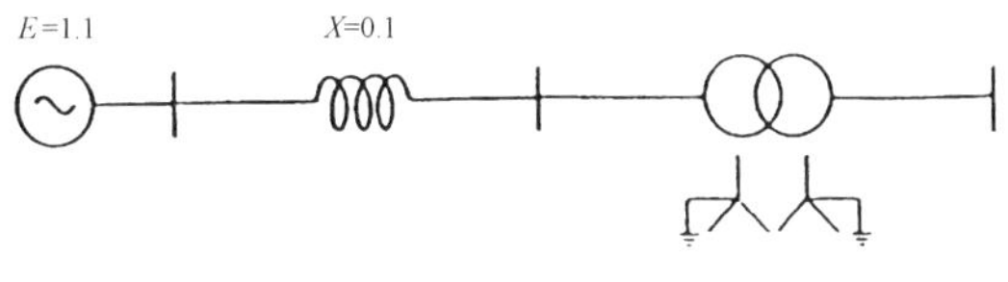

Figure 7.18 Test system

The magnetizing characteristics associated with the winding, yoke branches and zero-sequence flux paths are represented by curves 1, 2 and 3 in Figure 7.19.

For verification purposes, the test system was also analysed in the time domain, using the EMTDC program [15] with a 50 μs time step.

The results of the simulations involve the first 15 harmonics and convergence was achieved to a tolerance of 0.0001 (p.u.) for the harmonic voltages. These are shown in Figures 7.20(a) and 7.20(b) for the magnetizing current waveforms and their

$$
\begin{bmatrix} I_A - \alpha I_{n3} \\ I_B - \alpha I_{n2} \\ I_C - \alpha I_{n3} \\ I_a - \beta I_{n1} \\ I_b - \beta I_{n2} \\ I_c - \beta I_{n3} \\ I_N + \beta I_{n1} + \beta I_{n2} + \beta I_{n3} \\ I_N + \alpha I_{n1} + \alpha I_{n2} + \alpha I_{n3} \end{bmatrix} = \left[\begin{matrix} y_l + \alpha y_{11} & y'_l + \alpha y_{12} & y'_l + \alpha y_{13} & -y_l \\ y'_l + \alpha y_{21} & y_l + \alpha y_{22} & y'_l + \alpha y_{23} & y''_l \\ y'_l + \alpha y_{31} & y'_l + \alpha y_{32} & y_l + \alpha y_{33} & y''_l \\ -y_l & y''_l & y''_l & y_l + \beta y_{11} \\ y''_l & -y_l & y''_l & y'''_l + \beta y_{21} \\ y''_l & y''_l & -y_l & y'''_l + \beta y_{31} \\ \begin{matrix} -y_l - 2y'_l \\ -\alpha y_{11} - \alpha y_{21} \\ -\alpha y_{31} \end{matrix} & \begin{matrix} y_l - 2y'_l \\ -\alpha y_{12} - \alpha y_{22} \\ -\alpha y_{32} \end{matrix} & \begin{matrix} y_l - 2y'_l \\ -\alpha y_{13} - \alpha y_{23} \\ -\alpha y_{33} \end{matrix} & y_l - 2y''_l \\ y_l - 2y''_l & y_l - 2y''_l & y_l - 2y''_l & \begin{matrix} -y_l - 2y'''_l \\ -\beta y_{11} - \beta y_{21} \\ -\beta y_{31} \end{matrix} \end{matrix} \right.
$$

Connection: Star–Star

$$
\begin{bmatrix} I_A - \alpha I_{n1} \\ I_B - \alpha I_{n2} \\ I_C - \alpha I_{n3} \\ I_a - \beta I_{n1} + \beta I_{n3} \\ I_b + \beta I_{n1} - \beta I_{n2} \\ I_c + \beta I_{n2} - \beta I_{n3} \\ I_N + \alpha I_{n1} + \\ \alpha I_{n2} + \alpha I_{n3} \end{bmatrix} = \left[\begin{matrix} y_l + \alpha y_{11} & y'_l + \alpha y_{12} & y'_l + \alpha y_{13} \\ -y'_l + \alpha y_{21} & y_l + \alpha y_{22} & y'_l + \alpha y_{23} \\ y'_l + \alpha y_{31} & y'_l + \alpha y_{32} & y_l + \alpha y_{33} \\ -y_l - y''_l & 0 & y_l + y''_l \\ y_l + y''_l & -y_l - y''_l & 0 \\ 0 & y_l + y''_i & -y_l - y''_l \\ \begin{matrix} -y_l - 2y'_i \\ -\alpha y_{11} - \alpha y_{21} \\ -\alpha y_{31} \end{matrix} & \begin{matrix} -y_l - 2y'_l \\ -\alpha y_{12} - \alpha y_{22} \\ -\alpha y_{32} \end{matrix} & \begin{matrix} -y_l - 2y'_l \\ -\alpha y_{13} - \alpha y_{23} \\ -\alpha y_{33} \end{matrix} \end{matrix} \right.
$$

Connection: Star–Delta

$$
\left.\begin{array}{cccc}
y_l'' & y_l'' & \begin{array}{c}-y_l - 2y_l' \\ -\alpha y_{11} - \alpha y_{12} \\ -\alpha y_{13}\end{array} & y_l - 2y_l'' \\
-y_l & y_l'' & \begin{array}{c}-y_l - 2y_l' \\ -\alpha y_{21} - \alpha y_{22} \\ -\alpha y_{23}\end{array} & y_l - 2y_l'' \\
y_l'' & -y_l & \begin{array}{c}-y_l - 2y_l' \\ -\alpha y_{31} - \alpha y_{32} \\ -\alpha y_{33}\end{array} & y_l - 2y_l'' \\
y_l''' + \beta y_{12} & y_l''' + \beta y_{13} & y_l - 2y_l'' & \begin{array}{c}-y_l - 2y_l' \\ -\beta y_{11} - \beta y_{12} - \beta y_{13}\end{array} \\
y_l + \beta y_{22} & y_l''' + \beta y_{23} & y_l - 2y_l'' & \begin{array}{c}-y_l - 2y_l''' \\ -\beta y_{21} - \beta y_{22} - \beta y_{23}\end{array} \\
y_l + \beta y_{32} & y_l''' + \beta y_{33} & y_l - 2y_l'' & \begin{array}{c}-y_l - 2y_l''' \\ -\beta y_{31} - \beta y_{32} - \beta y_{33}\end{array} \\
y_l - 2y_l'' & y_l - 2y_l'' & \begin{array}{c}-3y_l + 6y_l'' \\ +\alpha y_{11} + \alpha y_{12} + \alpha y_{13} \\ +\alpha y_{21} + \alpha y_{22} + \alpha y_{23} \\ +\alpha y_{31} + \alpha y_{32} + \alpha y_{33}\end{array} & -3y_l + 6y'' \\
\begin{array}{c}-y_l - 2y_l''' \\ -\beta y_{12} - \beta y_{22} \\ -\beta y_{32}\end{array} & \begin{array}{c}-y_l - 2y_l''' \\ -\beta y_{13} - \beta y_{23} \\ -\beta y_{33}\end{array} & -3y_l + 6y_l''' & \begin{array}{c}+3y_l + 6y_l''' \\ +\beta y_{11} + \beta y_{12} + \beta y_{13} \\ +\beta y_{21} + \beta y_{22} + \beta y_{23} \\ +\beta y_{31} + \beta y_{32} + \beta y_{33}\end{array}
\end{array}\right]
\begin{bmatrix} V_A \\ V_B \\ V_C \\ V_a \\ V_b \\ V_c \\ V_N \\ V_n \end{bmatrix}
\qquad (7.72)
$$

$$
\left.\begin{array}{cccc}
-y_l - y_l'' & y_l + y_l'' & 0 & \begin{array}{c}-y_l - 2y_l' \\ -\alpha y_{11} - \alpha y_{12} \\ -\alpha y_{13}\end{array} \\
0 & -y_l - y_l'' & y_l + y_l'' & \begin{array}{c}-y_l - 2y_l' \\ -\alpha y_{21} - \alpha y_{22} \\ -\alpha y_{23}\end{array} \\
y_l + y_l'' & 0 & -y_l - y_l'' & \begin{array}{c}-y_l - 2y_l' \\ -\alpha y_{31} - \alpha y_{32} \\ -\alpha y_{33}\end{array} \\
\begin{array}{c}2y_l - 2y_l''' \\ +\beta y_{11} - \beta y_{31} \\ -\beta y_{13} + \beta y_{33}\end{array} & \begin{array}{c}-y_l + y_l''' \\ -\beta y_{11} + \beta y_{31} \\ +\beta y_{12} - \beta y_{32}\end{array} & \begin{array}{c}-y_l + y_l''' \\ +\beta y_{32} - \beta y_{33} \\ -\beta y_{12} + \beta y_{13}\end{array} & 0 \\
\begin{array}{c}2y_l - 2y_l''' \\ -\beta y_{11} + \beta y_{21} \\ +\beta y_{13} - \beta y_{23}\end{array} & \begin{array}{c}-y_l + y_l''' \\ +\beta y_{11} - \beta y_{21} \\ -\beta y_{12} + \beta y_{22}\end{array} & \begin{array}{c}-y_l + y_l''' \\ +\beta y_{12} - \beta y_{22} \\ -\beta y_{13} + \beta y_{23}\end{array} & 0 \\
\begin{array}{c}-y_l + y_l''' \\ -\beta y_{21} + \beta y_{31} \\ +\beta y_{23} - \beta y_{33}\end{array} & \begin{array}{c}2y_l - 2y_l''' \\ +\beta y_{21} - \beta y_{31} \\ -\beta y_{22} + \beta y_{32}\end{array} & \begin{array}{c}-y_l + y_l''' \\ \beta y_{22} - \beta_{32} \\ -\beta y_{23} + \beta y_{33}\end{array} & 0 \\
0 & 0 & 0 & \begin{array}{c}3y_l + 6y_l'' \\ +\alpha y_{11} + \alpha y_{12} + \alpha y_{13} \\ +\alpha y_{21} + \alpha y_{22} + \alpha y_{23} \\ +\alpha y_{31} + \alpha y_{32} + \alpha y_{33}\end{array}
\end{array}\right]
\begin{bmatrix} V_A \\ V_B \\ V_C \\ V_a \\ V_b \\ V_c \\ V_N \end{bmatrix}
\qquad (7.73)
$$

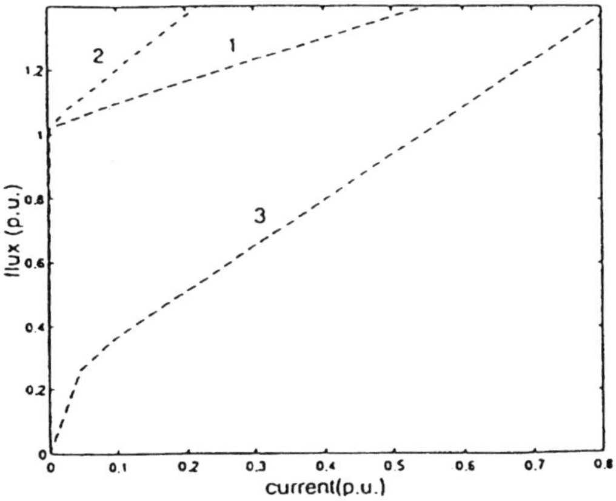

Figure 7.19 Assumed magnetizing characteristics

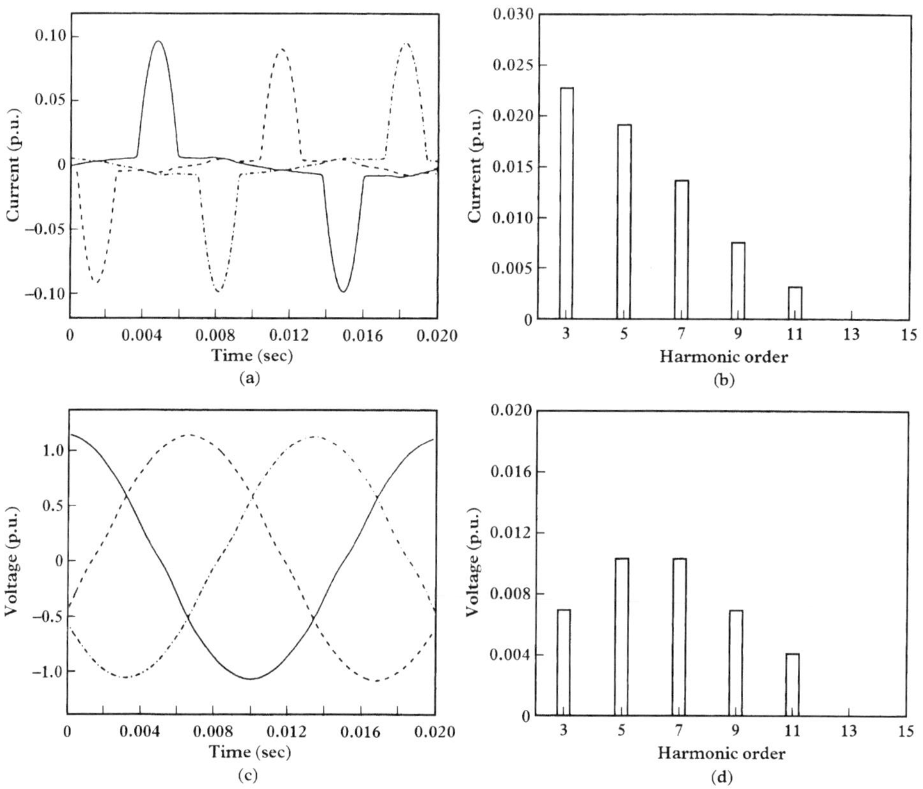

Figure 7.20 HDA simulations—Example

spectra, respectively. Due to the saturation and magnetic coupling between the three-phases, the three-limb transformer magnetizing currents have two maxima within one half-wave. The second and smaller one is not very significant in this example due to the slight transformer saturation considered. The corresponding voltage waveforms at the transformer primary side and their harmonic spectra are illustrated in Figure 7.18(c) and 7.18(d). Although the voltage distortions caused by the magnetizing currents are within 1.0% of the fundamental, it should be pointed out that the individual harmonic levels permitted by legislation are usually of this order.

Of course, the magnitude of the harmonic voltages are affected by the overvoltage levels and by the frequency-dependent network impedances. By increasing the infinite bus source to 1.2 (p.u.), or the series reactance to 1.0 (p.u.), distortion levels of up to 4% are calculated.

7.4 Summary

The harmonic domain, an alternative frame of reference to represent the process of frequency conversion in the presence of non-linearities, has been introduced. All the relevant harmonics are explicitly and simultaneously represented, including frequency-dependent factors such as skin effect and, therefore, this method provides greater accuracies than time domain simulation. It is also more efficient computationally since it avoids the long runs of time domain simulation needed to reach the steady state.

In the harmonic domain the order of the network matrix admittance is equal to the number of nodes times the number of phases multiplied by twice the number of harmonics (when positive and negative frequencies are used).

This chapter has only discussed the synchronous generator and transformer non-linearities. In the case of the synchronous machine, the harmonic Norton equivalents include the effect of rotor saliency and machine saturation. The Norton equivalent of the transformer represents the effect of saturation in the various magnetic branches of the multi-limb configuration.

Because of their greater contribution to the harmonic content, the static converters are given detailed consideration in the following chapters.

7.5 References

1. Park, R H, (1929). Two-Reaction Theory of Synchronous Machines; Generalized Method of Analysis — Part I, *AIEE Transactions*, **48**(3), 716–730.
2. Park, R H, (1933). Two-Reaction Theory of Synchronous Machines — Part II, *AIEE Transactions*, **52**(2), 352–355.
3. Clarke, E, (1950). *Circuit Analysis of A–C Power Systems — Vol II*, John Wiley & Sons, London.
4. Hwang, H H, (1965). Unbalanced Operations of A.C. Machines, *IEEE Transactions on Power Apparatus and Systems*, **PAS-84**(11), 1054–1066.
5. O'Kelly, D and Simmons, S, (1968). *Introduction to Generalized Electrical Machine Theory*, McGraw-Hill, London.
6. Semlyen, A, Eggleston, J F and Arrillaga, J, (1987). Admittance Matrix Model of a Synchronous Machine for Harmonic Analysis, *IEEE Transactions on Power Systems*, **PWRS-2**(4), 833–840.
7. Medina, A, (1992). Power Systems Modelling in the Harmonic Domain. PhD Thesis, University of Canterbury, New Zealand.

8. Anderson, P M and Fouad, A A, (1977). *Power System Control and Stability*, The IOWA State University Press, USA.
9. Brandwajn, V, (1980). Representation of Magnetic Saturation in the Synchronous Machine Model in an Electro-Magnetic Transients Program, *IEEE Transactions on Power Apparatus and Systems*, **PAS-99**(5,) 1996–2002.
10. Semlyen, A and Castro, A, (1975). A Digital Transformer Model for Switching Transient Calculations in Three-Phase Systems, *9th PICA Conference*, New Orleans, Lousiana, 121–126.
11. Acha, E, (1988). Modelling of Power System Transformers in the Complex Conjugate Harmonic Space. PhD Thesis, University of Canterbury, New Zealand.
12. Lisboa, M L V, Enright, W and Arrillaga, J, (1995). Harmonic and Time Domain Simulation of Transformer Magnetisation Non-Linearities, *Proc IPENZ Conf*, **2**, 72–77.
13. Dommel, H W (1975). Transformer Models in the Simulation of Electromagnetic Transients, *5th Power Systems Computation Conference*, Cambridge, England, **1**, 16.
14. Dick, E P and Watson, W, (1981). Transformer Models for Transient Studies Based on Field Measurements, *IEEE Transactions on Power Apparatus and Systems*, **PAS-100**(1), 106–110.
15. *The EMTDC Users Manual*, Manitoba Hydro, Canada, 1988.

8

AC–DC CONVERSION—HARMONIC DOMAIN

8.1 Introduction

Using the transfer function approach, Chapter 5 has described a general linearized solution of the converter for small levels of distortion. The transfer functions, by means of modulation theory, have been expressed in terms of switching instants that are themselves modulated as a result of applied distortions. The modulation of the switching instants and the transfer function shapes involve approximations valid for small levels of distortion, and low order harmonic, sub-harmonic, and interharmonic frequencies.

Most of the approximations made in previous chapters can be removed by modelling the converter in the time domain, but at the expense of solution speed, since time domain simulations must run until transients have decayed. Thus the motivation for this chapter is the need for greater accuracy while retaining computational efficiency.

In this chapter a general set of non-linear equations is derived to describe harmonic transfer through the ac–dc converter in the steady state. The proposed formulation convolves periodic sampled quantities in the harmonic domain with their sampling functions, so that no Fourier transform is required, resulting in substantial computational savings. The sampling functions are defined in terms of the exact switching instants, which are obtained as part of the over-all iterative procedure that accurately models the effect of ac voltage and dc current distortion on the valve conduction periods. As described here, the model takes one cycle of the ac voltage as the fundamental, and so only harmonics are analysed. However, an extension to the steady state over several cycles would allow inter-harmonics to be solved.

The following are important considerations that must be taken into account in a complete harmonic model of the converter:

- The converter terminal voltage may be unbalanced and include harmonic components.

- There is a voltage drop across the commutating impedance, due to phase current harmonics.
- The commutating impedances may be unbalanced.
- There is a conduction voltage drop across the valves.
- The firing instants are a function of the controller and periodic converter variables (for example the dc ripple and terminal voltage harmonics).
- The commutation current is affected by the ac voltage and dc current harmonics.
- Each commutation ends when the instantaneous commutation current is equal to the instantaneous dc current. The dc ripple therefore affects the overlap angles.
- The dc voltage is a function of the unbalanced and distorted ac voltages, the irregular firing and overlap angles, and the voltage drop across the commutating impedance.
- The ac phase currents are a function of the irregular firing and overlap angles, the commutation currents, and also of the dc current harmonics.

8.2 The Commutation Process

As explained in Chapter 5, central to the analysis of the six-pulse bridge is the commutation process, which must be modelled allowing for imbalance and harmonic distortion in the terminal voltage and dc current. The presence of resistance in the commutation circuit complicates the analysis, however when the interaction of the converter with the ac system is to be solved, the effect of commutation resistance can be accounted for by placing it between the ac system terminal and the converter transformer primary windings.

Two separate analyses of the commutation process are carried out in this section. One is for a bridge connected to a star-g/star transformer, and the other for a bridge connected to a star-g/delta transformer. The angle reference used is arbitrary, and possibly unrelated to the converter at all. For example, all angles may be referenced to the most recent positive zero crossing of the fundamental internal emf of the slack bus generator. At present, the first equidistant timing reference has arbitrarily been assigned an angle of zero. All firing angles and end of commutation angles are referenced to this angle, as opposed to the converter terminal voltage fundamental frequency component, as is the usual practice.

8.2.1 Star Connection Analysis

The commutation circuit to be analysed is that of Figure 8.1, where V_a, V_b, I_c and I_d are sums of harmonic phasors. In this diagram phase 'a' is commutating off, whilst phase 'b' is commutating on. The commutation ends when $I_c = I_d$. Note that I_c is always the commutating 'on' current waveform, not the current in phase 'c'. In this case it is the current commutating on in phase 'b', i.e. it is equivalent to i_b during the commutation period. Assuming the periodic steady state, and summing voltage drops around the commutation loop at harmonic order k:

$$V_{ak} + jkX_a I_{ck} - jkX_a I_{dk} + jkX_b I_{ck} - V_{bk} = 0. \tag{8.1}$$

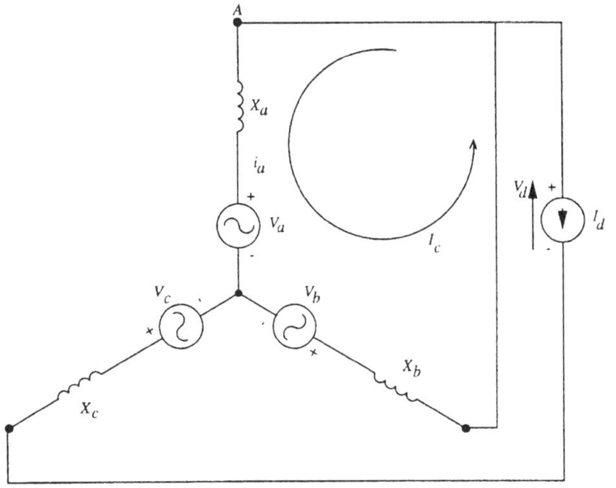

Figure 8.1 Circuit for star-g/star commutation analysis

Solving equation 8.1 for the commutation current:

$$I_{c_k} = \frac{\mathrm{j}kX_aI_{d_k} - V_{ab_k}}{\mathrm{j}k(X_a + X_b)}. \tag{8.2}$$

The periodic commutation current in the time domain is therefore

$$I_c(t) = D + \mathcal{I}\left\{\sum_{k=1}^{n_h} I_{c_k}\mathrm{e}^{\mathrm{j}k\omega t}\right\}, \tag{8.3}$$

where

$$D = -\mathcal{I}\left\{\sum_{k=1}^{n_h} I_{c_k}\mathrm{e}^{\mathrm{j}k\theta_i}\right\}. \tag{8.4}$$

D can be considered to be either a constant of integration, an initial condition, or equivalently a circulating dc current in the commutation loop. Assigning this value to D ensures that at the moment of firing a valve, θ_i, the current in it is zero.

The solution obtained for the commutation current is the steady state solution of the commutation circuit. Since there is no resistance in the circuit, the steady state is achieved instantaneously when the appropriate valve is fired. The commutation ends when the instantaneous commutation current is equal to the instantaneous dc current. This angle, the end of commutation ϕ_i, cannot be solved directly, as Equation 8.3 is transcendental. Instead, the end of each commutation is determined by the zero crossing of a differentiable mismatch equation, solvable by Newton's method. The mismatch equation is easily constructed by substituting $\omega t = \phi_i$ into the Fourier series for the dc and commutation currents, and taking the difference:

$$F_{\phi_i} = \mathcal{I}\left\{ j(I_{d_0} - D) + \sum_{k=1}^{n_h} (I_{d_k} - I_{ci_k}) e^{jk\phi_i} \right\}. \tag{8.5}$$

Equation 8.5 completes the commutation analysis for a bridge connected to a star connected source via an inductance. This equation is suitable for modelling the connection to an unbalanced star-g/star connected transformer, if the leakage reactances and terminal voltages are referred to the secondary side after scaling by off-nominal tap ratios on the secondary or primary windings. This issue is addressed fully in section 8.6.

8.2.2 Delta Connection Analysis

The circuit to be analysed is that of Figure 8.2, which corresponds to a particular commutation. The objective is to solve for the commutation current I_c, in terms of the sources and commutation reactances. Proceeding directly with a phasor analysis in the steady state, a series of loop and nodal equations can be obtained for this circuit at each harmonic k:

$$\begin{aligned} I_{dk} - I_{ck} + i_{bk} - i_{ak} &= 0 \\ I_{ck} + i_{ck} - i_{bk} &= 0 \\ i_{ak} - i_{ck} - I_{dk} &= 0 \\ V_{bk} - j i_{bk} X_{bk} &= 0 \\ V_{ck} - j i_{ck} X_{ck} + V_{dk} &= 0 \\ V_{ak} - j i_{ak} X_{ak} - V_{dk} &= 0 \end{aligned} \tag{8.6}$$

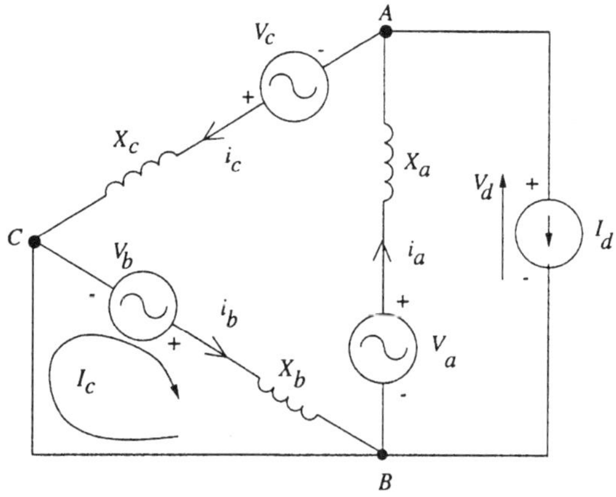

Figure 8.2 Circuit for star-g/delta commutation analysis

where for an inductance $X_k = kX_1$. This set of equations is readily solved to yield the dc voltage during the commutation, and the commutation current itself:

$$V_{dk} = \frac{V_{ak}X_{ck} - V_{ck}X_{ak} + jI_{dk}X_{ck}X_{ak}}{X_{ak} + X_{ck}} \tag{8.7}$$

$$I_{ck} = \frac{V_{bk}}{jX_{ck}} - \frac{V_{ck} + V_{dk}}{jX_{ck}} \tag{8.8}$$

A similar analysis holds for every separate commutation, with appropriate modifications to the phase subscripts, and the direction of the dc current. The end of commutation mismatch equation for the star-g/delta connection is obtained as for the star-g/star connection above.

8.3 The Valve Firing Process

There are two aspects to modelling the valve firing process; the firing controller, and the converter controller [1]. In modern schemes, the firing controller consists of a phase locked oscillator (PLO) [2] tracking the fundamental component of the terminal voltage, and generating essentially equi-spaced timing references. A well designed phase locked oscillator is unaffected by harmonics in the terminal voltage, since its time constant is of the same order as the fundamental. Consequently, the PLO is not modelled, and the timing pulses are assumed perfectly equidistant, spaced by 60°.

In general the effect of a nonideal PLO would be to introduce a coupling between terminal voltage harmonics, and the firing mismatch equation to be derived below. The analysis of the valve firing given here requires a frequency transfer function description of the controller: a simple Proportional–Integral (PI) controller is used, which can readily be extended to any other linear controller. The use of a nonlinear control characteristic would require a linearisation around an operating point.

A valve firing occurs when the elapsed angle from a timing pulse is equal to the instantaneous value of the alpha order. The alpha order is a command variable received from the converter controller. The constant current control is of the proportional integral type (Figure 8.3), and will respond to harmonics in the dc current, causing modulation of the firing angles. From Figure 8.3, the alpha order can be expressed as a sum of harmonic phasors:

$$\alpha = \mathcal{I}\left\{ j\alpha_0 + \sum_{k=1}^{n_h} \alpha_k e^{jk\omega t} \right\}, \tag{8.9}$$

where

$$\alpha_k = \frac{G}{1 + jk\omega T}\left(P + \frac{1}{jk\omega T_I}\right) I_{d_k}. \tag{8.10}$$

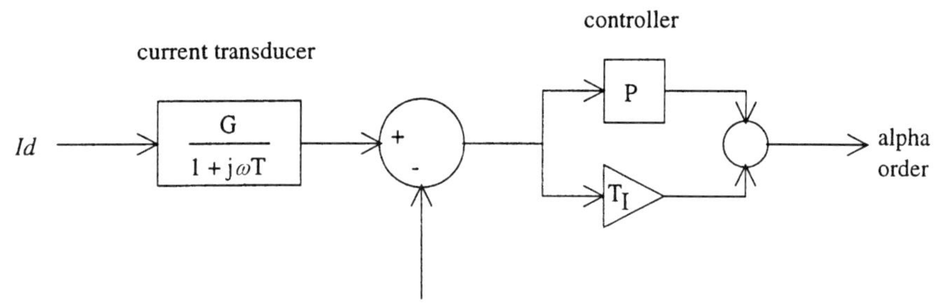

Figure 8.3 Current controller

With reference to Figure 8.4, it can be seen that firing occurs when the elapsed angle from the equidistant timing reference is equal to the instantaneous value of the alpha order, i.e. $\alpha = \theta_i - \beta_i$. The equidistant timing references are represented by $\beta_i = (i - 1)\pi/3$. The firing mismatch equation is therefore:

$$F_{\theta_i} = \mathcal{I}\left\{ \mathrm{j}(\beta_i + \alpha_0 - \theta_i) + \sum_{k=1}^{n_h} \alpha_{i_k} \mathrm{e}^{\mathrm{j}k\theta_i} \right\} = 0. \tag{8.11}$$

This analysis of the firing process is also valid for a bridge connected via a star-g/delta bridge to the ac system, in which case, the equidistant timing references should be advanced by 30°. The constant component of the alpha order, α_0, cannot be solved for directly, as the PI control has a pole at zero frequency. However in the steady state, the average delay angle, α_0, takes on a value that causes the dc component of the dc current to be equal to the current order. This requirement is easily expressed as a mismatch equation that has a zero crossing at the current order:

$$F_{\alpha 0} = (V_{d0} - 4V_{fwd} - E_d)Y_{d0} - I_{d0}, \tag{8.12}$$

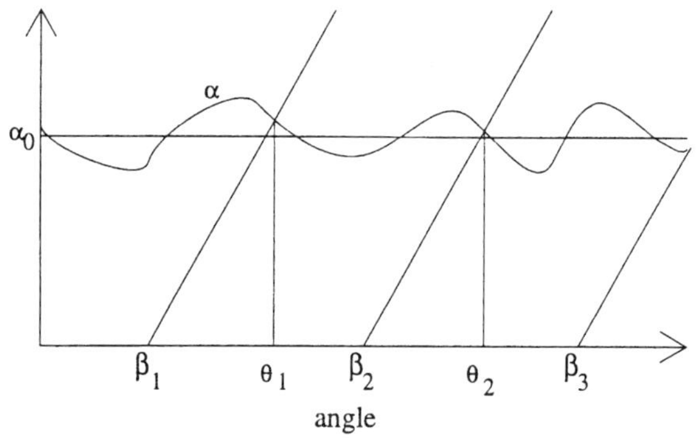

Figure 8.4 Method of finding the firing instants. The timing instants are assumed perfectly equidistant ($\pi/3$)

where V_{fwd} is the constant forward voltage drop through a group. This equation states that the dc voltage, when applied to the dc system, causes the current order to flow in the dc system. The dc voltage is obtained from Equation 8.3. Note that Equation 8.12 is not a function of α_0, the average delay angle does however feature in the firing angle mismatch Equations 8.11. When the converter is solved in Chapter 9, the average alpha order emerges from the Newton solution.

A similar equation to Equation 8.12 can also be written to describe a constant power control:

$$F_S = P_{specified} - (V_{d0} - 4V_{fwd})I_{d0}. \tag{8.13}$$

This equation has a zero crossing at a value of average dc current which causes the power order to be satisfied.

8.4 DC-Side Voltage

The six-pulse bridge passes through twelve states per cycle. Six of these are commutation states, and six are 'direct' conduction states. During direct conduction the positive and negative rails of the dc side are connected to the ac side via two conducting thyristors. As in the case of the commutation analysis, either state can be modelled by the immediate steady state of a simple linear circuit. The circuit consists of a star or delta connected ac voltage source with inductance, connected to a current source representing the dc system. The particular configuration of each circuit depends upon the conduction pattern of the valves in the bridge.

Although it is straightforward to solve the representative linear circuits, the outcome of each steady state solution is a harmonic spectrum, which when transformed into the time domain, matches the dc voltage during the appropriate conduction interval only. The objective is a single spectrum that is valid for one complete cycle of dc voltage, not twelve spectra each valid for only one twelfth of a cycle. The complete spectrum is obtained by convolving each of the twelve 'sample' spectra with the spectrum of a periodic square pulse that has value of one during the corresponding conduction interval, and a value of zero everywhere else. This yields twelve dc voltage sample spectra, the sum of which is the spectrum of the dc voltage across the bridge.

8.4.1 Star Connection Voltage Samples

During normal conduction the positive and negative rails of the dc side are directly connected to different phases of the ac terminal via the commutating reactance in each phase. The kth harmonic component of the dc voltage is therefore:

$$V_{dp_k} = V_+ - V_- - jk(X_+ + X_-)I_{d_k}, \tag{8.14}$$

Table 8.1 Construction of dc voltage and ac phase current samples

sample (p)	Phase Currents			DC voltage (V_{dp})					
	A	B	C	e	b	o	+	−	eqn
1	I_{c1}	$-I_d$	$I_d - I_{c1}$	A	C	B	.	.	(8.15)
2	I_d	$-I_d$	0	.	.	.	A	B	(8.14)
3	I_d	$-I_{c2} - I_d$	I_{c2}	C	B	A	.	.	(8.16)
4	I_d	0	$-I_d$	.	.	.	A	C	(8.14)
5	$I_d - I_{c3}$	I_{c3}	$-I_d$	B	A	C	.	.	(8.15)
6	0	I_d	$-I_d$	.	.	.	B	C	(8.14)
7	I_{c4}	I_d	$-I_{c4} - I_d$	A	C	B	.	.	(8.16)
8	$-I_d$	I_d	0	.	.	.	B	A	(8.14)
9	$-I_d$	$I_d - I_{c5}$	I_{c5}	C	B	A	.	.	(8.15)
10	$-I_d$	0	I_d	.	.	.	C	A	(8.14)
11	$-I_{c6} - I_d$	I_{c6}	I_d	B	A	C	.	.	(8.16)
12	0	$-I_d$	I_d	.	.	.	C	B	(8.14)

where the subscripts + and − refer to the phases connected to the positive and negative dc rails respectively. During a commutation on the positive rail, analysis of Figure 8.1 yields:

$$V_{dp_k} = \frac{X_e V_b + X_b(V_e - jkX_e I_{d_k})}{X_e + X_b} - V_o - jkX_o I_{d_k}, \tag{8.15}$$

and

$$V_{dp_k} = V_o - jkX_o I_{d_k} - \frac{X_e V_b + X_b(V_e + jkX_e I_{d_k})}{X_e + X_b} \tag{8.16}$$

for a commutation on the negative rail. In these equations e refers to phase ending conduction, b to a phase beginning conduction, and o to the other phase. The subscript p refers to the conduction interval being described.

From the known conduction pattern in each of the twelve states, Equations 8.14, 8.15 and 8.16 are used to assemble the twelve samples of the dc voltage. These samples are summarized in Table 8.1.

8.4.2 Delta Connection Voltage Samples

The dc voltage during a particular commutation has already been derived in section 8.2.2 with reference to Figure 8.2. The general result is

$$V_{dpk} = P_{apk} V_{ak} + P_{bpk} V_{bk} + P_{cpk} V_{ck} + P_{dpk} I_{dk}, \tag{8.17}$$

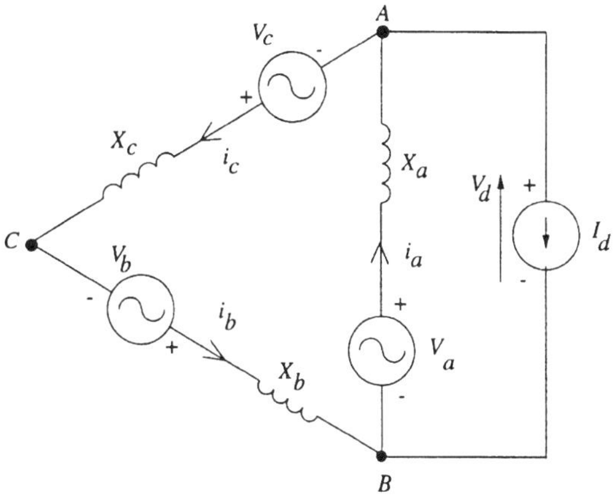

Figure 8.5 Representative linear circuit for a particular conduction period with a delta connected source

where $p = 1, 3, 5, 7, 9, 11$ refers to the conduction interval number. The coefficient matrix P is constant, and need only be calculated once. During a commutation on the positive rail

$$P_{epk} = \frac{X_{co}}{X_{co} + X_{ce}} \tag{8.18}$$

$$P_{opk} = \frac{-X_{ce}}{X_{co} + X_{ce}} \tag{8.19}$$

$$P_{bpk} = 0 \tag{8.20}$$

$$P_{dpk} = \frac{-jkX_{co}X_{ce}}{X_{co} + X_{ce}}, \tag{8.21}$$

where for the commutation periods (i.e. $p = 1, 3, 5, 7, 9, 11$) the subscripts $\{e, b, o\}$ are a permutation of $\{a, b, c\}$ according to which phases are involved in the commutation. A similar result holds for a commutation on the negative rail.

During a normal conduction period all three phases of the voltage source contribute to the dc voltage. Figure 8.5 shows the representative linear circuit of a particular conduction period. This circuit is analysed by first writing nodal and loop equations at harmonic k:

$$\begin{aligned} i_{ak} - i_{ck} - I_{dk} &= 0 \\ I_{dk} - i_{ak} + i_{bk} &= 0 \\ i_{ck} &= i_{bk} \\ V_{ak} - jki_{ak}X_a - V_{dk} &= 0 \\ V_{bk} - jki_{bk}X_b + V_{dk} + V_{ck} - jki_{ck}X_c &= 0. \end{aligned} \tag{8.22}$$

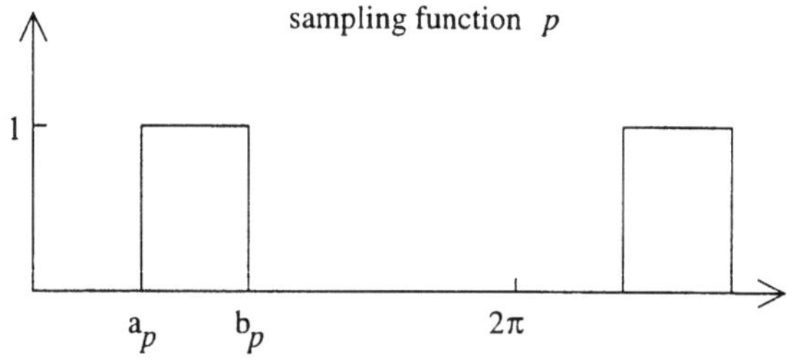

Figure 8.6 Sampling functions used for convolutions

This system is readily solved for the dc voltage sample at harmonic k:

$$V_{dk} = \frac{(V_{ak} - jkX_a I_{dk})(X_b + X_c) - X_a(V_b + V_{ck})}{X_a + X_b + X_c} \tag{8.23}$$

As for the star connected source, the solution for the dc voltage samples is generalised over all twelve conduction periods into a matrix of coefficients of the dc and ac sources, ie

$$V_{dpk} = P_{apk}V_{ak} + P_{bpk}V_{bk} + P_{cpk}V_{ck} + P_{dpk}I_{dk}, \tag{8.24}$$

where $p = 2, 4, 6, 8, 10, 12$.

8.4.3 Convolution of the Samples

Having obtained twelve dc voltage samples as a function of dc and ac side sources, the overall solution for the dc voltage spectrum is constructed by convolving each sample with a square pulse sampling function (Figure 8.6). The convolution described here uses positive frequencies only, and so generates phase conjugated terms [4] [3]. The square pulse is periodic at the fundamental frequency, and delimited alternately by the firing and end of commutation angles as listed in Table 8.2. The complex Euler coefficient for the sampling function at harmonic k is

$$\Psi_{p_k} = \frac{1}{k\pi}[\cos(ka_p) - \cos(kb_p)] + \mathrm{j}\frac{1}{k\pi}[\sin(kb_p) - \sin(ka_p)]. \tag{8.25}$$

$$\Psi_{p_0} = \begin{cases} \mathrm{j}\dfrac{b_p - a_p}{2\pi} & b_p > a_p \\ \mathrm{j}\left[1 - \dfrac{a_p - b_p}{2\pi}\right] & \text{otherwise} \end{cases} \tag{8.26}$$

Table 8.2 Limits of converter states for use in sampling function

sample (p)	a_p	b_p
1	θ_1	ϕ_1
2	ϕ_1	θ_2
3	θ_2	ϕ_2
4	ϕ_2	θ_3
5	θ_3	ϕ_3
6	ϕ_3	θ_4
7	θ_4	ϕ_4
8	ϕ_4	θ_5
9	θ_5	ϕ_5
10	ϕ_5	θ_6
11	θ_6	ϕ_6
12	ϕ_6	θ_1

Since the end of one conduction interval is the beginning of the next, all of the trigonometric evaluations are used in two consecutive sampling functions, thus halving the number of calculations. The dc voltage can now be written as:

$$V_d = \sum_{p=1}^{12} V_{dp} \otimes \Psi_p. \tag{8.27}$$

The convolution of two phasors is given by,

$$A_k \otimes B_l = \begin{cases} \frac{1}{2}\mathrm{j}(A_k B_l^*)_{(k-l)} - \frac{1}{2}\mathrm{j}(A_k B_l)_{(k+l)} & \text{if } k \geqslant l \\ \frac{1}{2}\mathrm{j}(A_k B_l^*)_{(l-k)}^* - \frac{1}{2}\mathrm{j}(A_k B_l)_{(k+l)} & \text{otherwise.} \end{cases} \tag{8.28}$$

The conjugate operator makes the convolution non-analytic, and so not differentiable in the complex form. It avoids the need for negative harmonics however, and it is still possible to obtain partial derivatives by decomposing into real and imaginary parts. Sum and difference harmonics are generated by the convolution, and since it is required to calculate voltage harmonics up to n_h, the sampling function spectra must be evaluated up to $2n_h$. Since the convolution operator is linear, the twelve convolutions in Equation 8.27 can be decomposed into convolutions of the component phasors:

$$V_{dp} \otimes \Psi_p = \sum_{k=1}^{n_h} \sum_{l=0}^{2n_h} V_{dp_k} \otimes \Psi_{p_l}. \tag{8.29}$$

This equation generates voltage harmonic components of order above n_h which are discarded. By using Equations 8.27, 8.28, and 8.29, the kth harmonic phasor component of V_d is

$$V_{d_k} = \tfrac{1}{2}\mathrm{j}\sum_{p=1}^{12}\left\{\sum_{l=k}^{n_h}(V_{dp_l}\Psi^*_{p_{l-k}}) + \sum_{l=1}^{n_h}(V_{dp_l}\Psi^*_{p_{l+k}})^* - \sum_{l=0}^{k}(V_{dp_l}\Psi_{p_{k-l}})\right\}, k > 0. \quad (8.30)$$

$$V_{d_0} = \tfrac{1}{2}\mathrm{j}\sum_{p=1}^{12}\left\{\sum_{l=0}^{n_h}(V_{dp_l}\Psi^*_{p_l}) - 2\mathrm{j}(V_{dp_0}\Psi_{p_0})\right\} \quad (8.31)$$

This completes the derivation of the dc voltage harmonics in terms of a dc side harmonic source, and a three phase ac side voltage source connected in star or delta, with source inductance. The process of sampling and convolution to obtain the dc voltage is demonstrated graphically in Figure 8.7. This sequence of plots was obtained by first simulating a six-pulse rectifier to the steady state in the time domain. The spectra of the terminal voltage and dc current were then used to calculate the dc voltage, allowing a validation against the time domain solution.

8.5 Phase Currents on the Converter Side

The derivation of the dc voltage involves the convolution of twelve different dc voltage samples, so by using the same sampling functions, 36 convolutions would be required to obtain the three phase currents. However referring to Table 8.1, and using the linearity of the convolution, the phase A secondary current can be written as:

$$\begin{aligned} I_a = I_d \otimes \{\Psi_2 + \Psi_3 + \Psi_4 + \Psi_5 - \Psi_8 - \Psi_9 - \Psi_{10} - \Psi_{11}\} \\ + I_{c_1} \otimes \Psi_1 - I_{c_3} \otimes \Psi_5 - I_{c_4} \otimes \Psi_7 + I_{c_6} \otimes \Psi_{11}, \end{aligned} \quad (8.32)$$

and similarly for one of the other two phases. The third phase must always be the negative sum of the first two, since there is no path for zero sequence into a bridge. This leads to a total of 8 convolutions to calculate the three phase currents. As evident in Equation 8.32, the periodic samples for the phase current calculation are just the dc side current, and the commutation currents derived in section 8.2. The calculation of the phase current flowing into the transformer primary is addressed in the next section. The derivation of a phase current is illustrated graphically in Figure 8.8.

8.6 Phase Currents on the System Side

Unbalance in the tap changer setting between the two six-pulse groups will lead to imperfect cancellation of six-pulse harmonics on the ac and dc sides of the converter. If the impedances in each phase of a three-phase bank are not all equal, the converter will generate positive and negative sequence odd triplen harmonics. The unbalanced

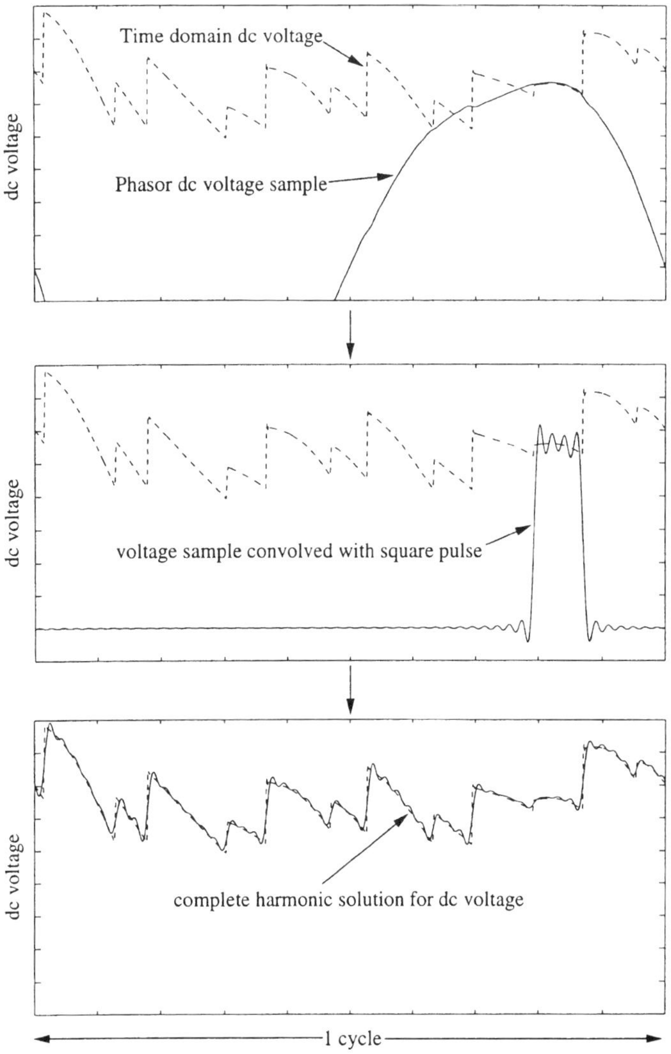

Figure 8.7 Construction of the dc voltage, and validation against time domain solution

star-g/delta connected transformer also acts as a sequence transformer, causing the converter to both respond to, and generate zero sequence harmonics.

In this section the transformer is modelled as a series connection of ideal tap changing transformers on the primary and secondary sides, a conduction resistance, a leakage reactance, and a star/delta connection; core saturation and hysteresis are not modelled. The resistance and reactance may be unbalanced, but the tap settings are assumed to be the same on all phases. The tap change controller is not modelled as it does not respond to harmonics. The magnetizing current injection is approximated by a shunt to ground at the primary terminal.

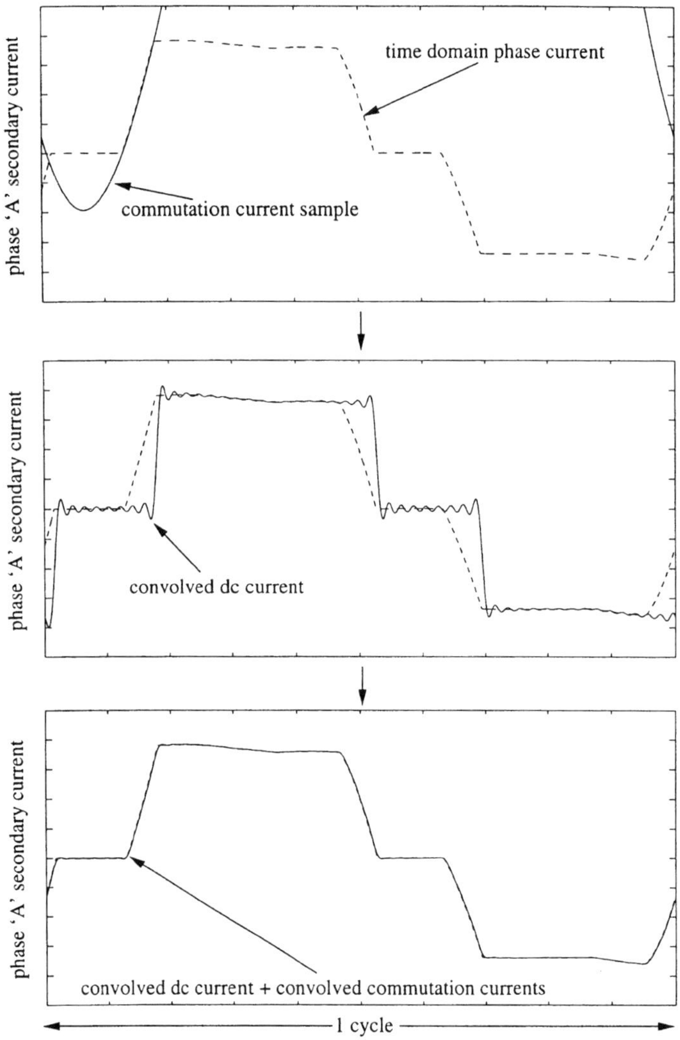

Figure 8.8 Construction of the phase current, and validation against time domain solution

The outcome of this analysis is a transfer model of the transformer; the primary currents are related directly to the secondary currents, and the secondary voltage to the primary voltage. This is easily achieved for the star-g/star connection, shown as a single line diagram in Figure 8.9.The transformer and thyristor resistances have been referred to an equivalent ac side resistance, R_{ac}:

$$R_{ac} = a_1^2\left(\frac{R_t}{a_2^2} + R_c\right). \tag{8.33}$$

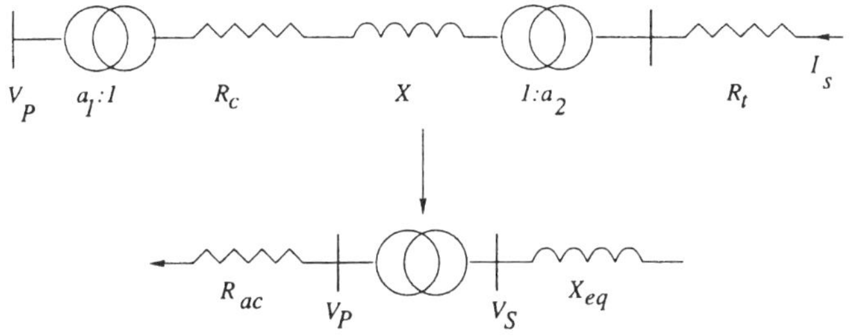

Figure 8.9 Equivalent circuit for star-g/star transformer

The leakage reactance has been referred to an equivalent on the secondary side:

$$X_{eq} = a_2^2 X. \tag{8.34}$$

Since all impedance has been removed from the transformer and incorporated into either the ac system or commutation circuit, the secondary voltage is now written as if it were independent of the current through the transformer:

$$V_S = \frac{a_2}{a_1} V_P, \tag{8.35}$$

and similarly for the phase current:

$$I_P = \frac{a_2}{a_1} I_S. \tag{8.36}$$

These equations are repeated over all harmonics, and all three phases.

The star-g/delta connected transformer is considerably more difficult to model, and in fact requires two separate analyses for transfers from the star to delta side and vice versa. As shown in Figure 8.10, the transfer from star to delta is primarily concerned with setting up the delta connected source for the voltage sampling and commutation analyses. The secondary side delta connected voltage source is scaled by:

$$V_S = \sqrt{3}\frac{a_2}{a_1} V_P, \tag{8.37}$$

and the equivalent reactance is:

$$X_{eq} = 3a_2^2 X. \tag{8.38}$$

The $\sqrt{3}$ scaling for the delta winding does not affect the transfer of thyristor resistance through the transformer, since it is not connected in delta. Thus, the referred ac system resistance is the same as Equation 8.33.

Calculation of the primary side phase currents in terms of the secondary currents is complicated by the circulating current in the delta winding. If the transformer is unbalanced, some of this appears as a positive or negative sequence current on the primary side.

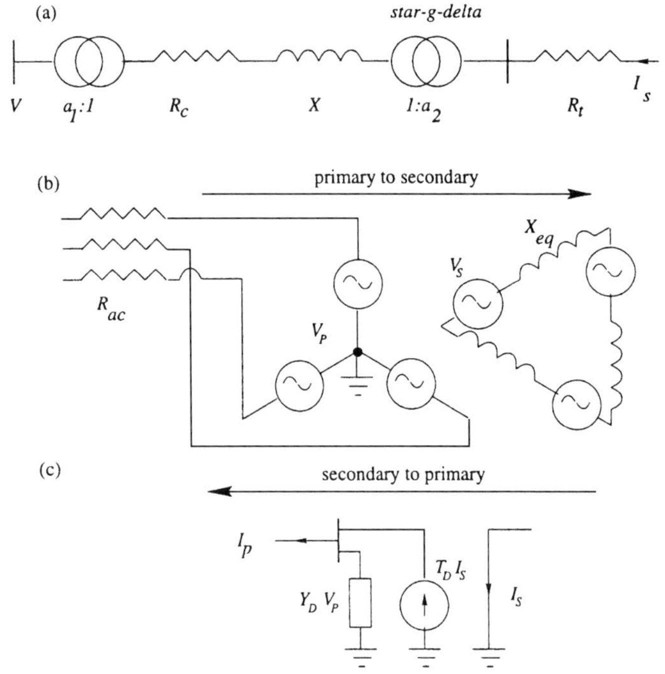

Figure 8.10 (a) equivalent circuit for star-g/delta transformer. (b) transfer from star to delta. (c) transfer from delta to star

The admittance matrix for an unbalanced star-g/delta transformer is readily obtained:

$$\begin{bmatrix} I_{Pa} \\ I_{Pb} \\ I_{Pc} \\ I_{Sa} \\ I_{Sb} \\ I_{Sc} \end{bmatrix} = \begin{bmatrix} \alpha^2 Y_a & 0 & 0 & -\alpha\beta Y_a & \alpha\beta Y_a & 0 \\ 0 & \alpha^2 Y_b & 0 & 0 & -\alpha\beta Y_b & \alpha\beta Y_b \\ 0 & 0 & \alpha^2 Y_c & \alpha\beta Y_c & 0 & -\alpha\beta Y_c \\ -\alpha\beta Y_a & 0 & \alpha\beta Y_c & \beta^2(Y_a + Y_c) & -\beta^2 Y_a & -\beta^2 Y_c \\ \alpha\beta Y_a & -\alpha\beta Y_b & 0 & -\beta^2 Y_a & \beta^2(Y_a + Y_b) & -\beta^2 Y_b \\ 0 & \alpha\beta Y_b & -\alpha\beta Y_c & -\beta^2 Y_c & -\beta^2 Y_b & \beta^2(Y_b + Y_c) \end{bmatrix} \begin{bmatrix} V_{Pa} \\ V_{Pb} \\ V_{Pc} \\ V_{Sa} \\ V_{Sb} \\ V_{Sc} \end{bmatrix} \tag{8.39}$$

where:

$$Y = \frac{1}{R_c + jX} \tag{8.40}$$

$$\alpha = \frac{1}{a_1} \tag{8.41}$$

$$\beta = \frac{1}{\sqrt{3}a_2}. \tag{8.42}$$

Equation 8.39 is used to calculate the primary current, by assuming that V_P and I_S are known, and eliminating V_S. The admittance matrix in Equation 8.39 is not invertible, as the delta winding is floating; there are an infinite number of possible potentials of the delta winding which are consistent with a given current injection into the transformer. One such potential is that obtained by grounding phase 'c' on the secondary so that $V_{Sc} = 0$. This permits the removal of the last row and column from Equation 8.39:

$$\begin{bmatrix} I_{Pa} \\ I_{Pb} \\ I_{Pc} \\ I_{Sa} \\ I_{Sb} \end{bmatrix} = \begin{bmatrix} A & B \\ C & D \end{bmatrix} \begin{bmatrix} V_{Pa} \\ V_{Pb} \\ V_{Pc} \\ V_{Sa} \\ V_{Sb} \end{bmatrix}, \tag{8.43}$$

where:

$$A = \begin{bmatrix} \alpha^2 Y_a & 0 & 0 \\ 0 & \alpha^2 Y_b & 0 \\ 0 & 0 & \alpha^2 Y_c \end{bmatrix}, \tag{8.44}$$

$$B = \begin{bmatrix} -\alpha\beta Y_a & \alpha\beta Y_a \\ 0 & -\alpha\beta Y_b \\ \alpha\beta Y_c & 0 \end{bmatrix}, \tag{8.45}$$

$$C = \begin{bmatrix} -\alpha\beta Y_a & 0 & \alpha\beta Y_c \\ \alpha\beta Y_a & -\alpha\beta Y_b & 0 \end{bmatrix}, \tag{8.46}$$

$$D = \begin{bmatrix} \beta^2(Y_a + Y_c) & -\beta^2 Y_a \\ -\beta^2 Y_a & \beta^2(Y_a + Y_b) \end{bmatrix}. \tag{8.47}$$

Eliminating V_s, the primary phase currents are:

$$I_P = [A - BD^{-1}C]V_P + BD^{-1}I_S, \tag{8.48}$$

$$\stackrel{\text{def}}{=} Y_D V_P + T_D I_S. \tag{8.49}$$

Y_D is a shunt admittance to ground at the converter terminal that is added to the filter shunt. T_D is a transfer matrix across the transformer, of size 3×2, indicating that there is no zero sequence current on the secondary side, and that only the phase 'a' and 'b' currents need be calculated. If the transformer is balanced, then Y_D is a zero sequence shunt, and so is not invertible. This also implies that if the transformer is nearly balanced, Y_D has a high condition number, and should not be inverted into an impedance without first being combined with an admittance that offers a path to positive and negative sequence currents.

8.7 Summary

A set of equations have been derived which describe steady-state relationships between variables relevant to a twelve-pulse controlled rectifier, with ac and dc system representation. The equations model the effect of unbalance in the ac and dc systems, harmonic sources in both systems, and unbalance in the converter transformers. The firing and commutation processes have also been modelled in detail, as have the current and power control constraints, and the current control response to harmonic ripple on the dc side.

Harmonic transfer through the converter has been analysed using a convolution method that avoids the problems of aliasing associated with numerical FFT calculations, or the complexity of Fourier analysis. An additional advantage of the convolution analysis described here is that all of the equations are differentiable when decomposed into real and imaginary components, a feature that enables a straighforward implementation of a Newtons method solution in Chapter 9.

8.8 References

1. Arrillaga, J, (1983). *High Voltage Direct Current Transmission, Vol. 6. Peter Peregrinus Ltd, London, UK.*
2. Ainsworth, JD, (1968). The phase locked oscillator—a new control system for controlled static converters, *IEEE Transactions on Power Apparatus and Systems*, **87**(3); 859–865.
3. Smith, BC, *et al.*, (1995). A steady state model of the ac–dc converter in the harmonic domain, *IEE Proceedings Generation, Transmission and Distribution*, **142**(2); 109–118.
4. Smith, BC, (1996). A harmonic domain model for the interaction of the HVdc convertor with ac and dc systems. PhD thesis, University of Canterbury, New Zealand.

9

ITERATIVE HARMONIC ANALYSIS

9.1 Introduction

It has already been explained in earlier chapters that the analysis of harmonic interactions between the network linear and non-linear components requires detailed models of these components as well as an iterative solution. The main non-linearities of synchronous generators and transformers have been analysed in Chapter 7 and those of the three-phase static converter in Chapters 5, 6 and 8. These models are used as the main components of the iterative algorithms discussed in this chapter. Considering the power ratings and complex controllability of HVdc converters much of this chapter is devoted to them, although the techniques discussed are equally relevant to other power electronic devices.

9.2 Fixed Point Iteration Techniques

The simplest iterative scheme used to simulate harmonic interactions uses the fixed point iteration concept. At each iteration the latest values of the distorted terminal voltages are used to derive updated information of the harmonic current injections. The direct analysis of Chapter 4 is then invoked to update the ac voltage harmonics for the next iteration. With reference to the three-phase static converter, the updated terminal voltages are used as the commutating voltages for the converter solution. The calculated dc voltage waveform is then impressed upon the dc side impedance to derive the dc current waveform, which in turn, together with the switching instants and commutation process, provides the ac side current harmonic injections. The latter are then used to derive the ac voltage harmonics in the frequency domain for the next iteration.

Early fixed point iteration methods involved the solution of the switching instants [1–3]. An alternative approach is to derive time domain waveforms for the direct voltage and ac side phase currents by evaluating analytical expressions for those quantities on a point by point basis, and then applying the FFT to yield the desired harmonic information [4].

In general, the solution diverges when the dc system harmonic impedance is large and the commutating reactance small [4]. Divergence can sometimes be avoided, or convergence improved by inserting a fictitious reactance pair between the converter transformer primary and the filter bus. The inserted pair consists of a series combination of a reactance, and its negative, with the midpoint voltage being the new voltage to be used as the commutating voltage [5]. The reactance value is chosen to cancel the ac system reactance and to increase the commutating reactance.

However, the reactance pair cannot cancel high impedance in the ac system due to parallel resonances; instead, a matched impedance pair can be chosen to mirror quite closely the frequency dependence of the ac system impedance, and consists of a simple RLC network [6,7].

The main problem with the matched impedance pair method is the added complexity. Selection of the RLC network is by no means straightforward, and the resulting commutation process is formidably difficult to solve. An additional requirement is a separate ac terminal for every six-pulse group, as the matched impedance pair can only be placed in one series path.

More recent work has been directed toward improving the solution method itself, rather than improving the fixed point iterative technique.

9.3 The Method of Norton Equivalents

In the fixed point iteration, the non-linear component is represented at each iteration by a constant current source. Far better convergence can be expected when using instead a Norton equivalent for the non-linear component, with the Norton admittance representing a linearization, possibly approximate, of the component response to variation in terminal voltage harmonics.

Full linearization requires either an admittance tensor (discussed in Chapter 10) or a complex conjugate cross harmonic admittance matrix representation (described in Chapter 7). The latter approach is best used with devices that can be described by a static (time invariant) voltage-current relationship,

$$i(t) = f(v(t)), \tag{9.1}$$

in the time domain. For such devices, both the current injection and the Norton admittance can be calculated by an elegant procedure involving an excursion into the time domain. At each iteration, the applied voltage harmonics are inverse Fourier transformed to yield the voltage waveshape. The voltage waveshape is then applied point by point to the static voltage/current characteristic, to yield the current waveshape. By calculating the voltage and current waveshapes at $2n$ equi-spaced points, a FFT is readily applied to the current waveshape, to yield the total harmonic injection.

To calculate the Norton admittance, the waveshape of the total derivative,

$$\frac{\mathrm{d}I}{\mathrm{d}V} = \frac{\mathrm{d}i(t)}{\mathrm{d}t}\frac{\mathrm{d}t}{\mathrm{d}v(t)} = \frac{\mathrm{d}i(t)/\mathrm{d}t}{\mathrm{d}v(t)/\mathrm{d}t}, \tag{9.2}$$

is calculated by dividing the point by point changes in the voltage and current waveshapes. Fourier transforming the total derivative yields columns of the Norton admittance matrix; in this matrix all the elements on any diagonal are equal, i.e. it has a Toeplitz structure. The Norton admittance calculated in this manner is actually the Jacobian for the source.

The complex conjugate solution, applied to the transformer magnetization non-linearity in Chapter 7, is used here as a basis for an iterative solution combining the linear and linearized components as illustrated in the flow diagram of Figure 9.1.

At each iteration the fluxes in all magnetic branches are required to adequately evaluate the harmonic Norton equivalents. The winding fluxes can be determined directly from the transformer terminal voltages, but the remaining fluxes can only be obtained by solving the magnetic circuit. Some of the equations are non-linear and, therefore, the solution has to be achieved through an iterative procedure. Matrix equation 7.65 expresses this set of equations in a linearized form, and a Newton-Raphson solution is possible with the Jacobian defined by matrix $[H]$.

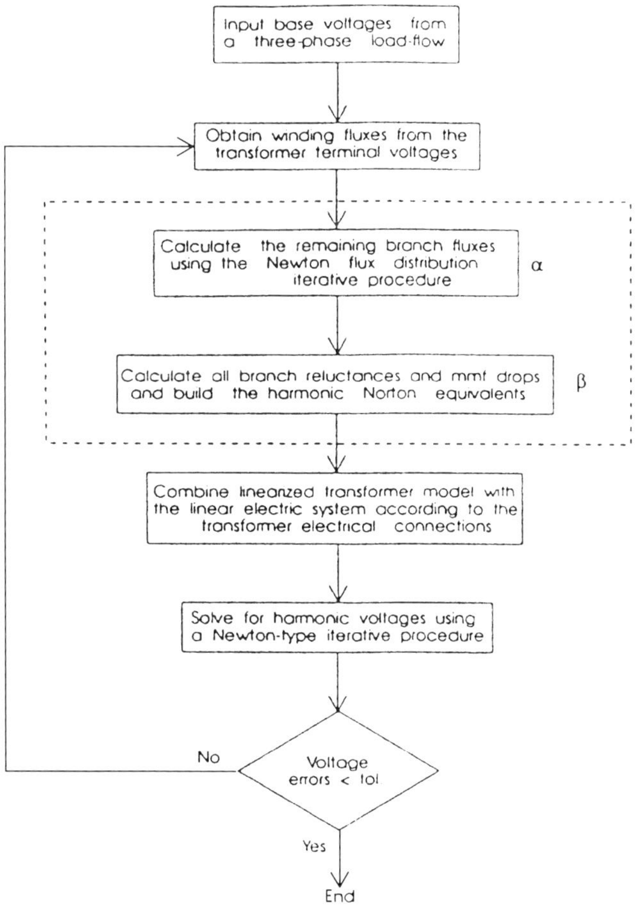

Figure 9.1 Structure of the algorithm to model transformer magnetization non-linearity

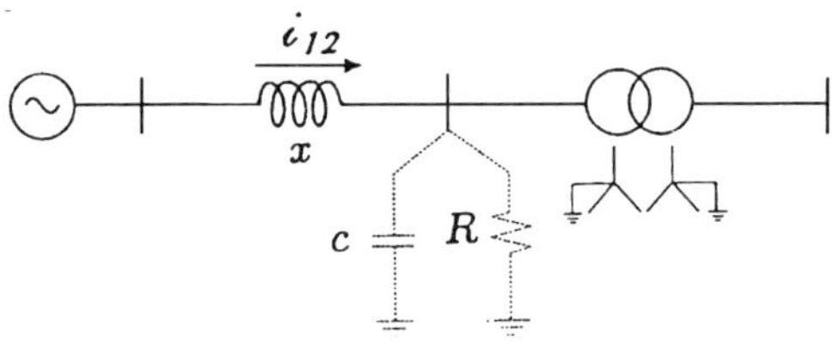

Figure 9.2 Test system to assess the effect of transformer magnetization characteristics. $x = 0.1$ p.u., $c = 0.9$ p.u., $R = 1.0$ p.u.

The starting values for each iteration can be those obtained from the flux distribution derived in the previous iteration, except at the first iteration when the starting values have to be derived from an approximation, such as the linear distribution suggested in the EMTP manual [8].

A simple three-phase system for the application of the linear and linearized algorithm is shown in Figure 9.2, with a multi-limb transformer.

Figure 9.3(a) illustrates the waveforms of the current flowing through the inductance and 9.3(b) their corresponding harmonic content. The transformer

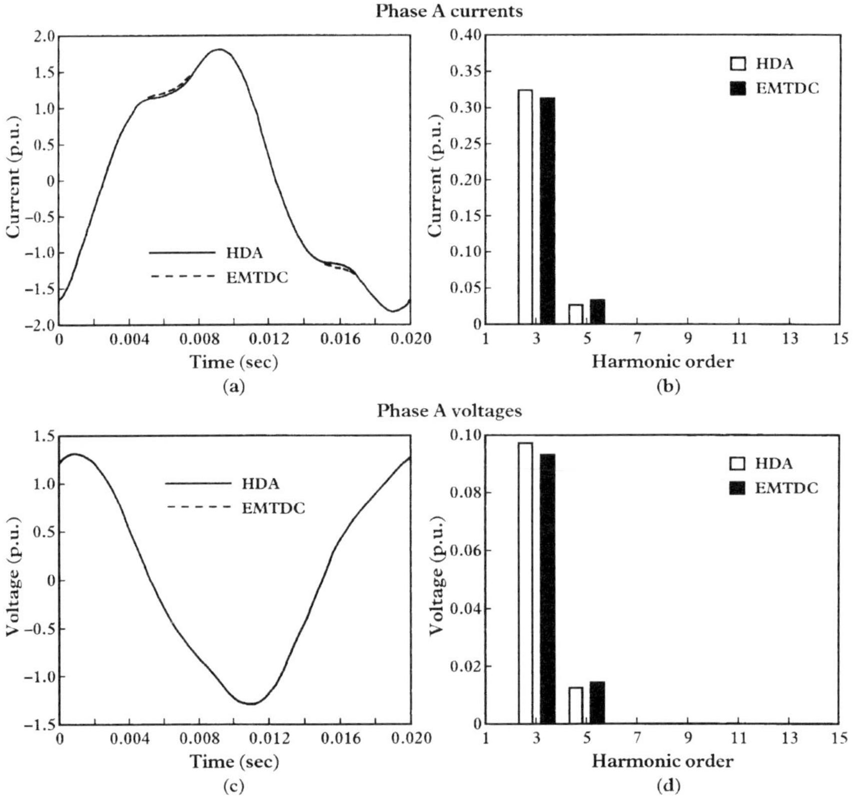

Figure 9.3 Comparison of time and harmonic domain solutions for the test system

primary side voltage waveforms and harmonic content are shown in Figures 9.3(c) and 9.3(d) (only phase A results are shown). Almost a perfect match is achieved for the voltage waveforms, the difference between the two solutions is indistinguishable in Figure 9.3(c). In this case, the third harmonic voltage is nearly 7%.

The results indicate very good agreement with the time domain solutions, especially considering the difficulty of the case analysed. The maximum magnitude differences are 0.003 (p.u.) for the voltages and 0.009 for the currents, both occurring at the third harmonic. The solution is achieved in four iterations.

However, the ac–dc converter does not fall in the above category defined by Equation 9.1. Instead, the V–I relationships result from many interdependent factors such as phase and magnitude of each of the ac voltage and dc current harmonic components, control system functions, firing angle constraints, etc.

The converter Norton admittance is not Toeplitz, and, in general, contains of the order of n^2 different elements, as opposed to the n elements obtained from a FFT. This is illustrated in Figure 9.4, for the ac side of a twelve pulse converter. The admittance clearly displays a lattice structure related to the twelve pulse switching action of the converter.

The problem with the method of Norton equivalents is that the converter is really an interface between the ac and dc systems, with only the ac system represented in the overall solution process. If the converter controller is modelled, a separate iterative procedure is required to solve the converter interaction with the dc system at each iteration.

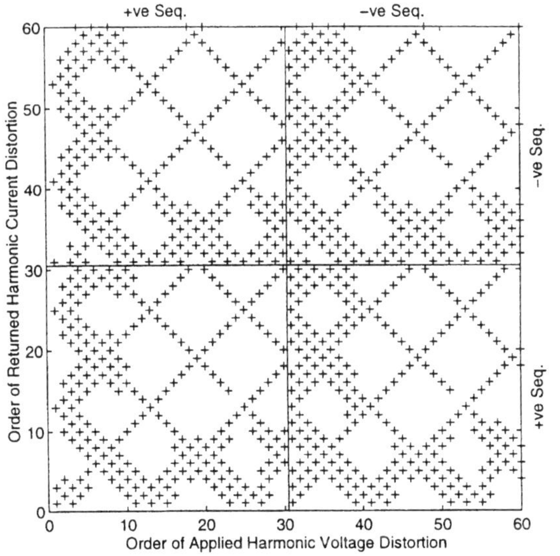

Figure 9.4 Structure of the Norton admittance at a solution with 0.85% negative sequence fundamental voltage distortion at the converter terminal

9.4 ABCD Parameters Model

The use of an *ABCD* parameters matrix has been proposed [9] to linearize the harmonic transfer across a converter. The matrix equation is

$$\begin{bmatrix} \Delta I \\ \Delta Vd \end{bmatrix} = \begin{bmatrix} A & B \\ C & D \end{bmatrix} \begin{bmatrix} \Delta V \\ \Delta Id \end{bmatrix} \tag{9.3}$$

where ΔI, ΔVd, ΔV, ΔId are vectors of harmonic perturbations and Vd, Id are dc side quantities. The *ABCD* matrix, therefore, links harmonics of different orders on both sides of the converter. Only positive order harmonics are used in the formulation and the *ABCD* matrix is obtained by harmonic perturbations of a time domain simulation. However, to be fully general, both positive and negative harmonics must be included, or the matrix should be of Cartesian components.

If the effects of control and commutation variations are neglected, the converter is linear in the harmonic domain, and the Δ symbol in Equation 9.3 can be dropped. An iterative approach is possible between a direct solution of the ac–dc system interaction described by an ABCD matrix and an update to the commutation duration [10,11]; at each iteration the *ABCD* matrix is updated by evaluating the harmonic domain convolutions of terminal voltage and direct current spectra with switching functions.

However, a decoupled solution similar to this model, iterating between a linear solution of the ac–dc system interaction, and a Newton solution for the switching angles has been found to diverge in cases where even order harmonic sources are present in the ac system [12].

There is thus a motivation to linearize switching angle variation with terminal harmonic variation, and develop a full Newton solution, incorporating more than just electrical equivalents. This approach is considered in the rest of the chapter.

9.5 Newton's Method

This section describes a sparse Newton solution for the steady state interaction of a controlled converter with the ac and dc systems. The test system, derived from the CIGRE benchmark model[13] is shown in Figure 9.5. The ac network is represented by a three-phase Thevenin equivalent and the dc system by a T-circuit equivalent and a dc source. This test system is operationally difficult in that it consists of a weak dc system resonant at the 2nd harmonic, coupled via the converter to a fundamental resonance on the dc side. Further details of the test system are given in Appendix VI The variables to solve in Figure 9.5 are the terminal voltage harmonic phasors, V^S and V^D, the dc current harmonics I_d, the twelve firing instants, θ_i, the twelve end of commutation instants, ϕ_i, the average delay angle, α_0, and, in the case of constant power control, the average dc current, I_{d0}. For a solution of every harmonic up to the 50th, there are therefore 726 unknowns, as each harmonic requires two variables to be specified. Being three phase and 2-port, a full specification of the terminal voltage requires six hundred variables. If the commutating resistance, R_c, is removed from the circuit, the number of unknowns reduces to 426, since the two six-pulse rectifiers

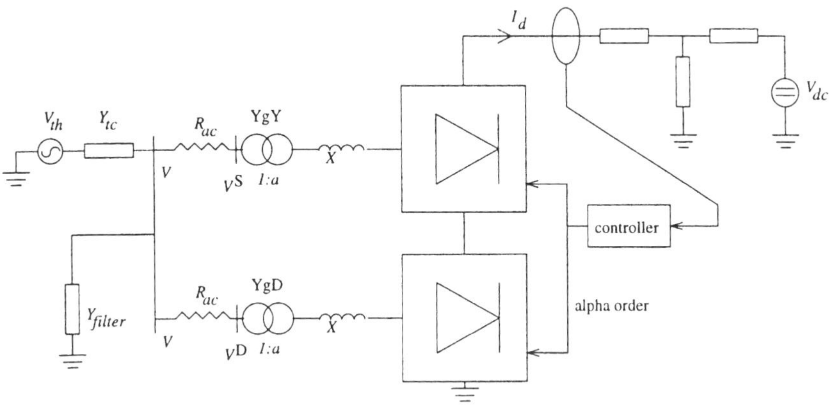

Figure 9.5 Twelve pulse converter model to be solved

share the same ac terminal voltage. A solution for this system is developed by using the inter-relationships of Chapter 8 to derive a set of simultaneous mismatch equations, the simultaneous zero of which corresponds to the desired steady state solution. This non-linear problem is solved by Newton's method with sparsity, and validated against a time domain simulation of a test system to the steady state.

The test system of Figure 9.5 will be described by the following quantities:

- V, the three phase harmonic series converter terminal voltage.
- V_P^D, the voltage at the equivalent star-g/delta transformer primary, on the converter side of the lumped commutation resistance. As shown in Figure 9.5 the transformer and thyristor conduction resistances have been lumped into an equivalent resistance between each bridge transformer and the ac terminal of the twelve pulse converter.
- V_P^S, as above for for the star-g/star transformer.
- I_S^S, I_S^D the phase currents flowing into the secondary windings of the star-g/star and star-g/delta transformers, respectively.
- I_P^S, I_P^D the primary phase currents for the two transformers.
- V_{dk}, V_{d0} the dc voltage harmonics and dc component, respectively.
- I_{dk}, I_{d0} the dc current harmonics and dc component, respectively.
- $F_{\alpha 0}$ the average delay angle mismatch equation.
- $F_{\theta i}^D$, $F_{\theta i}^S$, the firing angle mismatches of the thyristors in the bridges attached to the star-g/delta and star-g/star transformers, respectively.
- $F_{\phi i}^D$, $F_{\phi i}^S$, the end of commutation mismatches in the bridges attached to the star-g/delta and star-g/star transformers, respectively.

9.5.1 Functional Description of the Twelve-Pulse Converter

In this section the relationships developed in Chapter 8 are expressed in terms of variables relevant to the system of Figure 9.5. A functional notation is introduced to describe relationships between the quantities that describe the converter in the steady state. The starting point will be a relationship between the converter phase currents, and the converter terminal voltage.

Given the primary phase currents, the terminal voltage can be found by;

$$[V]_k = [Y_{cc}]_k^{-1}(-[I_P^S + I_P^D]_k + [Y_{ct}]_k[V_{th}]_k), \tag{9.4}$$

where the square brackets denote a three phase quantity. The ac system impedance, $[Y_{cc}]_k^{-1}$, is calculated by inverting the sum of the admittances attached to the converter bus. These are the filter admittance, Y_{filter}, and the Thevenin source admittance, Y_{ct}. This equation will be represented over all phases and harmonics by:

$$V = f_1(I_P^S, I_P^D) \tag{9.5}$$

The converter terminal voltage is then related to the equivalent transformer primary voltages by the voltage drops through the equivalent commutating resistances:

$$[V_P^S]_k = [V]_k - [R^S][I_P^S]'_k \tag{9.6}$$

$$[V_P^D]_k = [V]_k - [R^D][I_P^D]'_k. \tag{9.7}$$

The commutating resistance matrix is assumed frequency independent and diagonal, but possibly unbalanced. The use of phase currents from the previous iteration in Equations 9.6 and 9.7, leads to the derivation of a smaller nonlinear system in Section 9.5.2. This is denoted by the primed quantities in these equations. Using the function notation again, where phase currents from the previous iteration are treated as parameters, not variables:

$$V_P^S = f_2(V), \tag{9.8}$$

$$V_P^D = f_3(V). \tag{9.9}$$

Chapter 8 described how to calculate the dc voltage across a six pulse bridge attached to either a star or delta connected ac source, with inductive source impedance. Using the transformer models of Section 8.6, the primary voltages, V_P^S and V_P^D, can be transformed into equivalent star or delta connected inductive sources on the secondary side. The dc voltages across each group are then added to obtain the total dc voltage, and the constant forward voltage drop associated with each group is subtracted from the total dc side voltage direct component. The calculation of the dc voltage is again represented in functional notation:

$$V_{dl} = f_4(V_P^S, V_P^D, I_{dk}, \theta_i^S, \theta_i^D, \phi_i^S, \phi_i^D), \tag{9.10}$$

$$V_{d0} = f_5(V_P^S, V_P^D, I_{dk}, \theta_i^S, \theta_i^D, \phi_i^S, \phi_i^D). \tag{9.11}$$

The dc voltage is functionally dependent on the switching angles since they define the limits of the convolution analysis used to calculate the dc side voltage across each group. The dc current harmonics are present in the calculation of the voltage samples which are convolved with the sampling functions. The dc voltage, when applied to the dc system, yields the dc side current. For example

$$I_{dk} = V_{dk} Y_{dk}, \tag{9.12}$$

$$I_{d0} = (V_{d0} - 4V_{fwd} - E_d) Y_{d0}, \tag{9.13}$$

where V_{fwd} is the constant forward voltage drop through a group. Summarizing for any topology:

$$I_{dk} = f_6(V_{dk}) \tag{9.14}$$

$$I_{d0} = f_7(V_{d0}). \tag{9.15}$$

The dc current, switching angles, and primary transformer voltage can be used to calculate the transformer secondary phase currents by applying the analysis of Section 8.5 to each bridge:

$$I_S^S = f_8(V_P^S, I_{dk}, I_{d0}, \theta_i^S, \phi_i^S) \tag{9.16}$$

$$I_S^D = f_9(V_P^D, I_{dk}, I_{d0}, \theta_i^D, \phi_i^D). \tag{9.17}$$

The transformer analysis of Section 8.6 then describes how the primary currents are obtained:

$$I_P^S = f_{10}(I_S^S) \tag{9.18}$$

$$I_P^D = f_{11}(V_P^D, I_S^D). \tag{9.19}$$

Implicit equations have been derived in Chapter 8 for the converter switching angles, the power control, and the average delay angle. These equations, written in the form of mismatch equations which equal zero at the solution, are summarized below and in Table 9.1.

$$F_\phi^S = f_{12}(V_P^S, I_{dk}, \theta_i^S, \phi_i^S) \tag{9.20}$$

$$F_\phi^D = f_{13}(V_P^D, I_{dk}, \theta_i^D, \phi_i^D) \tag{9.21}$$

$$F_\theta^S = f_{14}(I_{dk}, I_{d0}, \theta_i^S, \alpha_0) \tag{9.22}$$

$$F_\theta^D = f_{15}(I_{dk}, I_{d0}, \theta_i^D, \alpha_0) \tag{9.23}$$

$$F_S = f_{16}(V_{d0}, I_{d0}) \tag{9.24}$$

$$F_{\alpha 0} = f_{17}(V_{d0}, I_{d0}) \tag{9.25}$$

A total of seventeen functional relationships have been described, representing 2328 equations in as many unknowns, for a solution up to the 50th harmonic. The functional relationships are summarized in Table 9.1

Table 9.1 Functional relationships between 12 pulse rectifier variables

Variable	No. of Vars.	Function of
V	300	$f_1(I_P^S, I_P^D)$
V_P^S	300	$f_2(V)$
V_P^D	300	$f_3(V)$
V_{dl}	100	$f_4(V_P^S, V_P^D, I_{dk}, \theta_i^S, \theta_i^D, \phi_i^S, \phi_i^D)$
V_{d0}	1	$f_5(V_P^S, V_P^D, I_{dk}, \theta_i^S, \theta_i^D, \phi_i^S, \phi_i^D)$
I_{dk}	100	$f_6(V_{dk})$
I_{d0}	1	$f_7(V_{d0})$
I_S^S	300	$f_8(V_P^S, I_{dk}, I_{d0}, \theta_i^S, \phi_i^S)$
I_S^D	300	$f_9(V_P^D, I_{dk}, I_{d0}, \theta_i^D, \phi_i^D)$
I_P^S	300	$f_{10}(I_S^S)$
I_P^D	300	$f_{11}(V_P^D, I_S^D)$
F_ϕ^S	6	$f_{12}(V_P^S, I_{dk}, \theta_i^S, \phi_i^S)$
F_ϕ^D	6	$f_{13}(V_P^D, I_{dk}, \theta_i^D, \phi_i^D)$
F_θ^S	6	$f_{14}(I_{dk}, I_{d0}, \theta_i^S, \alpha_0)$
F_θ^D	6	$f_{15}(I_{dk}, I_{d0}, \theta_i^D, \alpha_0)$
F_S	1	$f_{16}(V_{d0}, I_{d0})$
$F_{\alpha 0}$	1	$f_{17}(V_{d0}, I_{d0})$

9.5.2 Composition of Mismatch Functions

There is a great deal of redundancy in the system of equations summarized in Table 9.1. By a variety of substitutions of functions for variables, the number of simultaneous equations and variables can be reduced to 426. For example, taking f_1 and substituting functions for the variables I_P^S and I_P^D:

$$\begin{aligned}
V &= f_1(I_P^S, I_P^D) \\
&= f_1(f_{10}(I_S^S), f_{11}(V_P^D, I_S^D)) \\
&= f_1(f_{10}(f_8(V_P^S, I_{dk}, I_{d0}, \theta_i^S, \phi_i^S)), f_{11}(V_P^D, f_9(V_P^D, I_{dk}, I_{d0}, \theta_i^D, \phi_i^D))) \\
&= f_1(f_{10}(f_8(f_2(V), I_{dk}, I_{d0}, \theta_i^S, \phi_i^S)), f_{11}(f_3(V), f_9(f_3(V), I_{dk}, I_{d0}, \theta_i^D, \phi_i^D))) \\
&= f_{18}(V, I_{dk}, I_{d0}, \theta_i^D, \phi_i^D, \theta_i^S, \phi_i^S)
\end{aligned} \tag{9.26}$$

The new function, f_{18}, is a composition of several functions which describe how the dc current and terminal voltage, together with the switching angles are used to calculate the primary phase currents. The primary phase currents are then injected into the ac system impedance to yield the terminal voltage. The relationship

$$V = f_{18}(V, I_{dk}, I_{d0}, \theta_i^D, \phi_i^D, \theta_i^S, \phi_i^S) \tag{9.27}$$

can be written as a mismatch equation suitable for use in Newton's method:

$$F_V = V - f_{18}(V, I_{dk}, I_{d0}, \theta_i^D, \phi_i^D, \theta_i^S, \phi_i^S). \tag{9.28}$$

Equation 9.28 is called the *voltage mismatch equation*, and when decomposed into phases, harmonics, and rectangular components, yields 300 real equations. A similar type of mismatch equation can be constructed on the dc side:

$$\begin{aligned} I_{dk} &= f_6(V_{dk}) \\ &= f_6(f_4(V_P^S, V_P^D, I_{dl}, \theta_i^S, \theta_i^D, \phi_i^S, \phi_i^D)) \\ &= f_6(f_4(f_2(V), f_3(V), I_{dl}, \theta_i^S, \theta_i^D, \phi_i^S, \phi_i^D)) \\ &= f_{19}(V, I_{dk}, \theta_i^D, \phi_i^D, \theta_i^S, \phi_i^S) \end{aligned} \tag{9.29}$$

from which the following mismatch equation is constructed:

$$F_{Id} = I_{dk} - f_{19}(V, I_{dk}, I_{d0}, \theta_i^D, \phi_i^D, \theta_i^S, \phi_i^S). \tag{9.30}$$

Equation 9.30 yields 100 equations when decomposed into harmonic real and imaginary components, and is a composition of functions that describe the calculation of the dc voltage, and its application to the dc system model, to yield the dc current. A further 26 mismatch equations are obtained in a similar manner, related to the converter controller and the switching instants:

$$\begin{aligned} F_\phi^S &= f_{12}(V_P^S, I_{dk}, \theta_i^S, \phi_i^S) \\ &= f_{12}(f_2(V), I_{dk}, \theta_i^S, \phi_i^S) \\ &= f_{20}(V, I_{dk}, \theta_i^S, \phi_i^S) && (9.31) \\ F_\phi^D &= f_{13}(V_P^D, I_{dk}, \theta_i^D, \phi_i^D) \\ &= f_{13}(f_3(V), I_{dk}, \theta_i^D, \phi_i^D) && (9.32) \\ &= f_{21}(V, I_{dk}, \theta_i^D, \phi_i^D) && (9.33) \\ F_\theta^S &= f_{14}(I_{dk}, I_{d0}, \theta_i^S, \alpha_0) && (9.34) \\ F_\theta^D &= f_{15}(I_{dk}, I_{d0}, \theta_i^D, \alpha_0) && (9.35) \\ F_S &= f_{16}(V_{d0}, I_{d0}) \\ &= f_{16}(f_5(V_P^S, V_P^D, I_{dk}, \theta_i^S, \theta_i^D, \phi_i^S, \phi_i^D), I_{d0}) \\ &= f_{16}(f_5(f_2(V), f_3(V), I_{dk}, \theta_i^S, \theta_i^D, \phi_i^S, \phi_i^D), I_{d0}) \\ &= f_{22}(V, I_{dk}, I_{d0}, \theta_i^S, \theta_i^D, \phi_i^S, \phi_i^D) && (9.36) \\ F_{\alpha 0} &= f_{17}(V_{d0}, I_{d0}) \\ &= f_{23}(V, I_{dk}, I_{d0}, \theta_i^S, \theta_i^D, \phi_i^S, \phi_i^D) && (9.37) \end{aligned}$$

A smaller system of 426 simultaneous mismatch equations in 426 variables has now been developed, and is summarized in Table 9.2. The reduced set of variables to be solved for consists of the ac terminal voltage, the dc current, the switching angles, and the average delay angle.

Table 9.2 Mismatches and variables for the 12 pulse rectifier

Mismatch	No. of Vars.	Function of
F_V	300	$V - f_{18}(V, I_{dk}, I_{d0}, \theta_i^D, \phi_i^D, \theta_i^S, \phi_i^S)$
F_{Id}	100	$I_{dk} - f_{19}(V, I_{dk}, I_{d0}, \theta_i^D, \phi_i^D, \theta_i^S, \phi_i^S)$
F_ϕ^S	6	$f_{20}(V, I_{dk}, \theta_i^S, \phi_i^S)$
F_ϕ^D	6	$f_{21}(V, I_{dk}, \theta_i^D, \phi_i^D)$
F_θ^S	6	$f_{14}(I_{dk}, I_{d0}, \theta_i^S, \alpha_0)$
F_θ^D	6	$f_{15}(I_{dk}, I_{d0}, \theta_i^D, \alpha_0)$
F_S	1	$f_{22}(V, I_{dk}, I_{d0}, \theta_i^S, \theta_i^D, \phi_i^S, \phi_i^D)$
$F_{\alpha 0}$	1	$f_{23}(V, I_{dk}, I_{d0}, \theta_i^S, \theta_i^D, \phi_i^S, \phi_i^D)$

It would be possible to solve for a different set of variables, however those chosen have the advantage of being less distorted. In particular, the dc side terminal voltage is less distorted than the phase currents, and the dc side current is less distorted than the dc voltage. In fact, a fundamental frequency ac–dc load flow will give a very reasonable estimate of the fundamental voltage component on the ac side, and the dc current component on the dc side.

The interaction of the converter with the ac system has been specified in terms of terminal voltage mismatch. This requires the injection of phase currents into the ac system impedance to obtain a voltage to be compared with the estimated terminal voltage.

The ac system interaction can also be expressed in terms of a current mismatch. The estimated terminal voltage is applied to the ac system admittance to obtain phase currents that are compared with phase currents calculated by the converter model using the estimated voltage:

$$F_I = [Y_{cc}][V] - [Y_{ct}][V_{th}] - f_8(f_2(V), I_{dk}, I_{d0}, \theta_i^S, \phi_i^S) \\ - f_9(f_3(V), I_{dk}, I_{d0}, \theta_i^D, \phi_i^D) \tag{9.38}$$

Note that the current mismatch is still expressed in terms of the same variables as the voltage mismatch. The current mismatch has the advantage that it doesn't require the system admittance to be inverted, a possible difficulty if it has a high condition number. The current mismatch is also the preferred method of modelling the interaction with a purely inductive dc system, such as the unit connection, as the system admittance will decrease with increasing harmonic order.

Only the voltage mismatch is implemented here, as the ac system admittance is usually invertible, and the ac system impedance is typically much less than one per unit. A hybrid mixture of voltage and current mismatches at different harmonic orders would be the most robust and versatile approach to take; the voltage mismatch would be used at all harmonics where the system impedance is less than one per unit, otherwise the current mismatch would be used. In the hybrid mismatch method, the converter interacts with a reasonably strong ac system at all harmonic orders, and no admittance or impedance is larger than one per unit.

The dc current mismatch, F_{Id}, defining the interaction with the dc system, can also be written as a dc voltage mismatch, F_{Vd}. This mismatch is obtained by injecting the estimated dc current into the dc system impedance and comparing the resulting voltage with the calculated dc voltage:

$$F_{Vd} = \frac{I_{dk}}{Y_{dk}} - f_4(f_2(V), f_3(V), I_{dk}, \theta_i^S, \theta_i^D, \phi_i^S, \phi_i^D) \tag{9.39}$$

The dc voltage mismatch would be preferred in the unlikely instance of a capacitive dc system. Again, a hybrid current and voltage mismatch on the dc side is the most robust, as a series resonance can be modelled by a small impedance rather than a large admittance as in the dc current mismatch equation. Only the current mismatch has been implemented in this chapter, as the dc side admittance to harmonics is generally less than one per unit due to the presence of a smoothing reactor.

9.5.3 Solution Algorithm

The mismatch equations and variables of Section 9.5.1 are a mixture of real and complex valued. Newton's method is implemented here entirely in terms of real valued equations and variables. All complex quantities are therefore converted into real form by taking the real and imaginary components. A decomposition into real form is required, since the dc voltage and dc current mismatch equations are not differentiable in complex form. Newton's method is implemented by first assembling the variables to be solved for into a real vector X:

$$X = [\mathcal{R}\{V\}, \mathcal{I}\{V\}, \mathcal{R}\{I_d\}, \mathcal{I}\{I_d\}, \theta, \phi, \alpha_0, I_{d0}]^T. \tag{9.40}$$

The mismatch equations are likewise assembled into a real vector:

$$F(X) = [\mathcal{R}\{F_V\}, \mathcal{I}\{F_V\}, \mathcal{R}\{F_{Id}\}, \mathcal{I}\{F_{Id}\}, F_\theta, F_\phi, F_{\alpha 0}, F_{Id0}]^T. \tag{9.41}$$

Given an initial estimate of the solution, X^0, Newton's method is an iterative process for finding the solution vector, $X^\bullet$, that causes the mismatch vector to be zero:

$$F(X^\bullet) = 0. \tag{9.42}$$

The iterative process is defined by;

$$F(X^N) = J^N Y \tag{9.43}$$

$$X^{(N+1)} = X^N - Y, \tag{9.44}$$

with convergence deemed to have occurred when some norm of the residual vector $F(X^N)$ is less than a preset tolerance. Newton's method is not guaranteed to converge, but convergence is likely if the starting point is close to the solution. Central to Newton's method is a Jacobian matrix, J^N, of partial derivatives. For a system of 426 equations, the Jacobian is 426 elements square, as it contains the partial derivative of every mismatch function, with respect to every variable. This is illustrated for the twelve pulse converter functions and variables in Figure 9.6, for a system with constant current control.

$$
J = \begin{bmatrix}
\frac{\partial \mathcal{R}\{F_{V_k^a}\}}{\partial \mathcal{R}\{V_m^a\}} & \frac{\partial \mathcal{R}\{F_{V_k^a}\}}{\partial \mathcal{I}\{V_m^a\}} & \frac{\partial \mathcal{R}\{F_{V_k^a}\}}{\partial \mathcal{R}\{V_m^b\}} & \frac{\partial \mathcal{R}\{F_{V_k^a}\}}{\partial \mathcal{I}\{V_m^b\}} & \frac{\partial \mathcal{R}\{F_{V_k^a}\}}{\partial \mathcal{R}\{V_m^c\}} & \frac{\partial \mathcal{R}\{F_{V_k^a}\}}{\partial \mathcal{I}\{V_m^c\}} & \frac{\partial \mathcal{R}\{F_{V_k^a}\}}{\partial \mathcal{R}\{I_{d_m}\}} & \frac{\partial \mathcal{R}\{F_{V_k^a}\}}{\partial \mathcal{I}\{I_{d_m}\}} & \frac{\partial \mathcal{R}\{F_{V_k^a}\}}{\partial \phi_i} & \frac{\partial \mathcal{R}\{F_{V_k^a}\}}{\partial \theta_i} & 0 \\
\frac{\partial \mathcal{I}\{F_{V_k^a}\}}{\partial \mathcal{R}\{V_m^a\}} & \frac{\partial \mathcal{I}\{F_{V_k^a}\}}{\partial \mathcal{I}\{V_m^a\}} & \frac{\partial \mathcal{I}\{F_{V_k^a}\}}{\partial \mathcal{R}\{V_m^b\}} & \frac{\partial \mathcal{I}\{F_{V_k^a}\}}{\partial \mathcal{I}\{V_m^b\}} & \frac{\partial \mathcal{I}\{F_{V_k^a}\}}{\partial \mathcal{R}\{V_m^c\}} & \frac{\partial \mathcal{I}\{F_{V_k^a}\}}{\partial \mathcal{I}\{V_m^c\}} & \frac{\partial \mathcal{I}\{F_{V_k^a}\}}{\partial \mathcal{R}\{I_{d_m}\}} & \frac{\partial \mathcal{I}\{F_{V_k^a}\}}{\partial \mathcal{I}\{I_{d_m}\}} & \frac{\partial \mathcal{I}\{F_{V_k^a}\}}{\partial \phi_i} & \frac{\partial \mathcal{I}\{F_{V_k^a}\}}{\partial \theta_i} & 0 \\
\frac{\partial \mathcal{R}\{F_{V_k^b}\}}{\partial \mathcal{R}\{V_m^a\}} & \frac{\partial \mathcal{R}\{F_{V_k^b}\}}{\partial \mathcal{I}\{V_m^a\}} & \frac{\partial \mathcal{R}\{F_{V_k^b}\}}{\partial \mathcal{R}\{V_m^b\}} & \frac{\partial \mathcal{R}\{F_{V_k^b}\}}{\partial \mathcal{I}\{V_m^b\}} & \frac{\partial \mathcal{R}\{F_{V_k^b}\}}{\partial \mathcal{R}\{V_m^c\}} & \frac{\partial \mathcal{R}\{F_{V_k^b}\}}{\partial \mathcal{I}\{V_m^c\}} & \frac{\partial \mathcal{R}\{F_{V_k^b}\}}{\partial \mathcal{R}\{I_{d_m}\}} & \frac{\partial \mathcal{R}\{F_{V_k^b}\}}{\partial \mathcal{I}\{I_{d_m}\}} & \frac{\partial \mathcal{R}\{F_{V_k^b}\}}{\partial \phi_i} & \frac{\partial \mathcal{R}\{F_{V_k^b}\}}{\partial \theta_i} & 0 \\
\frac{\partial \mathcal{I}\{F_{V_k^b}\}}{\partial \mathcal{R}\{V_m^a\}} & \frac{\partial \mathcal{I}\{F_{V_k^b}\}}{\partial \mathcal{I}\{V_m^a\}} & \frac{\partial \mathcal{I}\{F_{V_k^b}\}}{\partial \mathcal{R}\{V_m^b\}} & \frac{\partial \mathcal{I}\{F_{V_k^b}\}}{\partial \mathcal{I}\{V_m^b\}} & \frac{\partial \mathcal{I}\{F_{V_k^b}\}}{\partial \mathcal{R}\{V_m^c\}} & \frac{\partial \mathcal{I}\{F_{V_k^b}\}}{\partial \mathcal{I}\{V_m^c\}} & \frac{\partial \mathcal{I}\{F_{V_k^b}\}}{\partial \mathcal{R}\{I_{d_m}\}} & \frac{\partial \mathcal{I}\{F_{V_k^b}\}}{\partial \mathcal{I}\{I_{d_m}\}} & \frac{\partial \mathcal{I}\{F_{V_k^b}\}}{\partial \phi_i} & \frac{\partial \mathcal{I}\{F_{V_k^b}\}}{\partial \theta_i} & 0 \\
\frac{\partial \mathcal{R}\{F_{V_k^c}\}}{\partial \mathcal{R}\{V_m^a\}} & \frac{\partial \mathcal{R}\{F_{V_k^c}\}}{\partial \mathcal{I}\{V_m^a\}} & \frac{\partial \mathcal{R}\{F_{V_k^c}\}}{\partial \mathcal{R}\{V_m^b\}} & \frac{\partial \mathcal{R}\{F_{V_k^c}\}}{\partial \mathcal{I}\{V_m^b\}} & \frac{\partial \mathcal{R}\{F_{V_k^c}\}}{\partial \mathcal{R}\{V_m^c\}} & \frac{\partial \mathcal{R}\{F_{V_k^c}\}}{\partial \mathcal{I}\{V_m^c\}} & \frac{\partial \mathcal{R}\{F_{V_k^c}\}}{\partial \mathcal{R}\{I_{d_m}\}} & \frac{\partial \mathcal{R}\{F_{V_k^c}\}}{\partial \mathcal{I}\{I_{d_m}\}} & \frac{\partial \mathcal{R}\{F_{V_k^c}\}}{\partial \phi_i} & \frac{\partial \mathcal{R}\{F_{V_k^c}\}}{\partial \theta_i} & 0 \\
\frac{\partial \mathcal{I}\{F_{V_k^c}\}}{\partial \mathcal{R}\{V_m^a\}} & \frac{\partial \mathcal{I}\{F_{V_k^c}\}}{\partial \mathcal{I}\{V_m^a\}} & \frac{\partial \mathcal{I}\{F_{V_k^c}\}}{\partial \mathcal{R}\{V_m^b\}} & \frac{\partial \mathcal{I}\{F_{V_k^c}\}}{\partial \mathcal{I}\{V_m^b\}} & \frac{\partial \mathcal{I}\{F_{V_k^c}\}}{\partial \mathcal{R}\{V_m^c\}} & \frac{\partial \mathcal{I}\{F_{V_k^c}\}}{\partial \mathcal{I}\{V_m^c\}} & \frac{\partial \mathcal{I}\{F_{V_k^c}\}}{\partial \mathcal{R}\{I_{d_m}\}} & \frac{\partial \mathcal{I}\{F_{V_k^c}\}}{\partial \mathcal{I}\{I_{d_m}\}} & \frac{\partial \mathcal{I}\{F_{V_k^c}\}}{\partial \phi_i} & \frac{\partial \mathcal{I}\{F_{V_k^c}\}}{\partial \theta_i} & 0 \\
\frac{\partial \mathcal{R}\{F_{Id_k}\}}{\partial \mathcal{R}\{V_m^a\}} & \frac{\partial \mathcal{R}\{F_{Id_k}\}}{\partial \mathcal{I}\{V_m^a\}} & \frac{\partial \mathcal{R}\{F_{Id_k}\}}{\partial \mathcal{R}\{V_m^b\}} & \frac{\partial \mathcal{R}\{F_{Id_k}\}}{\partial \mathcal{I}\{V_m^b\}} & \frac{\partial \mathcal{R}\{F_{Id_k}\}}{\partial \mathcal{R}\{V_m^c\}} & \frac{\partial \mathcal{R}\{F_{Id_k}\}}{\partial \mathcal{I}\{V_m^c\}} & \frac{\partial \mathcal{R}\{F_{Id_k}\}}{\partial \mathcal{R}\{I_{d_m}\}} & \frac{\partial \mathcal{R}\{F_{Id_k}\}}{\partial \mathcal{I}\{I_{d_m}\}} & \frac{\partial \mathcal{R}\{F_{Id_k}\}}{\partial \phi_i} & \frac{\partial \mathcal{R}\{F_{Id_k}\}}{\partial \theta_i} & 0 \\
\frac{\partial \mathcal{I}\{F_{Id_k}\}}{\partial \mathcal{R}\{V_m^a\}} & \frac{\partial \mathcal{I}\{F_{Id_k}\}}{\partial \mathcal{I}\{V_m^a\}} & \frac{\partial \mathcal{I}\{F_{Id_k}\}}{\partial \mathcal{R}\{V_m^b\}} & \frac{\partial \mathcal{I}\{F_{Id_k}\}}{\partial \mathcal{I}\{V_m^b\}} & \frac{\partial \mathcal{I}\{F_{Id_k}\}}{\partial \mathcal{R}\{V_m^c\}} & \frac{\partial \mathcal{I}\{F_{Id_k}\}}{\partial \mathcal{I}\{V_m^c\}} & \frac{\partial \mathcal{I}\{F_{Id_k}\}}{\partial \mathcal{R}\{I_{d_m}\}} & \frac{\partial \mathcal{I}\{F_{Id_k}\}}{\partial \mathcal{I}\{I_{d_m}\}} & \frac{\partial \mathcal{I}\{F_{Id_k}\}}{\partial \phi_i} & \frac{\partial \mathcal{I}\{F_{Id_k}\}}{\partial \theta_i} & 0 \\
\frac{\partial F_{\phi_i}}{\partial \mathcal{R}\{V_m^a\}} & \frac{\partial F_{\phi_i}}{\partial \mathcal{I}\{V_m^a\}} & \frac{\partial F_{\phi_i}}{\partial \mathcal{R}\{V_m^b\}} & \frac{\partial F_{\phi_i}}{\partial \mathcal{I}\{V_m^b\}} & \frac{\partial F_{\phi_i}}{\partial \mathcal{R}\{V_m^c\}} & \frac{\partial F_{\phi_i}}{\partial \mathcal{I}\{V_m^c\}} & \frac{\partial F_{\phi_i}}{\partial \mathcal{R}\{I_{d_m}\}} & \frac{\partial F_{\phi_i}}{\partial \mathcal{I}\{I_{d_m}\}} & \frac{\partial F_{\phi_i}}{\partial \phi_i} & \frac{\partial F_{\phi_i}}{\partial \theta_i} & 0 \\
\frac{\partial F_{\theta_i}}{\partial \mathcal{R}\{V_m^a\}} & \frac{\partial F_{\theta_i}}{\partial \mathcal{I}\{V_m^a\}} & \frac{\partial F_{\theta_i}}{\partial \mathcal{R}\{V_m^b\}} & \frac{\partial F_{\theta_i}}{\partial \mathcal{I}\{V_m^b\}} & \frac{\partial F_{\theta_i}}{\partial \mathcal{R}\{V_m^c\}} & \frac{\partial F_{\theta_i}}{\partial \mathcal{I}\{V_m^c\}} & \frac{\partial F_{\theta_i}}{\partial \mathcal{R}\{I_{d_m}\}} & \frac{\partial F_{\theta_i}}{\partial \mathcal{I}\{I_{d_m}\}} & 0 & \frac{\partial F_{\theta_i}}{\partial \theta_i} & \frac{\partial F_{\theta_i}}{\partial \alpha_i} \\
\frac{\partial \mathcal{I}\{F_{\alpha 0}\}}{\partial \mathcal{R}\{V_m^a\}} & \frac{\partial \mathcal{I}\{F_{\alpha 0}\}}{\partial \mathcal{I}\{V_m^a\}} & \frac{\partial \mathcal{I}\{F_{\alpha 0}\}}{\partial \mathcal{R}\{V_m^b\}} & \frac{\partial \mathcal{I}\{F_{\alpha 0}\}}{\partial \mathcal{I}\{V_m^b\}} & \frac{\partial \mathcal{I}\{F_{\alpha 0}\}}{\partial \mathcal{R}\{V_m^c\}} & \frac{\partial \mathcal{I}\{F_{\alpha 0}\}}{\partial \mathcal{I}\{V_m^c\}} & \frac{\partial \mathcal{I}\{F_{\alpha 0}\}}{\partial \mathcal{R}\{I_{d_m}\}} & \frac{\partial \mathcal{I}\{F_{\alpha 0}\}}{\partial \mathcal{I}\{I_{d_m}\}} & \frac{\partial \mathcal{I}\{F_{\alpha 0}\}}{\partial \phi_i} & \frac{\partial \mathcal{I}\{F_{\alpha 0}\}}{\partial \theta_i} & 0
\end{bmatrix}
$$

Figure 9.6 Assembly of the Jacobian matrix from partial derivatives

There are two methods for obtaining the Jacobian elements; numerical partial differentiation, and the evaluation of analytically derived expressions for the partial derivatives. The numerical method is used here to validate the analytic expressions for the Jacobian elements. Numerical calculation of the Jacobian has the advantage of ease of coding, but is quite slow. Each column of the Jacobian requires an evaluation of all the mismatch equations, and the resulting calculation is only an approximation to the partial derivative. The *numerical Jacobian* is obtained by sequentially perturbing each element of X, and calculating the change in all the mismatches: $J_{ij} = \Delta F_i / \Delta X_j$. Provided ΔX_j is small enough, this gives a good approximation to the Jacobian.

The analytical method of calculating the Jacobian matrix requires considerable effort to obtain all the partial derivatives in analytic form, but is exceptionally fast. Frequently, the amount of computation required to calculate the *analytic Jacobian* is of the same order as that required to calculate the complete set of mismatches just once. For the converter system of 426 mismatch equations, the analytic Jacobian can be calculated about twenty times faster than the numerical Jacobian. The numerical Jacobian for the test system has been plotted in Figure 9.7, for a solution up to the thirteenth harmonic. This Jacobian was calculated at the solution, and so represents a linearization of the system of equations in Table 9.2 around the converter operating point.

Referring to Figure 9.7, the elements of the Jacobian have been ordered in blocks corresponding to the three phases of terminal voltage, the dc current, the end of commutation angles, the firing angles, and the average delay angle. The blocks associated with interactions between the dc current harmonics, and the ac voltage harmonics comprise the *ac–dc partition*, which is 104 elements square. All other parts of the Jacobian are called the *switching terms*. Within the ac–dc partition, elements have been arranged within each block in ascending harmonic order, with the real and imaginary parts of each harmonic alternating. Each block in the ac–dc partition is therefore 26 elements square in Figure 9.7, but 100 elements square for a solution to the fiftieth harmonic.The Jacobian displays several important structural features:

A: The test system contains a parallel resonance in the ac system at the second harmonic. This leads to rows of large terms in the Jacobian aligned with the second harmonic terminal voltage mismatch (the *resonance terms*).

B: A change in harmonic k on one side of the converter affects harmonics $k+1$ and $k-1$ on the other side of the converter, causing the double diagonal structures in the ac to dc, and dc to ac blocks. These are the *three port terms*.

C: The end of commutation mismatch is very sensitive to harmonics in the terminal voltage and dc current.

D: The individual firing instants are sensitive to harmonics in the dc current.

E: The average delay angle mismatch, since it relates to the average dc current, is extremely sensitive to changes in the fundamental terminal voltage. There is also some sensitivity to harmonics coupled to the fundamental; i.e. the 11th and 13th harmonics on the ac side.

F: As would be expected, there is strong coupling between the switching angles and the switching mismatches.

G: The Jacobian contains a strong diagonal, approximately equal to one in the ac/dc partition. A large diagonal is often beneficial when the linear system corresponding to the Jacobian is solved.

There is also apparently little dependence of the terminal voltage mismatches on the end of commutation angles, even though there is a sensitivity to the firing angles. This seemingly anomalous result is due to the formulation of the mismatch equations. A change in θ moves the entire commutation current curve, whereas a change in ϕ moves only the end of commutation, consequently having a negligible effect. This is demonstrated in Figure 9.8, and it is apparent that different behaviour would be observed with a different formulation of the mismatch equations.

A very useful feature of the Jacobian, is the large number of small elements. Since the Jacobian is only an estimate of the behaviour of the nonlinear system in response to small perturbations, it is acceptable to approximate elements in the

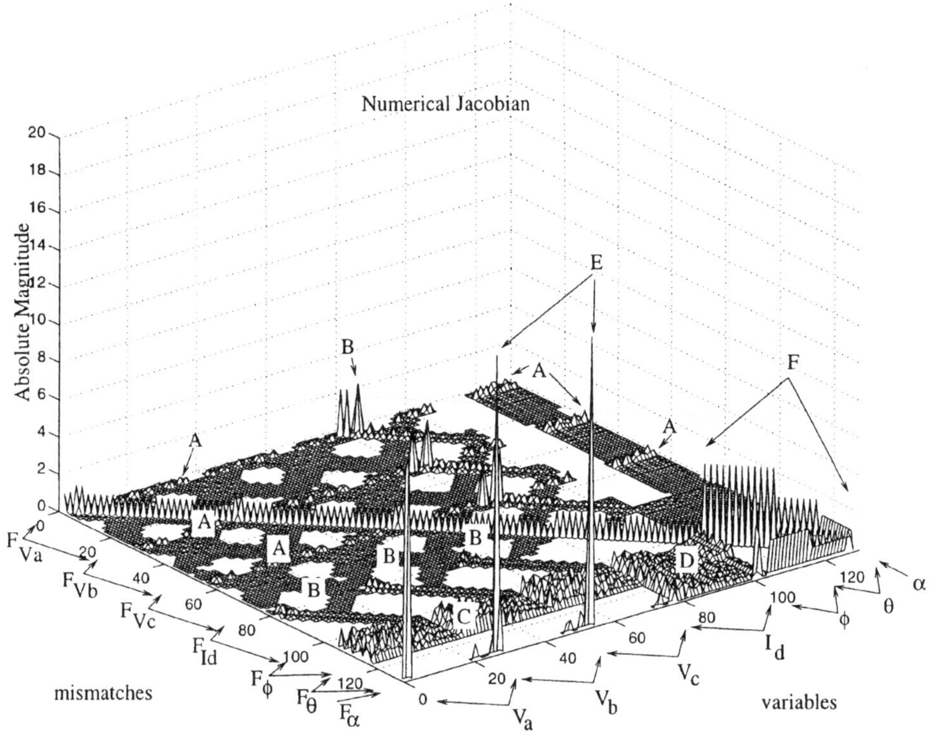

Figure 9.7 Numerically calculated Jacobian for the test system; 13 harmonics

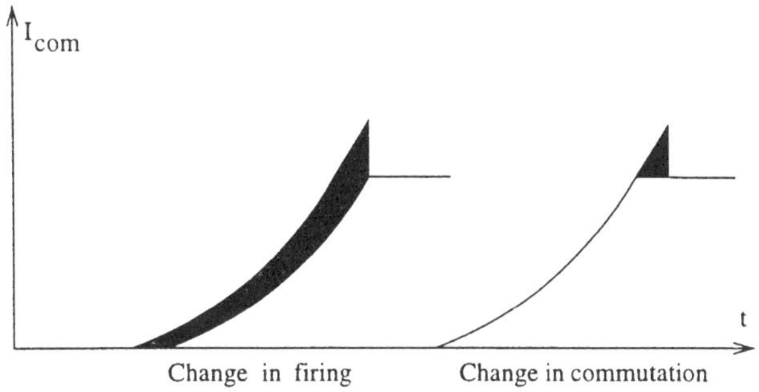

Figure 9.8 Effect of variation in firing angle versus end of commutation angle

Jacobian, without affecting the convergence of Newton's method. Small elements may therefore be approximated by zero, making the Jacobian sparse. The sparse structure of the Jacobian is illustrated in Figure 9.9, and again the structural elements described above are evident. For a solution up to the fiftieth harmonic, the Jacobian is typically 96% sparse. The structure of Figure 9.9 was obtained by scanning through the Jacobian, analytically calculating selected elements, and retaining those elements in the ac–dc partition larger than 0.05, and switching elements larger than 0.02.

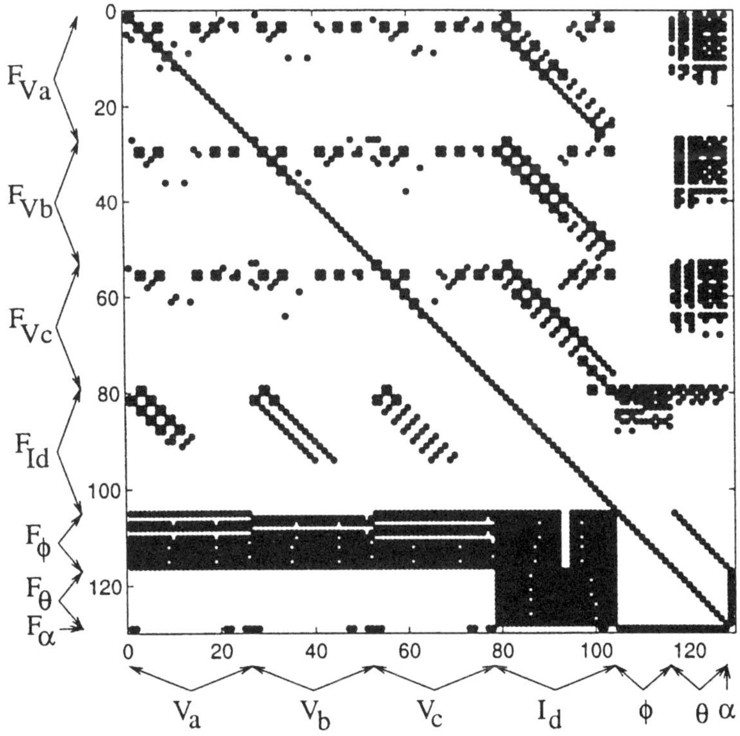

Figure 9.9 Sparsity structure of the Jacobian; 13 harmonics

In scanning the Jacobian, it is not necessary to calculate all elements. On the ac or dc side of the converter, an odd harmonic never couples to an even harmonic, while for transfers across the converter, an odd harmonic only couples to an even harmonic. The total time to scan the ac–dc partition can therefore be halved by scanning with the checkerboard pattern of Figure 9.10. The difference between Figures 9.9 and 9.10 shows that with prior system knowledge, a more sophisticated scanning method could be derived.

Analytic expressions for the partial derivatives that comprise the Jacobian elements are derived in Appendix IV. In order to simplify the derivation and notation, partial derivatives for a six pulse current controlled rectifier with a star-g/star connected transformer are obtained. The extension to 12-pulse partial derivatives is relatively straightforward, the 12-pulse ac–dc Jacobian partition being the sum of the ac–dc partitions for the two six pulse star and delta converters, minus the identity matrix. The switching terms associated with each group are the same.

The partial derivative analytic expressions have been verified by calculating the sparse matrix of differences between the analytic Jacobian and the numerical Jacobian. The difference matrix is sparse, since it is only calculated at the sparse locations of the analytic Jacobian. As illustrated in Figure 9.11, the two Jacobians agree to within 5×10^{-4}. Since convergence is not affected by a scan tolerance of 0.02, these differences are inconsequential, and are inherent to a numerical calculation of the Jacobian.

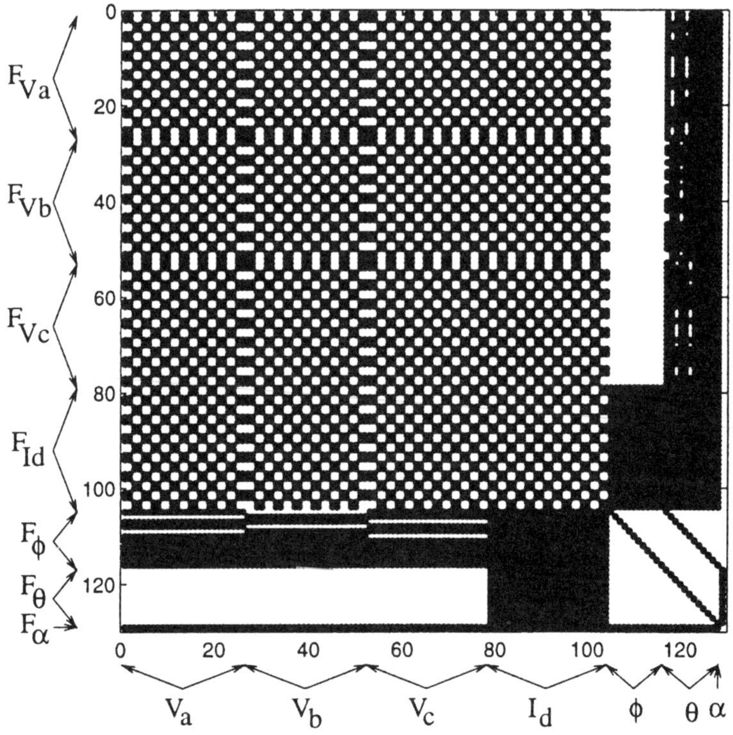

Figure 9.10 Scan structure of the sparse Jacobian; 13 harmonics

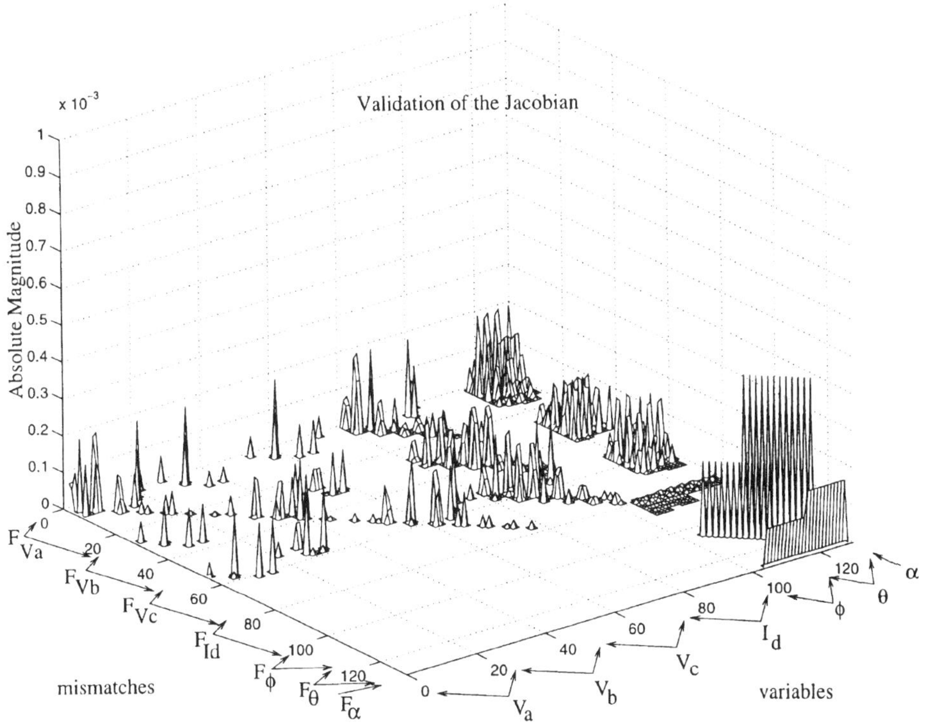

Figure 9.11 Difference between numerical and analytic Jacobians; 13 harmonics

9.5.4 Computer Implementation

In this section the application of Newton's method to the case at hand is described in detail. Several issues are addressed that have not yet been discussed. Of particular importance is the method of determining a suitable starting point for the Newton method, the updating of the Jacobian matrix, the sparse solution of the linear Jacobian system, and the stopping criteria for the iterative process. These points are illustrated in the flow diagram for the solution (Figure 9.12), where it can be seen that a two stage process is employed to calculate the starting point. A first estimate of the converter is obtained by using a classical analysis, followed by a Newton solution of the *switching system*, with no harmonics. If the switching system converges, a full harmonic solution follows, after which the results are printed to output files.

Initialization An initial estimate for the converter delay angle is obtained from the equation:

$$V_{d0} = \frac{2 \cdot 3\sqrt{2 \cdot 3}}{\pi} |V_{th1}| \cos \alpha - \frac{3X}{\pi} I_{d0}, \tag{9.45}$$

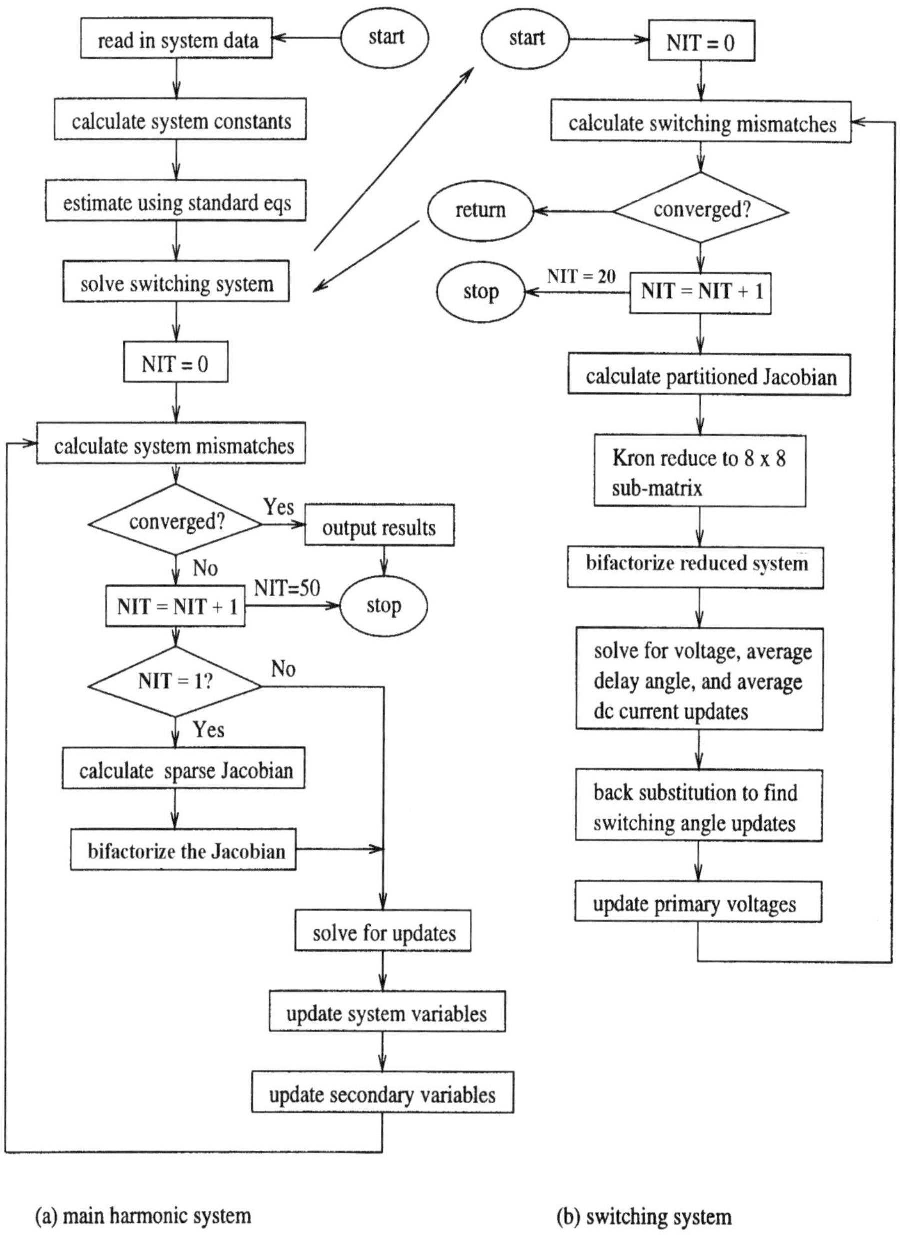

Figure 9.12 Flow chart for the sparse Newton solution

ignoring voltage magnitude drop through the ac system impedance. The dc voltage is estimated from the voltage drop through the dc system and the dc source:

$$V_{d0} = E + \frac{I_{d0}}{Y_{d0}} \tag{9.46}$$

The average commutation angle is obtained from:

$$V_{d0} = \frac{3\sqrt{2}}{\pi}|V_{th}|[\cos\alpha + \cos(\alpha + \mu)] \tag{9.47}$$

These angles are then used to assemble the individual firing and end of commutation angles:

$$\theta_i = \beta_i + \alpha \tag{9.48}$$

$$\phi_i = \theta_i + \mu \tag{9.49}$$

These calculations yield a very rough estimate of the converter switching angles, which is subsequently improved substantially by a Newton solution of the switching system.

The Switching System The purpose of the switching system is to solve the relationships between the fundamental terminal voltage, the dc current, and the switching angles for both bridges. The switching system is thus a complete model of the 12-pulse converter in the presence of constant terminal voltage and dc current harmonics, since the harmonic quantities appear as constant parameters. The mismatch equations and partial derivatives for the twelve pulse switching system have all been derived previously. The set of equations to be solved in the switching system (for constant current control) are:

$$\begin{aligned}
F^S_{\phi_i}(\phi^S_i, \theta^S_i, V, I_d) &= 0\\
F^D_{\phi_i}(\phi^D_i, \theta^D_i, V, I_d) &= 0\\
F^S_{\theta_i}(\theta^S_i, V, I_d) &= 0\\
F^D_{\theta_i}(\theta^D_i, I_d, \alpha_0) &= 0\\
F_{V_1}(\phi^S_i, \theta^S_i, V, I_d) &= 0\\
F_{\alpha_0}(\phi^S_i, \theta^S_i, V, I_d) &= 0.
\end{aligned} \tag{9.50}$$

A flow diagram for the switching system is shown in Figure 9.12, part (b), while the structure of the switching Jacobian can be seen in Figure 9.13. Those elements corresponding to the partial derivatives of voltage mismatch with respect to end of commutation angle have been set equal to zero, since they are always insignificant.

This Jacobian is quite sparse, and always has the same sparsity structure, however it is of an intermediate size, being too small for a general purpose sparse solution, and yet large enough to be significant. In the case of an interharmonic model, the switching Jacobian would be of size $24n + 8$, where n is the number of cycles over which the steady state is defined. It is therefore worthwhile to develop an *ad hoc* sparse solution of the switching Jacobian, and the best way to do this is to employ a partitioning method.

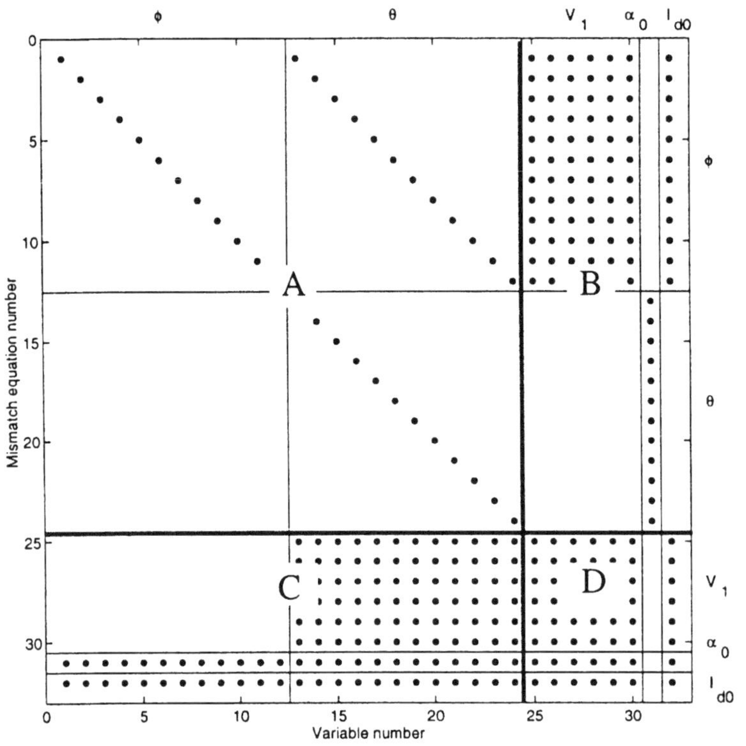

Figure 9.13 Sparsity structure of the switching Jacobian matrix (power control)

With reference to Figure 9.13, the partitioning method exploits the fact that the top left hand part of the switching Jacobian, A, is almost diagonal, and can easily be reduced to the identity matrix. It is then straightforward to create a reduced system of size 8 × 8 by multiplying the rectangular cross coupling partitions, C and B, to give $D - CB$. In this case the cross coupling matrix multiplication is very fast, since the variables have been ordered so that most of the nonzero elements of each of the cross coupling partitions correspond to the zeros of the other. Using this method, the linear system can be solved in approximately 1000 flops, instead of some 10 000. Indexing overheads have been virtually eliminated by storing partial derivatives directly into specific vectors, and then using *ad hoc* code for the partitioning method. The reduced 8 × 8 system is solved by LU decomposition for the terminal voltage, average delay angle, and dc current order updates. These are then backsubstituted to find the switching angle updates.

The switching Jacobian is updated every iteration of the Newton method, and convergence has been found to be rapid and robust. The convergence criteria for the switching system is:

$$\frac{|F_{V_1}|}{|V_1|} < 0.001$$
$$\frac{|F_{I_1}|}{|I_1|} < 0.001$$
$$\frac{|F_{\alpha_0}|}{|I_{d0}|} < 0.001$$
$$\frac{|F_{Id0}|}{|P|} < 0.001$$
$$|F_{\theta_i}| < 5 \times 10^{-8}$$
$$|F_{\phi_i}| < 5 \times 10^{-8}. \qquad (9.51)$$

Convergence typically occurs in 4 to 10 iterations, for starting terminal voltages ranging from 0.3 to 7 p.u. A case with a system impedance of 1.3 p.u. at the fundamental has been solved (which required 11 iterations). A failure to converge in 20 iterations has so far always implied that the system has no solution. An invalid solution (i.e. negative firing angle) has never occurred, nor have multiple valid solutions been observed. If the dc voltage source is negative, the system will solve as for a current controlled inverter.

The solution obtained with the switching system is an excellent starting point for the full harmonic solution, as the switching angles are largely determined by interactions at the fundamental frequency. Since the switching system is three phase, it includes the effect of any unbalance.

The Harmonic Solution The system of harmonic phasor and switching angle equations is quite large (426 elements square), and at each iteration of Newton's method, a linear system this size must be solved for the update vector: $F(X^N) = J^N Y^N$. This step represents the bulk of the computation required in Newton's method, and so techniques for speeding up the overall solution are concerned with details of the Jacobian linear system, and its solution method. The Jacobian has been made sparse, and it is essential that this sparsity is exploited in an efficient manner.

Three types of sparse linear solver have been implemented. One of these, the sparse symmetric bifactorization [14] method was found to be unsuitable, as it requires the Jacobian to be diagonally row dominant. Although the Jacobian has a large diagonal, it is not diagonally row dominant. The method of Zollenkopf pivots for sparsity, not numerical stability, and does not yield the correct solution when applied to the Jacobian system. The Zollenkopf method is essentially optimized for solving admittance matrix systems, which are necessarily symmetric in structure, and diagonally row dominant. The two other sparse solvers that have been implemented are an asymmetric sparse bifactorization that pivots for a compromise between numerical stability and sparsity, and the iterative conjugate gradient method [15]. The sparse bifactorization employed is the *y12m* solver from the *netlib* [16]. Both methods have been found to be satisfactory, but suited to different types of solution algorithm.

Frequently, the Newton method can be improved by calculating the Jacobian matrix only once, on the first iteration, and keeping it constant throughout the solution. In this case, the sparse bifactorization method is fastest, as the bifactorization need only be calculated once. On subsequent iterations the linear system is solved using the factorised Jacobian from the first iteration. This method also avoids many of the indexing overheads associated with the sparse bifactorization, since the sparsity structure is constant. Holding the Jacobian constant, leads to a larger number of faster iterations to obtain the overall solution.

Another method is to update 'important' parts of the Jacobian, holding the bulk of the Jacobian constant. Convergence in the least number of iterations has been obtained by holding the ac–dc partition constant, and updating the switching terms, since they can be recalculated quickly. In this case the Jacobian must be refactorized at each iteration, and the conjugate gradient method is almost as fast. The conjugate gradient method typically requires 100 iterations to converge to an accuracy in the update that does not slow the Newton solution. On the whole, the sparse bifactorization has been found to be the most versatile.

There are better iterative methods than the conjugate gradient, such as the preconditioned biconjugate gradient, that have not been implemented. The preconditioned biconjugate gradient method can be used efficiently if a sparse approximation to the inverse of the Jacobian can be created. This is very likely, as the inverse of the Jacobian contains many small elements. A final advantage of the conjugate gradient methods is that they can be easily implemented on a parallel processor, as the algorithm is based on successive sparse multiplications of the Jacobian by a vector.

Convergence Tolerance The basic requirement of convergence is that all of the mismatches $F(X)_i$ are small enough. The mismatches however, are of several types, entailing a different convergence requirement for each type of mismatch. The convergence tolerances are listed below:

$$
\begin{aligned}
&\frac{|F_{Vk}|}{|V_k|} < 0.001 \\
&\frac{|F_{Ik}|}{|I_k|} < 0.001 \\
&\frac{|F_{Id}|}{|Id|} < 0.001 \\
&\frac{|F'_{\alpha 0}|}{|I_{d0}|} < 0.001 \\
&\frac{|F_{Id0}|}{|P|} < 0.001 \\
&|F_{\theta_i}| < 5 \times 10^{-8} \\
&|F_{\phi_i}| < 5 \times 10^{-8}
\end{aligned}
\qquad (9.52)
$$

Note that convergence tolerance for the complex mismatches, F_V, F_I, F_{Id} is expressed in terms of the magnitude of the mismatch. This means that the error in an estimated value for a variable , for example V_{11}, is smaller than 0.1% of its own length. The advantage of a relative mismatch of this type, is that it treats all harmonics equally. However to prevent an attempt to converge to an absolute error of zero for harmonics that are not present (e.g. even harmonics in some cases), this convergence test is only applied to harmonics that have a size larger than 10^{-5} per unit. These convergence tolerances can be made tighter to obtain more accurate solutions if necessary. A relative convergence tolerance of 2×10^{-6} has been used in the impedance calculations of chapter 10.

The tolerance set for the switching mismatches of 5×10^{-8} corresponds to 1.4×10^{-4} degrees at the 50th harmonic, or 0.1 nsec. Another type of convergence tolerance that can be used, is to calculate a norm of the real mismatch vector; for example, the 1-norm:

$$|X|_1 = \sum_{i=1}^{n} |X_i| \tag{9.53}$$

A tolerance of $|X|_1 < 10^{-5}$ is suitable for general purpose use. This type of convergence test is fast and easy to apply, but does not imply that all harmonics have converged to a satisfactory accuracy.

9.5.5 Validation and Performance

The model has been verified against time domain simulation of the test system described in Appendix VI, using the program PSCAD/EMTDC. The steady state solution was obtained by simulating for one second, with a time step of 20 μs, and then obtaining waveforms over one cycle for subsequent comparison with the harmonic domain solution. The results of four tests are given here, comparing the dc voltage and ac phase current waveforms and spectra, since these quantities are the most distorted. The tests are designed to highlight any modelling, and convergence deficiencies. The tests carried out were:

Test 1: A base case solution with no harmonic sources in the ac system.

Test 2: The Thevenin voltage source in the ac system was distorted by 5% positive sequence second harmonic. This excites the complementary ac–dc system resonance, leading to noncharacteristic harmonics, and a high degree of interaction between converter switching angles and the ac–dc harmonics.

Test 3: The leakage reactance of the phase 'b' star-g/delta transformer was increased from 0.18 to 0.3 per unit. This imbalance causes the generation of odd harmonics, and a relatively large coupling to the zero sequence, which is illustrated in Figure 9.18.

Test 4: A 0.1 per unit resistance was placed in series with the star-g/delta transformer, and the secondary tap changer of that transformer was set to 1.1 p.u. Convergence with such a large series resistance indicates that the effect on convergence of not representing commutation resistance in the Jacobian is acceptable. Setting the tap changer on one transformer but not the other introduces six pulse unbalance. It was necessary to increase the ac system fundamental source from 1.10976 to 1.20976 p.u. to enable the current order to be satisfied.

The results of Test 1, the base case, are shown in Figures 9.14 and 9.15. It can be seen from the spectra that only the characteristic harmonics are present, and that

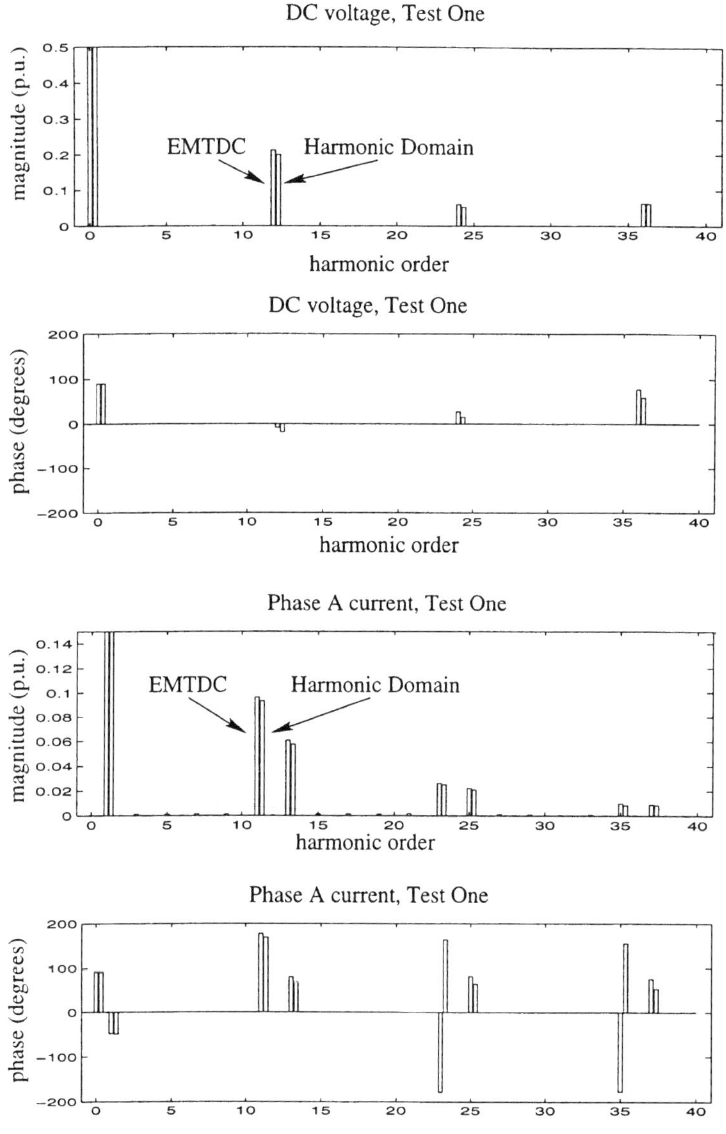

Figure 9.14 Comparison of time and harmonic domain solutions for phase currents and dc voltage spectra: Base Case

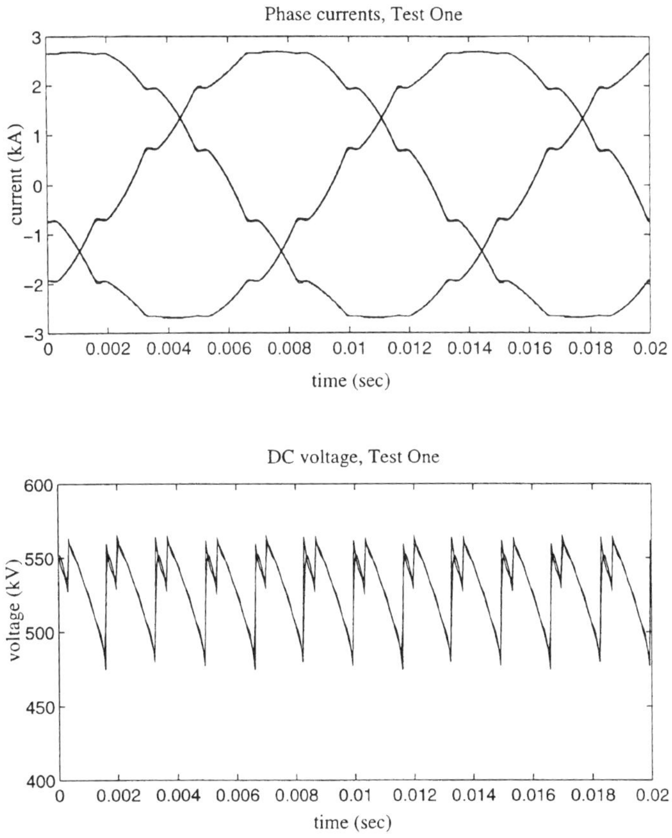

Figure 9.15 Comparison of time and harmonic domain solutions for phase current and dc voltage waveforms: Base Case

there is a close match between the time and harmonic domain solutions. The time domain solution yielded small residual non-characteristic harmonics in the spectra, which were suppressed in the phase graphs. The dc voltage waveform was generated from the harmonic domain solution by plotting the Fourier series for each dc voltage sample during the appropriate interval, rather than inverse transforming the dc voltage spectra. This eliminates Gibbs phenomena associated with the step changes in voltage, but gives overly sharp voltage spikes. These are not present in the time domain solution due to modelling of the snubber circuits, which limit the dV/dt. The time domain derived dc voltage waveform is therefore more rounded. Clearly, if an accurate time domain waveform was required from the harmonic domain solution, it would be necessary to post process the waveshape using knowledge of the snubber circuit time domain response. The comparison of the waveshapes indicates that all the switching angles are correct.

When a second order harmonic voltage source was placed in the ac system, the composite resonance was excited, resulting in non-characteristic harmonics. Referring to Figures 9.16 and 9.17, odd harmonics are present on the dc side, and

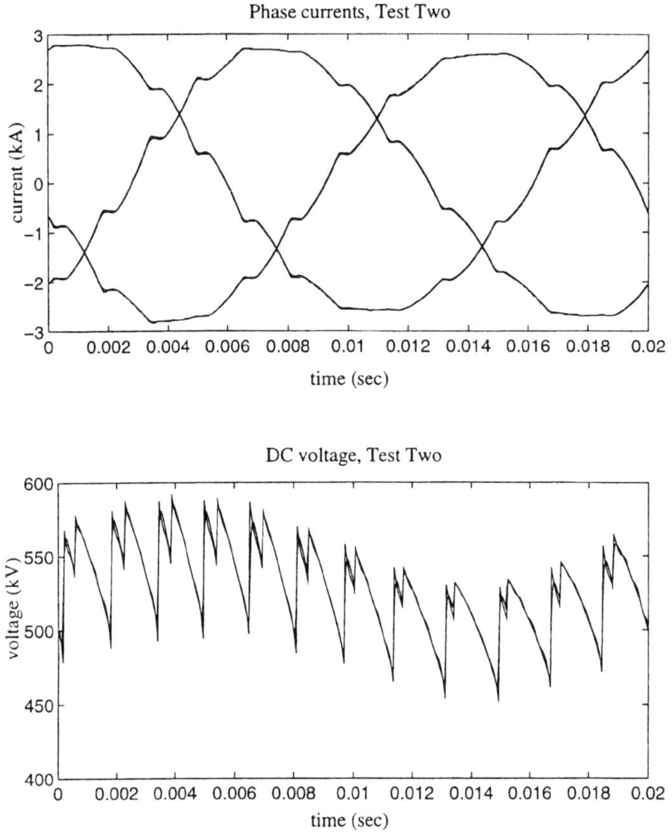

Figure 9.16 Comparison of time and harmonic domain solutions for phase current and dc voltage waveforms: Test 2

even harmonics on the ac side. In particular, the fundamental resonance on the dc side is excited, and there is a large fundamental component in the dc voltage.

Unbalancing the star-g/delta transformer leakage reactance (Figures 9.18 and 9.19), caused the generation of many harmonics, and a large zero sequence on the dc primary side. This is due to sequence transformation by the transformer, irrespective of its connection to a converter. The zero sequence has been plotted in Figure 9.18, and the close agreement with the time domain solution is evident. The final set of results, Figures 9.20 and 9.21, shows the expected generation of six pulse harmonics due to imbalance between the two six pulse groups.

Convergence of the sparse Newton solution is fast and robust, even if the Jacobian matrix is held constant. A convenient measure of the convergence is the 1-norm, the sum of the magnitudes of all the mismatches. At each iteration the 1-norm is reduced by an approximately constant *convergence factor*, λ, depending upon the difficulty of the system, and whether the switching terms are updated. The constant convergence factor is evident in Figures 9.22 and 9.23, which show the decreasing 1-norm as a function of iteration number for all five test cases. Note that the final 1-norm at

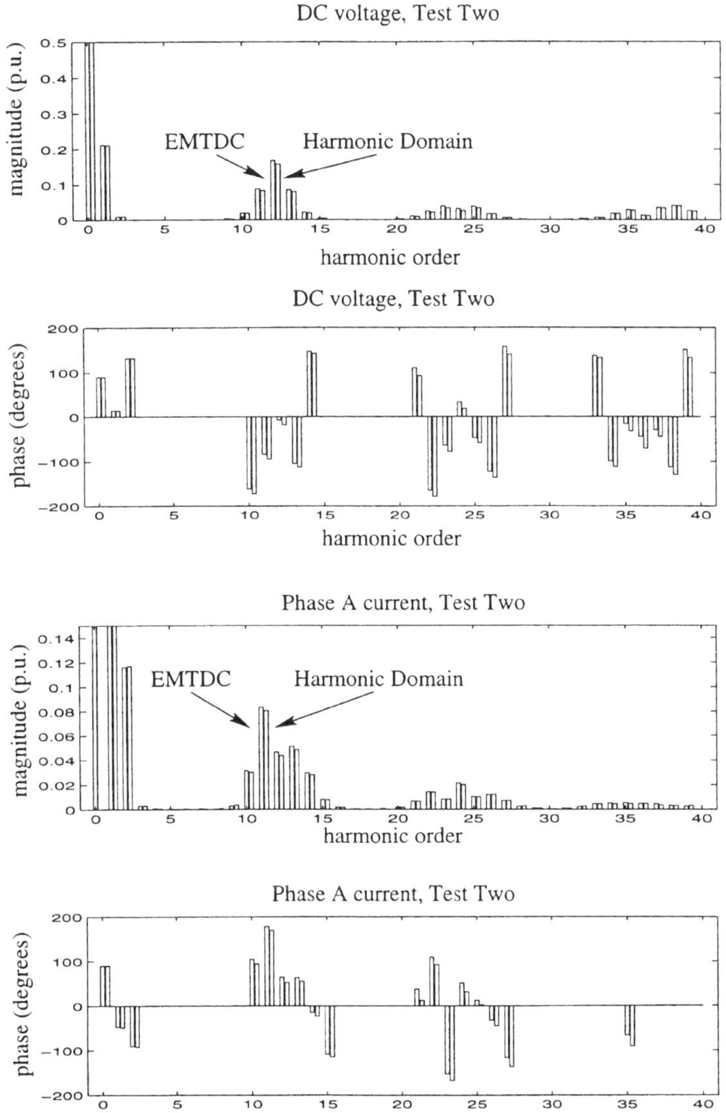

Figure 9.17 Comparison of time and harmonic domain solutions for phase currents and dc voltage spectra: Test 2

convergence varies according to the test, as the required convergence of a particular small noncharacteristic harmonic may require more iterations. An additional test has also been plotted, test five, where a 20% second order harmonic source was placed in the ac system. As indicated in Table 9.3, a smaller convergence factor (and hence fewer iterations to convergence) is obtained if the switching terms are updated (rows with subscript 'a'). However the constant Jacobian method is always faster as each iteration takes approximately 0.6 seconds as opposed to 2.3 seconds. The constant Jacobian method is only slower if it requires more than four times as many iterations

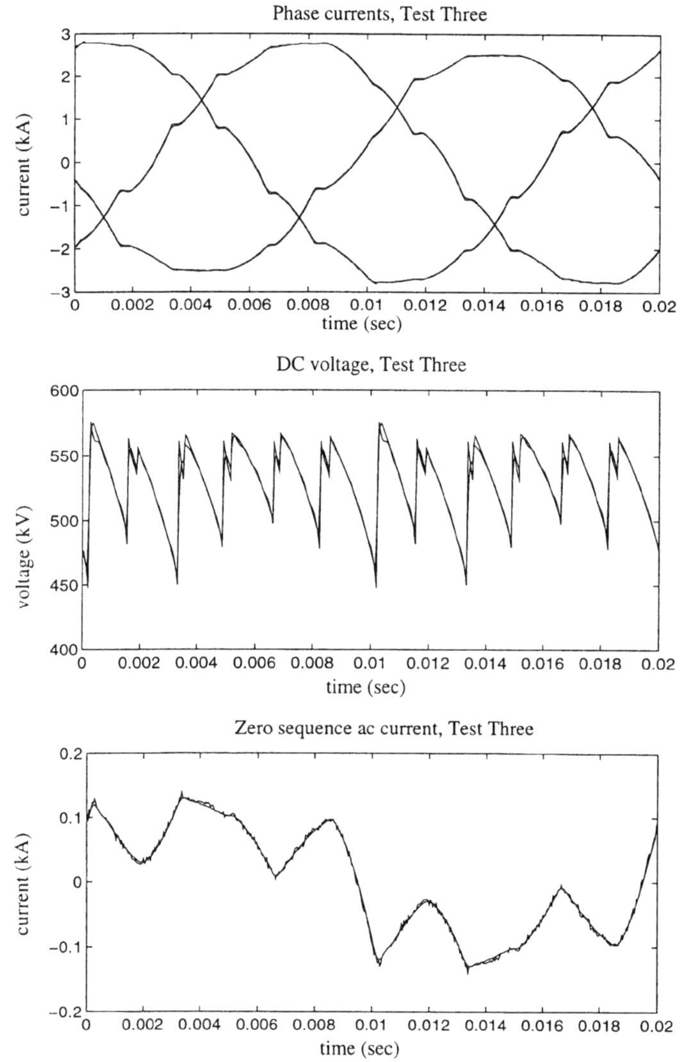

Figure 9.18 Comparison of time and harmonic domain solutions for phase current and dc voltage waveforms: Test 3

to converge. The only situation in which this has been observed is when the composite resonance is excited by a very large (0.3 p.u.) second harmonic source. The constant Jacobian method is therefore likely to be faster in any realistic case.

It is evident from Table 9.3 that convergence is slowed by the low harmonic order composite resonance, and by the presence of a large commutating resistance. The second harmonic composite resonance is particularly difficult for the constant Jacobian method, as there is a higher coupling between low order harmonics and the switching angles. This is evident in the convergence of Test 5, which required 21 iterations. For more realistic systems, convergence in eight iterations using a

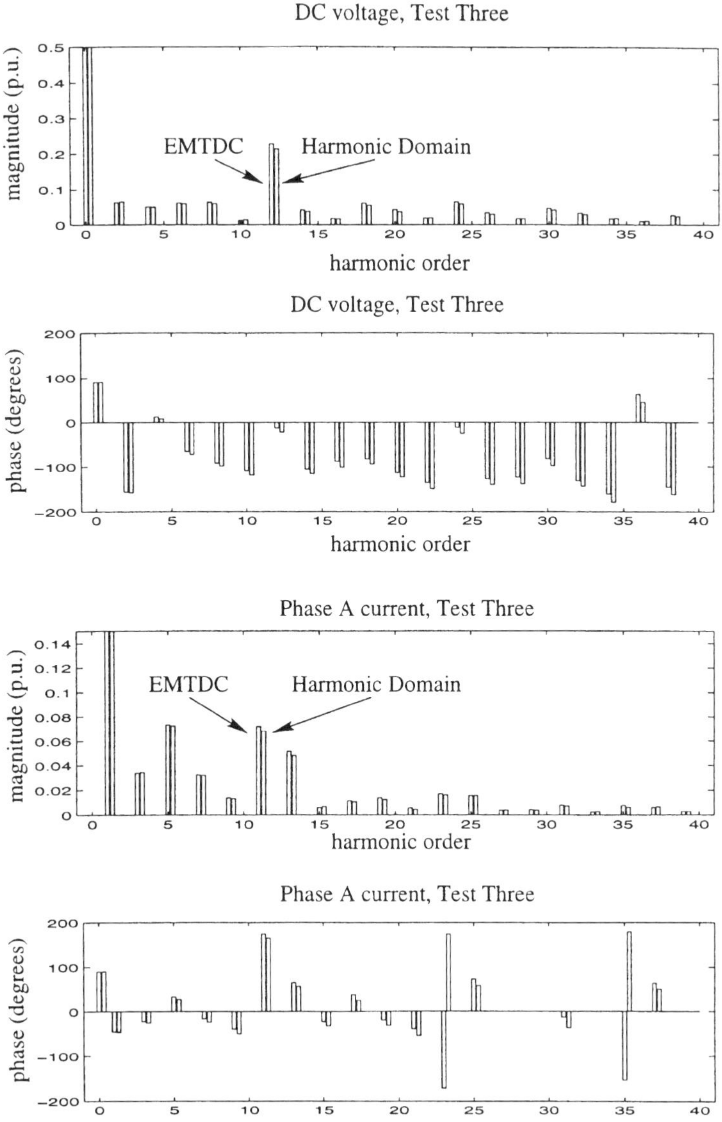

Figure 9.19 Comparison of time and harmonic domain solutions for phase currents and dc voltage spectra: Test 3

constant Jacobian might be expected. The execution times listed in Table 9.3 are for a Sun Sparcstation IPX.

9.6 Diagonalizing Transforms

A useful improvement to the performance of the converter model is obtained by applying linear transformations to the vector of mismatches and variables at each

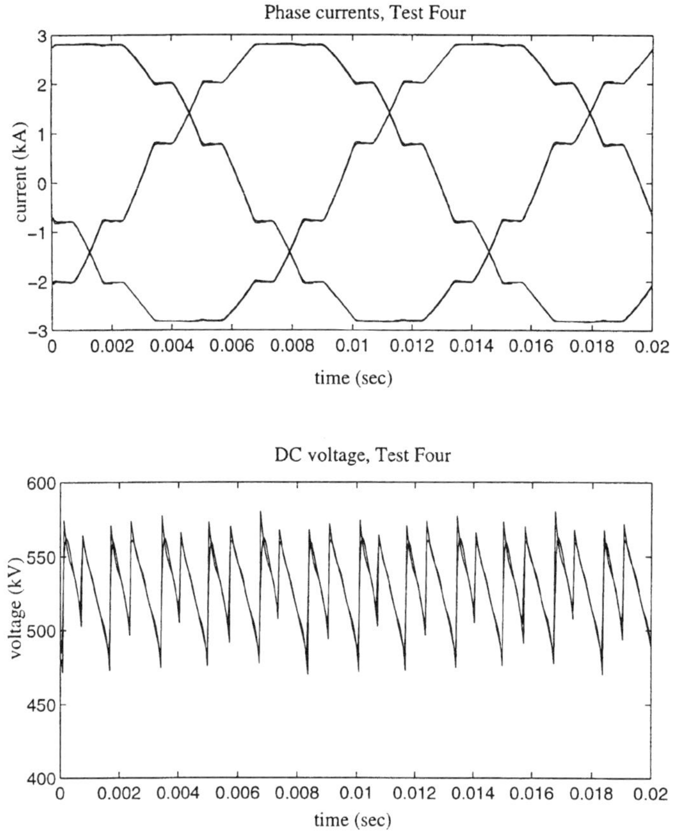

Figure 9.20 Comparison of time and harmonic domain solutions for phase current and dc voltage waveforms: Test 4

iteration of the Newton solution. The objective is to improve the sparsity of the Jacobian matrix by transforming to a new system of mismatches and variables which is more diagonal.

Balanced against the improved Jacobian sparsity of this method, is the extra calculation overhead associated with calculating the transform at each iteration. It is therefore not feasible to completely diagonalize the Jacobian matrix by calculating its eigenvalues, primarily because the matrix of eigenvectors which describes the diagonalizing transform is the same size as the Jacobian matrix, but full.

Applying either the sequence or dq0 transforms to quantities on the ac side of the converter yields a considerable improvement in Jacobian sparsity, with insignificant transformation overheads. Under either of these transformations, the zero sequence component is completely diagonalized, unless the star-g/delta transformer is unbalanced. Additional frequency coupling terms between the ac and dc sides are also removed in the sequence transform, since a harmonic k on the dc side couples mainly to harmonics $k+1$ in positive sequence, and $k-1$ in negative sequence on the ac side.

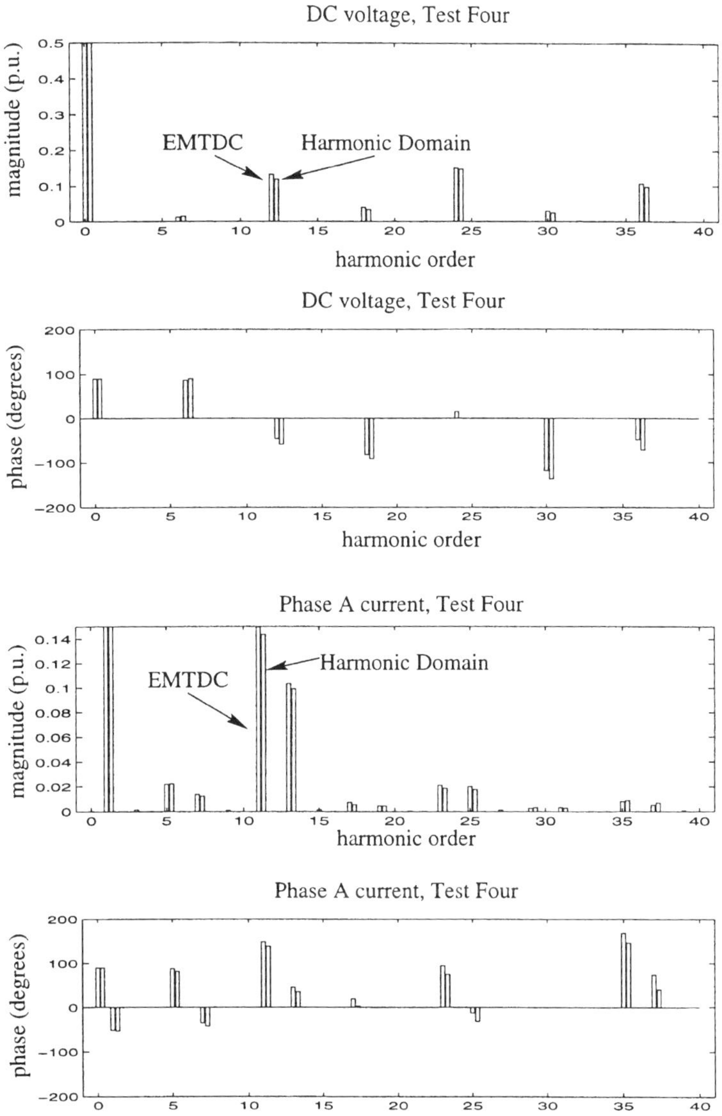

Figure 9.21 Comparison of time and harmonic domain solutions for phase currents and dc voltage spectra: Test 4

Application of the dq0 transform is not warranted since in the steady state, the dq0 transform is just a sequence transform followed by a rotation of the positive sequence into the direct and quadrature axes, and a rotation of the negative sequence into the conjugate of these axes. Unlike the synchronous machine, where the direct and quadrature axes are aligned with the rotor, there is no such preferred phase reference for the converter. It would therefore be necessary to choose a direct axis at every harmonic, which introduced the least coupling between the direct and quadrature axes quantities and other variables.

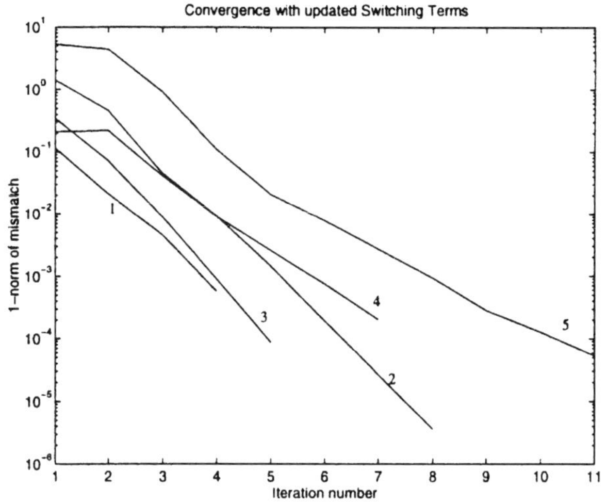

Figure 9.22 Convergence with the switching terms updated each iteration

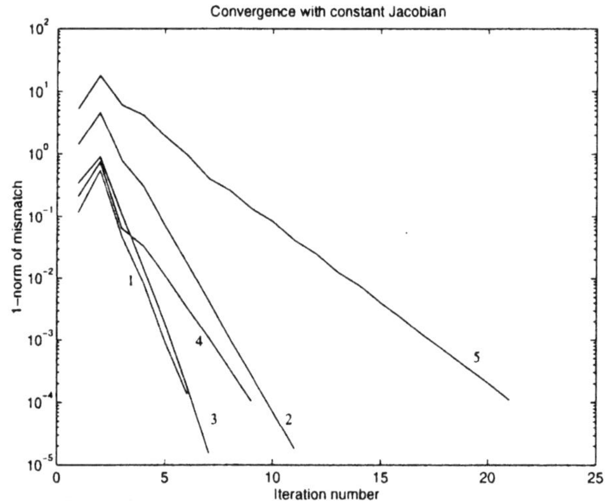

Figure 9.23 Convergence with the Jacobian held constant

Solving power system elements in sequence components affords a common frame of reference within which to model the system as a whole. In such a case it is necessary to integrate machine models, the load flow, HVDC, FACTS and transmission system components into a single iterative solution. In addition to being more computationally efficient, the sequence components frame of reference leads to greater insight into the interactions between component nonlinearities, communications interference, and power quality.

Table 9.3 Convergence and performance of the solution a) updating switching terms, b) constant Jacobian

Test No.	CPU time (seconds)	Switch Iterations	Main Iterations	λ
1a	15.5	6	4	0.1314
1b	11.1	6	6	0.1712
2a	24.9	6	8	0.1440
2b	14.8	6	11	0.2562
3a	17.7	6	5	0.1069
3b	12.3	6	7	0.1130
4a	22.7	9	7	0.2498
4b	14.1	9	9	0.3129
5a	31.5	6	11	0.3068
5b	20.8	6	21	0.5382

In the existing model, interaction with the dc system is specified by summing the dc voltage harmonics across each bridge, and applying the resulting voltage harmonics to the dc system linear model. The resulting harmonic current,

$$I'_{dk} = f_{19}(V, I_{dk}, \theta_i^D, \phi_i^D, \theta_i^Y, \phi_i^Y) \tag{9.54}$$

should be equal to the dc current harmonic ripple, I_{dk}. The dc side current mismatch equations can therefore be written;

$$F_{Idk} = I_{dk} - I'_{dk} \tag{9.55}$$

A similar equation is applied on the ac side, by summing the phase currents from each bridge and injecting them into the ac system. This yields voltage harmonics,

$$V'_k = f_{18}(V, I_{dk}, \theta_i^D, \phi_i^D, \theta_i^Y, \phi_i^Y) \tag{9.56}$$

that should equal the estimated converter terminal voltage harmonics, V_k. The ac side voltage mismatch equations can therefore be written

$$F_{Vk} = V_k - V'_k. \tag{9.57}$$

A sequence components solution is implemented by using the sequence transform matrix, T, to interface complex phase component calculations to the Newton solution in sequence components. For complex phasors, the complex 3×3 sequence transform is

$$\begin{bmatrix} V^0 \\ V^- \\ V^+ \end{bmatrix} = \begin{bmatrix} 1 & 1 & 1 \\ 1 & a^2 & a \\ 1 & a & a^2 \end{bmatrix} \begin{bmatrix} V^a \\ V^b \\ V^c \end{bmatrix}, \tag{9.58}$$

where $a = e^{j2\pi/3}$. Setting

$$W = \begin{bmatrix} V^0 \\ V^- \\ V^+ \end{bmatrix}$$

as the sequence components voltage yields $W = TV$, and $V = T^{-1}W$. If a three phase quantity has been decomposed in real rectangular components, as in the Jacobian matrix, a real components sequence transform matrix, $\bar{T}$, can be constructed:

$$\bar{W} = \bar{T}\bar{V}, \text{ which in full is} \tag{9.59}$$

$$\begin{bmatrix} V_R^0 \\ V_I^0 \\ V_R^- \\ V_I^- \\ V_R^+ \\ V_I^+ \end{bmatrix} = \begin{bmatrix} 1 & 0 & 1 & 0 & 1 & 0 \\ 0 & 1 & 0 & 1 & 0 & 1 \\ 1 & 0 & \mathcal{R}a^2 & -\mathcal{I}a^2 & \mathcal{R}a & -\mathcal{I}a \\ 0 & 1 & \mathcal{I}a^2 & \mathcal{R}a^2 & \mathcal{I}a & \mathcal{R}a \\ 1 & 0 & \mathcal{R}a & -\mathcal{I}a & \mathcal{R}a^2 & -\mathcal{I}a^2 \\ 0 & 1 & \mathcal{I}a & \mathcal{R}a & \mathcal{I}a^2 & \mathcal{R}a^2 \end{bmatrix} \begin{bmatrix} V_R^a \\ V_I^a \\ V_R^b \\ V_I^b \\ V_R^c \\ V_I^c \end{bmatrix},$$

where $\mathcal{R}, \mathcal{I}$ denote the real and imaginary parts of a quantity. The existing phase components mismatch equations summarized in Table 9.2 can now be written as sequence component mismatches, in terms of the sequence components terminal voltage, using the sequence transform matrix. Sequence component mismatches for the 12-pulse controlled rectifier are listed in Table 9.4 below.

The Jacobian matrix for a sequence components solution is readily obtained from the phase component partial derivatives by application of the chain rule:

$$\left[\frac{\partial F_W}{\partial W}\right] = I - \bar{T}\left[\frac{\partial V'}{\partial V}\right]\bar{T}^{-1} \tag{9.60}$$

$$\left[\frac{\partial F_W}{\partial x}\right] = -\bar{T}\left[\frac{\partial V'}{\partial x}\right], \quad x \in \{I_{dk}, \theta_i^D, \phi_i^D, \theta_i^Y, \phi_i^Y\}$$

$$\left[\frac{\partial F_x}{\partial W}\right] = \left[\frac{\partial F_x}{\partial V}\right]\bar{T}^{-1}, \quad x \in \{I_{dk}, \theta_i^D, \phi_i^D, \theta_i^Y, \phi_i^Y\}$$

The resulting Jacobian matrix is plotted in the bottom of Figure 9.24, and shows a greater degree of sparsity than that for the phase components, shown in Figure 9.7. This is primarily due to the absence of any coupling between the zero sequence ac voltage harmonics and any of the other converter variables.

Table 9.4 Mismatches and variables for the 12-pulse rectifier with the terminal voltage in sequence components

Eqn.	functional notation
F_{Wk}	$W - TV'_k(T^{-1}W, I_{dk}, \theta_i^D, \phi_i^D, \theta_i^Y, \phi_i^Y)$
F_{Idk}	$I_{dk} - I'_{dk}(T^{-1}W, I_{dk}, \theta_i^D, \phi_i^D, \theta_i^Y, \phi_i^Y)$
F_ϕ^Y	$f_{20}(T^{-1}W, I_{dk}, \theta_i^Y, \phi_i^Y)$
F_ϕ^D	$f_{21}(T^{-1}W, I_{dk}, \theta_i^D, \phi_i^D)$
F_θ^Y	$f_{14}(I_{dk}, I_{d0}, \theta_i^Y, \alpha_0)$
F_θ^D	$f_{15}(I_{dk}, I_{d0}, \theta_i^D, \alpha_0)$
$F_{\alpha 0}$	$F_{23}(T^{-1}W, I_{dk}, I_{d0}, \theta_i^Y, \theta_i^D, \phi_i^Y, \phi_i^D)$

Convergence of the sequence components solution is compared with that of the constant Jacobian phase components solution in Table 9.5, for three of the test cases. Convergence of the sequence components solution was similar to that for phase components, indicating that the converter model can be directly combined with other components (such as the synchronous machine) modelled in sequence components.

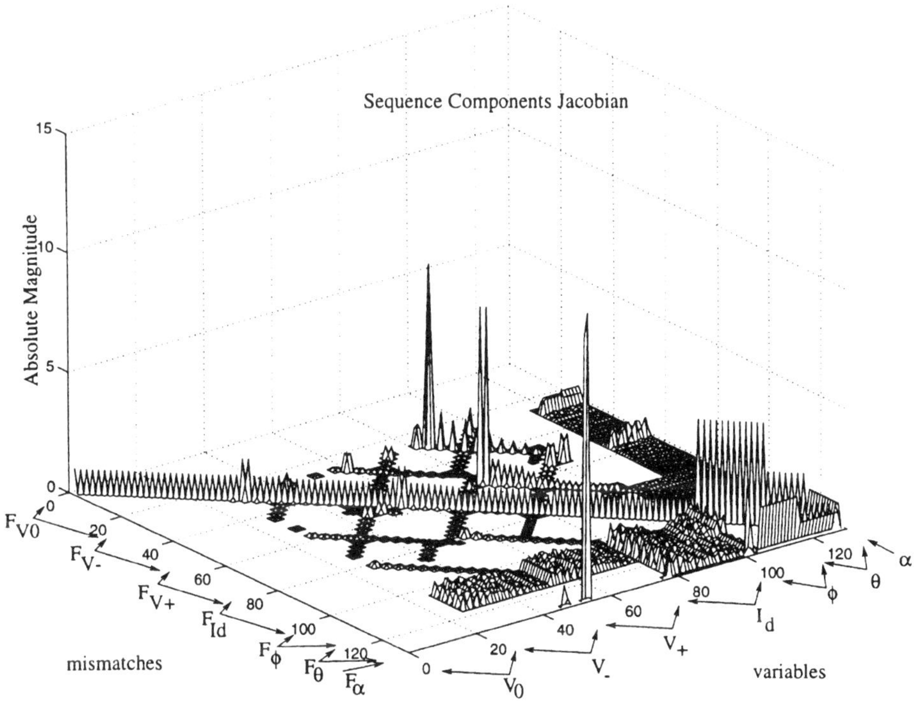

Figure 9.24 Sequence components Jacobian matrix; 13 harmonics

Table 9.5 Convergence and performance of the converter model, b) constant unified Jacobian, d) Constant sequence components Jacobian

Test No.	CPU time (seconds)	Main Iterations	λ
1b	11.1	6	0.1712
1d	10.9	5	0.0851
2b	14.8	11	0.2562
2d	15.0	12	0.2973
3b	12.3	7	0.1130
3d	11.5	6	0.1056

9.7 Integrated Converter and Load Flow Solution

A significant advantage of steady state modelling is the relative ease with which the load flow constraints can be accounted for. This is particularly important in the case of the HVdc converter, which responds strongly to conditions on the ac side at fundamental frequency. Apart from the influence of the positive sequence component magnitude on the average delay angle, the converter will respond to negative sequence at fundamental by injecting triplen harmonic currents into the ac system, and other noncharacteristic even harmonics on the dc side. An accurate solution of the converter harmonics therefore requires that it be interfaced to a three phase load flow on the ac side.

From the stand-point of the load flow, the reactive power and voltage at the converter bus are all unspecified, so the converter bus cannot be classified as PQ, PV or slack. An additional problem is the interaction between triplen harmonics on the ac side, and negative sequence fundamental current injected by the converter, which directly affects the three-phase load flow. The above points indicate the need for a simultaneous solution for both the converter and the load flow, preferably by Newton's method because of the strong interaction between triplen harmonics and negative sequence fundamental.

The Newton solution for the converter described in this chapter is readily augmented by a load flow at the fundamental frequency. The Thevenin equivalent of the ac system is retained at all harmonic frequencies except the fundamental, which is the point of coupling to the load flow equations. Similarly, three-phase load flow equations are retained at every bus except the converter bus. It is assumed that were it not for the converter, the converter bus would be PQ. Since the load-flow is three-phase, six real mismatch equations are required at the converter bus at fundamental frequency. A suitable set of equations is obtained from the requirement that Kirchoff's current law is satisfied, that is, the sum of the currents in every phase into the converter bus is zero. Taking real and imaginary parts of the current mismatch yields the required six equations.

Representing the ac system as a Thevenin equivalent at harmonic frequencies reduces the size of the problem substantially, since all system buses except the converter bus are eliminated in calculating the Thevenin impedance. The unknowns in the three-phase load flow are the fundamental frequency complex voltages in every phase, at every bus in the system. A total of $6(n_b - 1)$ real unknowns are therefore contributed by the load flow, in addition to the unknowns already specified for the converter model.

Unlike the converter mismatch equations, the load flow mismatch equations are very easy to calculate, as is the load flow partition of the Jacobian matrix. Since the processing required at every iteration is dominated by the converter equations, it is essential that convergence of the load flow part of the system be faster than that of the converter model, which typically converges in seven iterations. The decoupled load flow method is therefore not suitable, as it requires more iterations for convergence. A decoupled load flow, being framed in polar coordinates, is also incompatible with the converter model in Cartesian components, unless a non-linear polar transform is applied at every iteration at the converter bus. This additional non-linearity would be likely to degrade convergence, especially if the

Jacobian is held constant. Integration of the converter model with a load flow therefore requires that the load flow be reformulated in Cartesian components, with no decoupling in the Jacobian matrix.

In summary, the complete set of mismatch equations are;

- PQ bus. The real and reactive power in each phase are as specified.
- PV bus. The positive sequence voltage magnitude, and real power in the positive sequence are as specified. Kirchoff's current law is applied for the negative and zero sequence current, so that the current shunted by the machine balances that flowing out of the network.
- Slack bus. The real and imaginary parts of the positive sequence voltage are as specified. Kirchoffs current law is applied for the zero and negative sequence currents.
- Converter ac side at fundamental. Current mismatches are applied in all three phases,
- Converter ac side at harmonics. Voltage mismatch at every harmonic in all three sequences.
- Converter dc side at all frequencies. Current mismatches at every harmonic, including fundamental frequency.
- Converter switchings. Zero crossing mismatches.
- Converter control. The controlled quantity is as specified. For example the average dc current is as specified in the case of constant current control.

A sparsity diagram for the Jacobian of this system of equations is shown in Figure 9.25. The test system in this case consists of the rectifier end of the CIGRE benchmark, with the ac system replaced by a 9 bus load flow representing the lower part of the New Zealand South Island primary transmission system. This system has also been solved with a more detailed 70 bus load flow. The Jacobian of Figure 9.25 contains no more terms than the separate Jacobians of the load flow and converter, and yet convergence has been found to be faster and more robust than a fixed point iteration between separate load flow and converter model updates, which usually diverges.

9.8 Summary

After a brief review of currently available iterative methods for harmonic analysis, the inter-relationships of Chapter 8 have been used in this chapter to describe a Newton solution of a 12-pulse rectifier with ac and dc system representation. A functional description of inter-dependent quantities has been used to assemble a

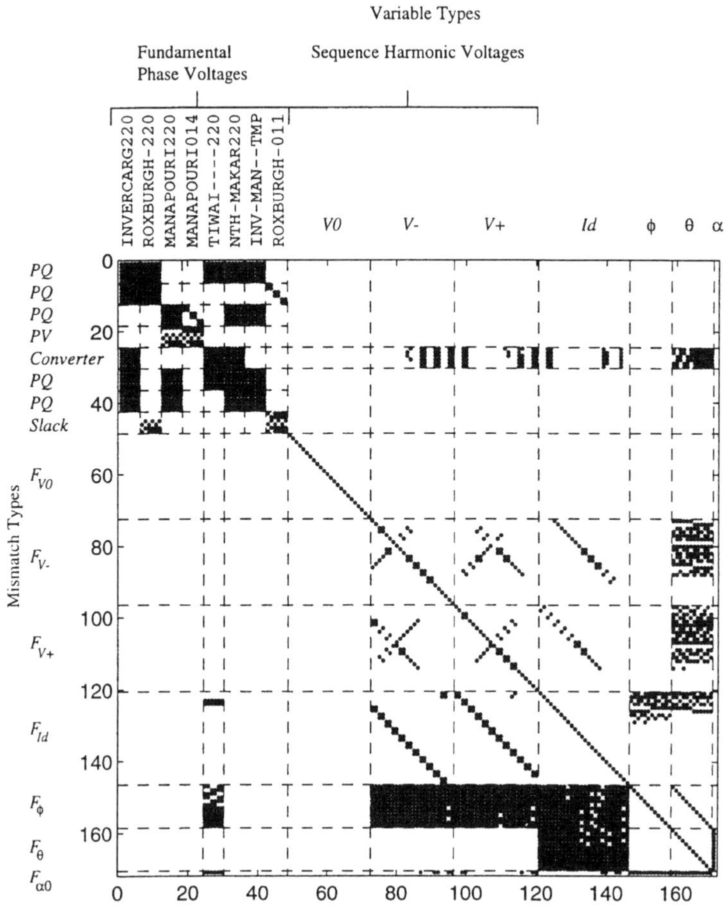

Figure 9.25 Sparsity of the Jacobian matrix for an integrated load flow and converter model

reduced set of mismatch equations suitable for use in Newton's method. The Newton's method solution has been described, and the Jacobian matrix of partial derivatives used in this method analysed.

The Jacobian matrix can be made sparse by setting small elements to zero. Among the methods of solving the sparse Jacobian system, a direct sparse bifactorization method is preferred when the Jacobian is held constant. Despite requiring more iterations, fastest convergence is obtained when the Jacobian is held constant.

Additional sparsity and compatibility with other component models, is obtained by solving for sequence components voltages on the ac side. Integration of the converter model with a reformulation of the three-phase load flow has been described.

The description of the system in terms of functions affords a modular implementation of the model. For example, the functions that describe transfer through the converter transformer could readily be extended to cover other transformer connections. The Jacobian matrix itself retains the same structure, even in multiconverter systems or when interfaced to a load flow. In either case, the Jacobian matrix described here would be a block in a block diagonal system Jacobian with linear coupling terms to other nonlinearities.

9.9 References

1. Yacamini, R and de Oliveira, JC, (1980). Harmonics in multiple convertor systems: a generalized approach, *IEE Proceedings Pt.B*, **127**(2), 96–106.
2. Yacamini, R and de Oliveira, JC, (1986). Comprehensive calculation of convertor harmonics with system impedances and control representation, *IEE Proceedings PT. B*, **133**(2), 95–102.
3. Reeve, J and Baron, JA, (1971). Harmonic interaction between hvdc convertors and ac power systems, *IEEE Transactions on Power Apparatus and Systems*, **90**(6), 2785–2793.
4. Callaghan, C and Arrillaga, J, (1989). A double iterative algorithm for the analysis of power and harmonic flows at ac–dc terminals, *Proc. IEE*, **136**(6), 319–324.
5. Callaghan, CD and Arrillaga, J, (1990). Convergence criteria for iterative harmonic analysis and its application to static convertors, *Proc. Intl. Conf. on Harmonics in Power Systems (IEEE), Budapest*, 38–43.
6. Carbone, R, *et al.*, (1992). Some considerations on the iterative harmonic analysis convergence, *Proc. Intl. Conf. on Harmonics in Power Systems (IEEE)*, Atlanta.
7. Carpinelli, G, Gagliardi, F, Russo, M and Villacci, D, (1994). Generalised converter models for iterative harmonic analysis in power systems, *Proceedings of the IEE Gener. Transm. Distrib.*, **141**(5), 445–451.
8. Dommel, HW (ed.), (1986). *Electromagnetic Transients Program Manual/EMTP Theory Book*. Bonneville Power Administration.
9. Larson, EV, Baker, DH and McIver, JC, (1989). Low order harmonic interaction on ac/dc systems, *IEEE Transactions on Power Delivery*, **4**(1), 493–501.
10. Jalali, SG and Lasseter, RH, (1994). A study of nonlinear harmonic interaction between a single phase line-commutated converter and a power system, *IEEE Transactions on Power Delivery*, **9**(3), 1616–1624.
11. Rajagopal, N and Quaicoe, JE, (1993). Harmonic analysis of three-phase ac/dc converters using the harmonic admittance method, *1993 Canadian Conference on Electrical and Computer Engineering, Vancouver, BC, Canada*, **1**, 313–316.
12. Smith, BC, (1996). A harmonic domain model for the interaction of the HVdc convertor with ac and dc systems. PhD thesis, University of Canterbury, New Zealand.
13. Szechtman, M, Weiss, T and Thio, CV, (1991). First benchmark model for hvdc control studies, *Electra*, **135**, 55–75.
14. Zollenkopf, K, (1970). Bi-factorization–basic computational algorithm and programming techniques, In *Confernce on Large Sets of Sparse Linear Equations*, 75–96 Oxford.
15. Press, WH, *et al.*, (1992). *Numerical Recipes in Fortran, The Art Of Scientific Computing*. Cambridge University Press, 2nd edition.
16. Accessible on the World Wide Web at http://netlib.att.com/netlib/y12m/index.html

10

CONVERTER HARMONIC IMPEDANCES

10.1 Introduction

The representation of ac–dc converters as frequency-dependent equivalents has been discussed in Chapter 5 (Section 5.4), as well as its application to filter design. A more accurate approach, based on the converter models described in Chapter 8, is used here to derive the converter harmonic impedances.

A widely used and effective technique for analysing nonlinear devices is to linearise their response around an operating point. If the nonlinearity is a voltage-current relationship, the linearization yields an impedance. This type of linearization is particularly relevant to the converter, as the converter impedance can be combined with the ac and dc system impedances to analyse resonances, harmonic transfers, and harmonic magnification factors.

An example of its application to the calculation of the dc side impedance of a converter, including the effect of ac side impedance at coupled harmonics, has been described by Bahrman [1]. Another reported application relates to a GIC induced 5th harmonic resonance [2], which could only be explained by combining the impedance of the converter with that of the ac system.

When linearizing general nonlinear devices to an equivalent impedance, representation by a single complex number is not possible. Instead the use of either complex or real valued matrices is necessary [3]. The fact that the complex value of an impedance can depend upon the phase angle of the current flowing through it, and still be linear, is not widely appreciated. The linear transfer from firing angle modulation to direct current at the 6th harmonic has been shown to be phase dependent [4], a fact that also applies to the converter linearized impedance [5].

Central to Newton's method for solving nonlinear systems is the Jacobian matrix, which represents a linearization of the system of equations at every iteration. Although the Jacobian is typically held constant during the solution, it can be re-calculated at convergence. It is therefore possible to use the Jacobian matrix to calculate equivalent impedances for the converter, or indeed any other linearized relationship. The advantage of calculating impedances in this way is that the effect of control, switching instant variation, unbalance, and system impedances are automatically taken into account.

10.2 Calculation of the Converter Impedance

In general the converter impedance for small distortions will be phase dependent, and therefore best represented by a tensor. In this case, the impedance tensor, described in Appendix V can be directly calculated from the converter Jacobian, since the Jacobian is a linearization of the converter around an operating point. Firstly however, the phase and magnitude dependence of the converter complex impedance to an applied harmonic voltage distortion is investigated by applying a variety of small harmonic distortions to the converter terminal.

10.2.1 Perturbation Analysis

An approximation to the converter impedance for small harmonic distortions applied to the ac terminal can be determined by solving the converter model for perturbations around an initial solution. For example, to derive the 7th harmonic positive sequence impedance, the converter base case is first solved, and a sequence transformation is then applied to the 7th harmonic voltage and phase current at the ac terminal to yield base values:

$$V_{7b}^{+} = [TV_{7b}]^{+}, \tag{10.1}$$

$$I_{7b}^{+} = [TI_{7b}]^{+}, \tag{10.2}$$

where T is the sequence transform matrix. Next, a small positive sequence 7th harmonic voltage source is added to the ac system Thevenin voltage source, and the converter model is re-converged. The perturbed positive sequence 7th harmonic ac terminal quantities are then calculated, as above:

$$V_{7p}^{+} = [TV_{7p}]^{+}, \tag{10.3}$$

$$I_{7p}^{+} = [TI_{7p}]^{+}. \tag{10.4}$$

A value of 7th harmonic impedance is then obtained from the ratio of change in voltage, to change in current:

$$Z_7^{+} = \frac{V_{7p}^{+} - V_{7b}^{+}}{I_{7p}^{+} - I_{7b}^{+}}. \tag{10.5}$$

The variation of the impedance with magnitude or phase, is determined by applying several different magnitudes or phase of voltage to the ac system Thevenin equivalent voltage source. For most harmonics, the series combination of filters and system impedance acts as a voltage divider to any applied harmonic voltage in the ac system, so that it is difficult to obtain a large voltage distortion at the converter terminal. Accordingly, for the purpose of determining the response of the converter impedance to a large harmonic distortion, the ac system and filter impedances at every harmonic were made equal to their impedances at the fundamental. The effect of harmonic magnitude was then determined by applying a voltage source in the ac system that resulted in approximately 0.001 p.u. change in voltage distortion at the converter terminal, and then a source that resulted in 0.12 p.u. voltage distortion.

The small source yields a good approximation to the 'slope' converter impedance, whereas the 0.12 p.u. distortion resulting from the large source is larger than would be encountered in practice.

Forty phase angles, from 0 to 2π, of voltage distortion were applied at both magnitudes, and at every harmonic order from 2 to 37, in positive and negative sequence. The resulting impedance loci have been plotted in Figure 10.1.

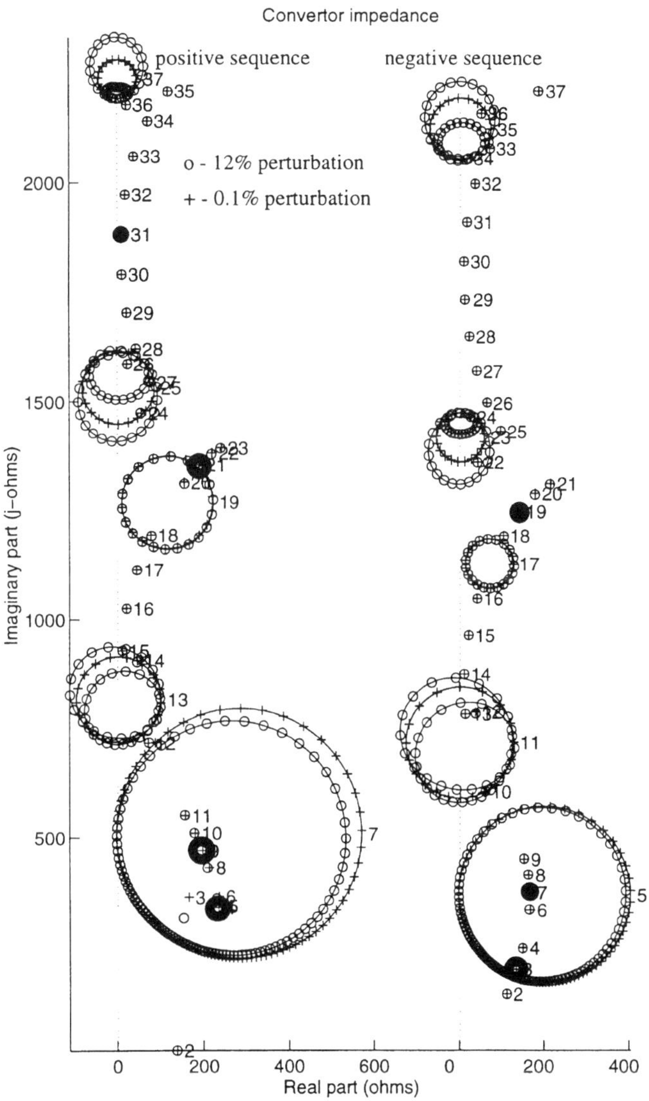

Figure 10.1 Dependence of converter impedances on magnitude of applied voltage perturbation

To obtain accurately the value of small current and voltage perturbations, the convergence tolerance for the model was reduced to:

$$
\begin{aligned}
&\frac{|F_V|}{|V|} < 2 \times 10^{-6} \\
&\frac{|F_I|}{|I|} < 2 \times 10^{-6} \\
&\frac{|F_{Id}|}{|Id|} < 2 \times 10^{-6} \\
&\frac{|F_{\alpha_0}|}{|I_{d0}|} < 2 \times 10^{-6} \\
&|F_{\theta_i}| < 5 \times 10^{-8} \\
&|F_{\phi_i}| < 5 \times 10^{-8},
\end{aligned}
\tag{10.6}
$$

and thresholding was removed for the impedance harmonic under consideration. Since the system was the same for every perturbation, the Jacobian was held constant for all 5760 runs. The main features to be observed in Figure 10.1 are the phase dependence of the converter impedance at six-pulse characteristic harmonics, and the independence of impedance to the magnitude of applied voltage. At the 12-pulse characteristic harmonics, a large value of distortion leads to an epicyclic phase dependent locus, due to the coupling of these harmonics to the average dc current, and hence control action. At all other harmonics and sequences, the converter impedance is essentially completely linear over the range of magnitudes likely to be encountered. In practice, there will also be filters at the characteristic and high order harmonics, and high levels of voltage distortion will not be encountered.

Despite a uniform progression of voltage phase angles applied to the 5th and 7th harmonics, the associated circular loci display a clustering of impedance points near the origin. This is a consequence of the perturbation method used, rather than the impedance itself, and is best explained by considering the locus of the returned current perturbations, as plotted in Figure 10.2 for the positive sequence 7th harmonic.

As the applied voltage phase progresses uniformly through 2π, the returned current is determined by a uniform double rotation around the admittance locus. The admittance locus is not concentric with the origin, and so the returned current is small only for those few points on the admittance locus close to the origin. This results in the clustering of points on the extrema of the elliptical current locus. When points on the impedance locus are calculated, this clustering effect is compounded by thc radial distancc from the origin on the impedance plane, since the current locus is clustered at large values of current, which correspond to small impedances. If the voltage perturbation method is used to calculate the admittance locus instead, the points are equi-spaced.

In relation to the earlier discussion on phase dependence, it is evident that the converter impedance can be described by complex numbers at all harmonics, apart from $h = 6n \pm 1$. At these harmonics the impedance should be a tensor representation, or a coupling to the conjugate of the applied voltage. The generation

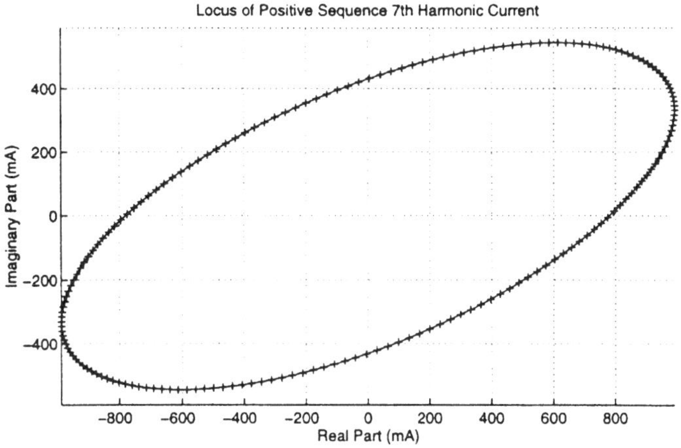

Figure 10.2 Locus of the perturbed fifth harmonic current

of harmonic currents by the converter related to the conjugate of an applied voltage distortion is predicted by the modulation theory analysis of Chapter 5. The impedance of the converter is calculated in this model by superimposing the effect of several small amplitude waveform distortion transfers, each calculated by modulation theory. Although this model predicts many relationships involving multiples of the modulation frequency, these higher order terms are very small, and it suffices for a qualitative understanding of the converter impedance to consider only the relationships described below.

Modulation Theory of the Impedance The application of either a positive sequence harmonic voltage of order $(k+1)$ angle δ, or a negative sequence harmonic $(k-1)\angle\delta$ on the ac side of the converter is directly transferred to a dc side voltage $k\angle(a_1+\delta)$ and a small $(12-k)\angle(a_2-\delta)$, where a_i is a constant angle. If $k=6$ then the first reflected term, $(12-k)$, is at the same harmonic, but phase conjugated. This mechanism of phase conjugation occurs quite generally for all the distortion transfers through the converter. The dc ripple at harmonic k, resulting from the voltage ripple, will be transferred to the ac side currents $(k+1)\angle(a_3+\delta)$ in positive sequence, and $(k-1)\angle(a_4+\delta)$ in negative sequence. There will also be ac side currents $(11-k)\angle(a_5-\delta)$ in positive sequence, and $(13-k)\angle(a_6-\delta)$ in negative sequence. The reflected terms, $(11-k)$ and $(13-k)$, are very small, but demonstrate one mechanism whereby there is a phase conjugation of applied voltage to returned current on the ac side.

Commutation Period Modulation The application of harmonic voltage orders of either $(k+1)\angle\delta$ in positive sequence, or $(k-1)\angle\delta$ in negative sequence on the ac side will directly modulate the end of commutation angles at harmonic $k\angle(a_7+\delta)$. This has a very significant effect on ac side impedances, since the full dc current will be modulated. The end of commutation modulation leads to a range of modulation products on the ac side, including the harmonic orders $(k+1)\angle(a_8+\delta)$ in positive sequence, $(k-1)\angle(a_9+\delta)$ in negative sequence, $(11-k)\angle(a_{10}-\delta)$ in negative sequence, and $(13-k)\angle(a_{11}-\delta)$ in positive sequence. The reflected terms in this case are significant, and are scaled by the product of two sinc functions: $\text{sinc}(k\mu/2)\text{sinc}(m\mu/4)$, where $m = 1, 11, 13, 23\ldots$ Setting $k = 6$, indicates that the 5th negative, and 7th positive sequence impedances of the converter are substantially phase dependent. If a 6th harmonic positive sequence voltage is applied to the converter there will be a negative sequence current returned, and vice versa. A similar analysis holds for all six-pulse order harmonics, reflecting as appropriate from higher order multiples of the modulating frequency.

Having established that the converter impedance is magnitude independent, and best represented by a tensor at particular harmonics, it is not necessary to calculate the complex impedance for a full 2π of applied voltage distortions. In fact, just two perturbations yield four equations in the four unknown tensor elements. However the perturbation method is essentially a numerical partial differentiation, rife with potential problems. For example, it is difficult to know if the perturbation is definitely small enough, and whether the convergence tolerance specified has yielded suitably accurate data points. In the next section, the Jacobian matrix is modified to yield directly any desired tensor relationship. Anticipating this result, Figure 10.13 is a comparison of the CIGRE rectifier impedance loci calculated analytically from the Jacobian (continuous circles), and by perturbation (data points on circles), for every harmonic order from 2 to 37 in positive and negative sequence.

10.2.2 The Lattice Tensor

Several authors, most notably Larson [6], have noted that from the point of view of the ac system, the converter presents a stable and quite linear set of interrelationships between harmonics and sequences. For example, the application of a positive sequence voltage distortion at harmonic order $k+1$, leads to the injection of harmonic currents $(11 \pm k)$, $(23 \pm k)$, etc. in negative sequence, and $(1 \pm k)$, $(13 \pm k)$, $(25 \pm k)$ etc. in positive sequence. The difference terms are always phase conjugated, unless the harmonic would be negative, in which case the sequence is reversed. In fact the analysis of Wood [7] predicts many other multiple reflections that usually, but not always, decay with order (for example $12m \pm nk \pm 1$). This 'numerology of the converter' is succinctly summarized by a lattice-like diagram of connections between harmonics and sequences, as shown in Figure 10.3, for $n = 1$.

Points on Figure 10.3 marked with a '+', can be represented by a complex admittance. Points marked by a 'O' represent couplings between positive and negative harmonics, and can be represented either by a circular admittance locus centered on the origin in the complex admittance plane, or by a tensor (the impedance tensor concept is described in Appendix V). Points with both markers, at lattice vertices, indicate that the total current returned consists of two components, related to the voltage applied, and its conjugate. Such points correspond to a circular locus in the complex plane shifted away from the origin. Figure 10.13 is a plot of impedances along the diagonal of Figure 10.3, with circular loci occurring at the six-pulse characteristic harmonics. Also marked on Figure 10.3 is the special case of a

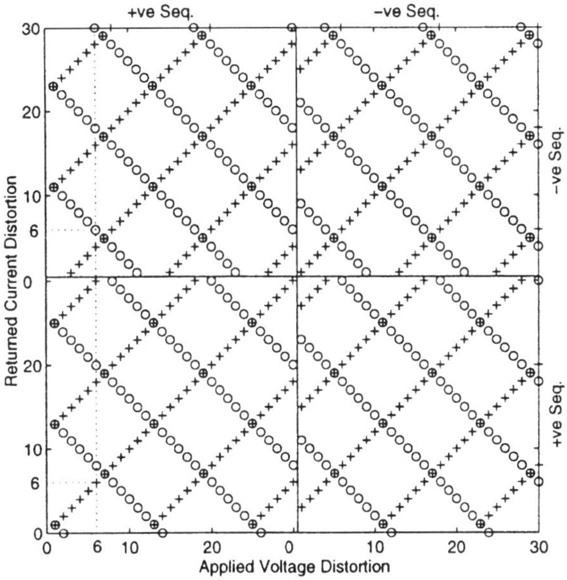

Figure 10.3 Principal harmonic currents returned by a twelve pulse converter in response to an applied voltage distortion. '+'- current phase related to phase of applied voltage, 'O' - current phase related to conjugate of applied voltage

sixth harmonic positive sequence distortion. A current is returned at the same harmonic in both positive and negative sequence, and it can be seen that the converter transforms between positive and negative sequence for all harmonics $k = 6n$.

If the converter is unbalanced, or there are large harmonic distortions, the cross coupling lattice will fill in to some extent, especially at low order harmonics. There will also be current injections related to the conjugate of applied harmonic voltages at most harmonics. In general then, the lattice is most accurately represented by a large *lattice tensor*, of second rank, but of size $4n_h \times 4n_h$ (unless there is coupling to the zero sequence). The lattice tensor can be calculated from the Jacobian matrix by first noting that partial derivatives of the voltage mismatch are of the form:

$$\frac{\partial F_V}{\partial X} = \frac{\partial V}{\partial X} - \frac{\partial f_{18}}{\partial X} \tag{10.7}$$

where f_{18} is the calculated terminal voltage, obtained by injecting the phase currents into the ac system impedance:

$$f_{18k}(X) = [Y_{cc}]_k^{-1}([I_P^S + I_P^D]_k + [Y_{ct}]_k[V_{th}]_k) \tag{10.8}$$

consequently

$$\frac{\partial f_{18}}{\partial X} = [Y_{cc}]_k^{-1} \frac{\partial [I_P^S + I_P^D]}{\partial X} \tag{10.9}$$

Those rows of the Jacobian that contain partial derivatives of the voltage mismatch are therefore easily modified to contain partial derivatives of the phase currents with respect to the system variables (terminal voltage, dc current, switching angles). The matrix equation defined by the new matrix, J', is now:

$$\begin{bmatrix} \Delta I \\ \Delta F_{Id} \\ \Delta F_{\phi} \\ \Delta F_{\theta} \\ \Delta F_{\alpha 0} \end{bmatrix} = \begin{bmatrix} \frac{\partial I}{\partial V} & \frac{\partial I}{\partial I_d} & \frac{\partial I}{\partial \phi} & \frac{\partial I}{\partial \theta} & \frac{\partial I}{\partial \alpha_0} \\ \frac{\partial F_{Id}}{\partial V} & \frac{\partial F_{Id}}{\partial I_d} & \frac{\partial F_{Id}}{\partial \phi} & \frac{\partial F_{Id}}{\partial \theta} & \frac{\partial F_{Id}}{\partial \alpha_0} \\ \frac{\partial F_{\phi}}{\partial V} & \frac{\partial F_{\phi}}{\partial I_d} & \frac{\partial F_{\phi}}{\partial \phi} & \frac{\partial F_{\phi}}{\partial \theta} & \frac{\partial F_{\phi}}{\partial \alpha_0} \\ \frac{\partial F_{\theta}}{\partial V} & \frac{\partial F_{\theta}}{\partial I_d} & \frac{\partial F_{\theta}}{\partial \phi} & \frac{\partial F_{\theta}}{\partial \theta} & \frac{\partial F_{\theta}}{\partial \alpha_0} \\ \frac{\partial F_{\alpha 0}}{\partial V} & \frac{\partial F_{\alpha 0}}{\partial I_d} & \frac{\partial F_{\alpha 0}}{\partial \phi} & \frac{\partial F_{\alpha 0}}{\partial \theta} & \frac{\partial F_{\alpha 0}}{\partial \alpha_0} \end{bmatrix} \begin{bmatrix} \Delta V \\ \Delta Id \\ \Delta \phi \\ \Delta \theta \\ \Delta \alpha_0 \end{bmatrix} \tag{10.10}$$

At the converter solution, all the mismatches are equal to zero. When a harmonic voltage perturbation is applied, it is required that all the mismatches in equation 10.10 remain zero, i.e. $\Delta F = 0$. Partitioning J' around I, and V, equation 10.10 can be written:

$$\begin{bmatrix} \Delta I \\ 0 \end{bmatrix} = \begin{bmatrix} A & B \\ C & D \end{bmatrix} \begin{bmatrix} \Delta V \\ \Delta X' \end{bmatrix}. \tag{10.11}$$

Eliminating $\Delta X'$:

$$\Delta I = (A - BD^{-1}C)\Delta V \tag{10.12}$$

The large matrix D is not actually inverted, instead columns, y_i of $D^{-1}C$ are obtained from solutions to the linear system

$$Dy_i = c_i, \tag{10.13}$$

where c_i are columns of C. This procedure is much faster as it avoids the matrix multiplication and storage of D^{-1}, and the LU decomposition of D is only calculated once.

Since I and V have been decomposed into real valued components, the matrix $\bar{Y}_{cp} = A - BD^{-1}C$ is a second rank tensor. It is actually a phase components version of the lattice network, connecting harmonics and phases, rather than harmonics and sequences. The phase components cross coupling tensor has been plotted in Figure 10.4, for harmonic interactions up to the 21st harmonic. Only admittances larger than 0.005 p.u. have been retained, and only magnitudes have been plotted. The tensor is 126 elements square ($2 \times 3 \times 21$).

Application of the sequence components transform yields the same tensor in sequence components, plotted in Figure 10.5. In sequence components the lattice tensor clearly contains far fewer terms, and the zero sequence part has been

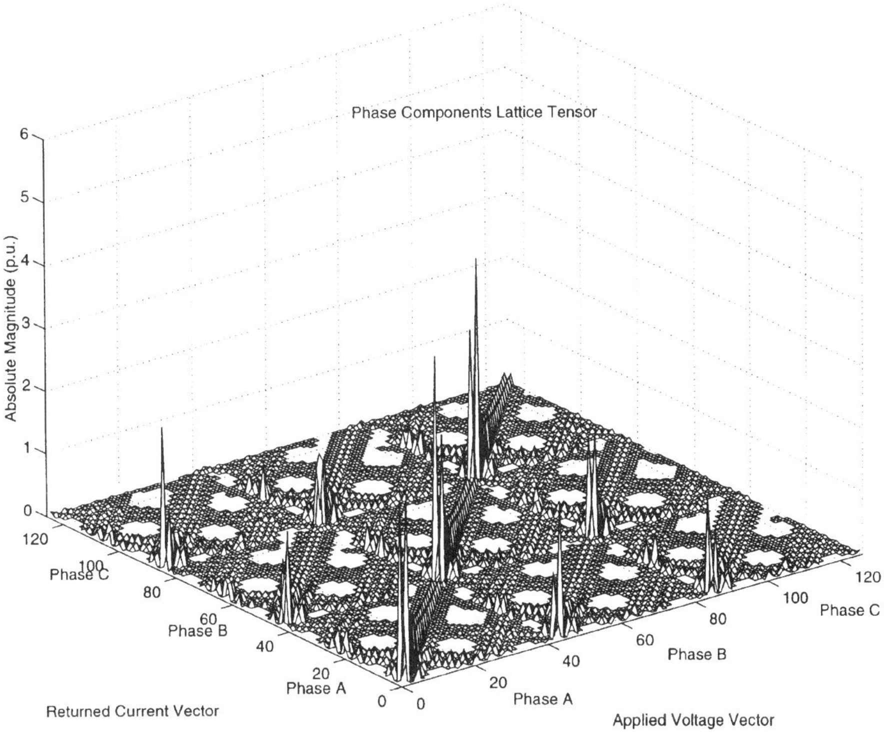

Figure 10.4 Phase components lattice tensor calculated at the solution of Test1

diagonalized by the transform. The lattice tensor has also been calculated around the operating points of Test 2 in Figure 10.6, and Test 3 in Figure 10.7. The presence of 2nd harmonic terminal voltage distortion in Figure 10.6 has increased the number of significant cross harmonic couplings. As expected, unbalance of the star-g/delta transformer in Figure 10.7 has resulted in substantial coupling to the zero sequence, due to sequence transformation by that transformer. As in Test 2, the unequal commutation periods and voltage samples lead to many more significant admittance terms.

Ignoring now the zero sequence, since there is usually little or no coupling to it in the converter, the lattice diagram of Figure 10.3 is verified against the lattice tensor calculated above. Each '+' or 'O' in Figure 10.3 represents four real elements in the lattice tensor, corresponding to variations in the real and imaginary parts of the returned current distortion in response to variations in the real and imaginary parts of the applied voltage distortion (which may be at a different harmonic and sequence). These four elements constitute a tensor cross coupling term that can be represented by a circular locus in the complex admittance plane. By considering sequentially every such cross coupling tensor, Figure 10.8 is the result of applying the following plotting rules to the lattice tensor calculated at the solution of Test 1:

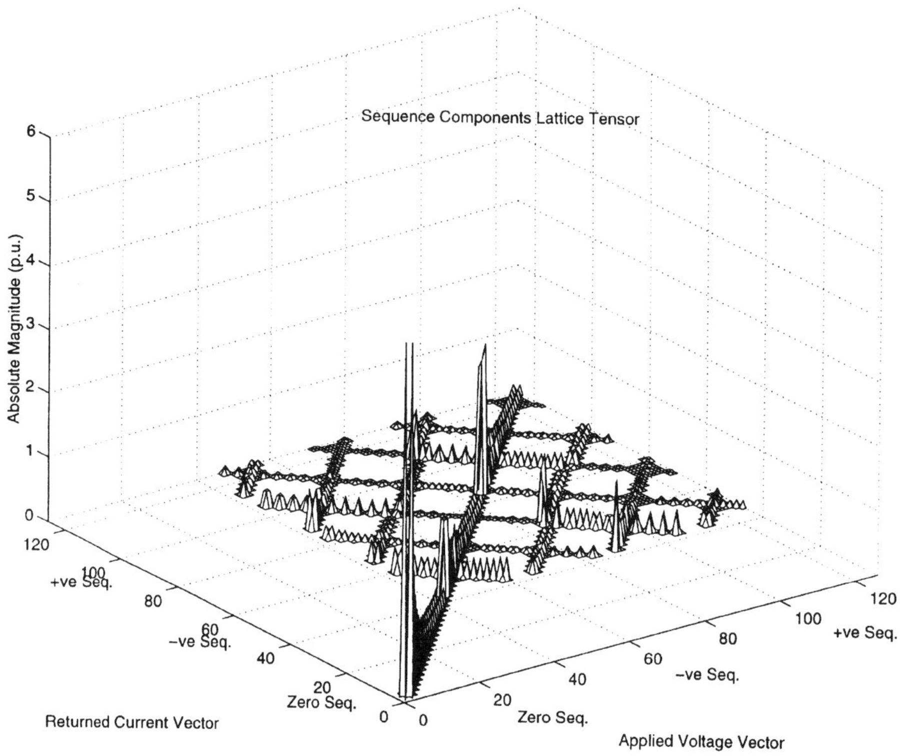

Figure 10.5 Sequence components lattice tensor calculated at the solution of Test 1

1. If the center of the locus is farther than 0.005 p.u. from the origin, plot a '+'.
2. If the radius of the circular locus is greater than 0.005, plot a 'O'.
3. If a '+' has been plotted, and the radius is greater than 1% of the distance of the center of the locus from the origin, plot a 'O' as well.
4. If an 'O' has been plotted, and the distance of the center of the locus from the origin is greater than 1% of the radius of the locus, plot a '+' as well.

The purpose of the plotting rules is to sift out very small admittances, but to still show the nature of admittances that have been retained. The result is a classification of the crosscoupling tensors into direct '+', and phase conjugating 'O' admittances, that essentially recreates the lattice diagram of Figure 10.3 in Figure 10.8. However, if the same process is applied to the lattice tensor calculated at the solution of Test 2, the result is a lattice diagram, Figure 10.9, with a considerable amount of fill-in. A more realistic case is plotted in Figure 10.10, where the ac system source was distorted by 1% negative sequence fundamental, resulting in 0.8% negative sequence

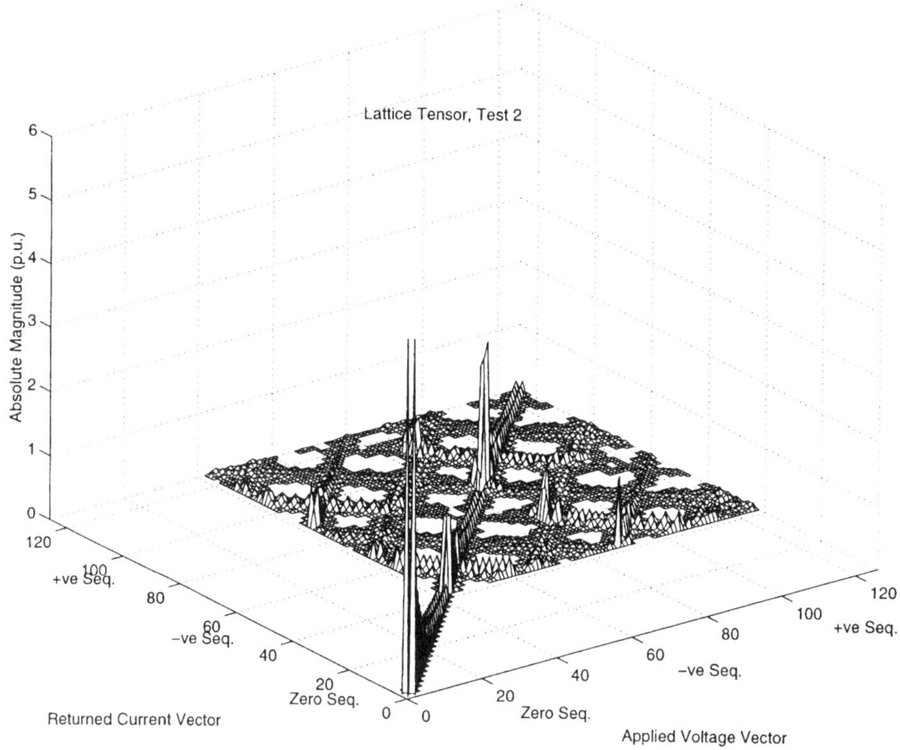

Figure 10.6 Sequence components lattice tensor calculated at the solution of Test 2

fundamental at the converter ac terminal. The conclusion to be drawn is that the lattice diagram is only valid in balanced, undistorted conditions.

The lattice tensor can be used to form a Norton equivalent for the converter. However given that the converter model developed in Chapters 8 and 9 converges rapidly, it is not necessary to approximate the converter by a linearized equivalent in order to solve ac system interactions.

Several authors have proposed the use of ABCD parameter matrices to characterize the harmonic interaction of the ac and dc systems, i.e.

$$\begin{bmatrix} I \\ V_d \end{bmatrix} = \begin{bmatrix} A & B \\ C & D \end{bmatrix} \begin{bmatrix} V \\ I_d \end{bmatrix} \tag{10.14}$$

with the ABDC parameters calculated by means of the perturbation of time domain simulations, or perturbation of a switching function approach. The ABCD matrix (or tensor in positive frequency analysis), can be calculated directly from the Jacobian matrix by modifying the dc mismatch derivatives to be derivatives of the dc voltage, and then performing a Kron reduction to eliminate the switching angle perturbations, as for the lattice tensor above. Once again however, this is of limited

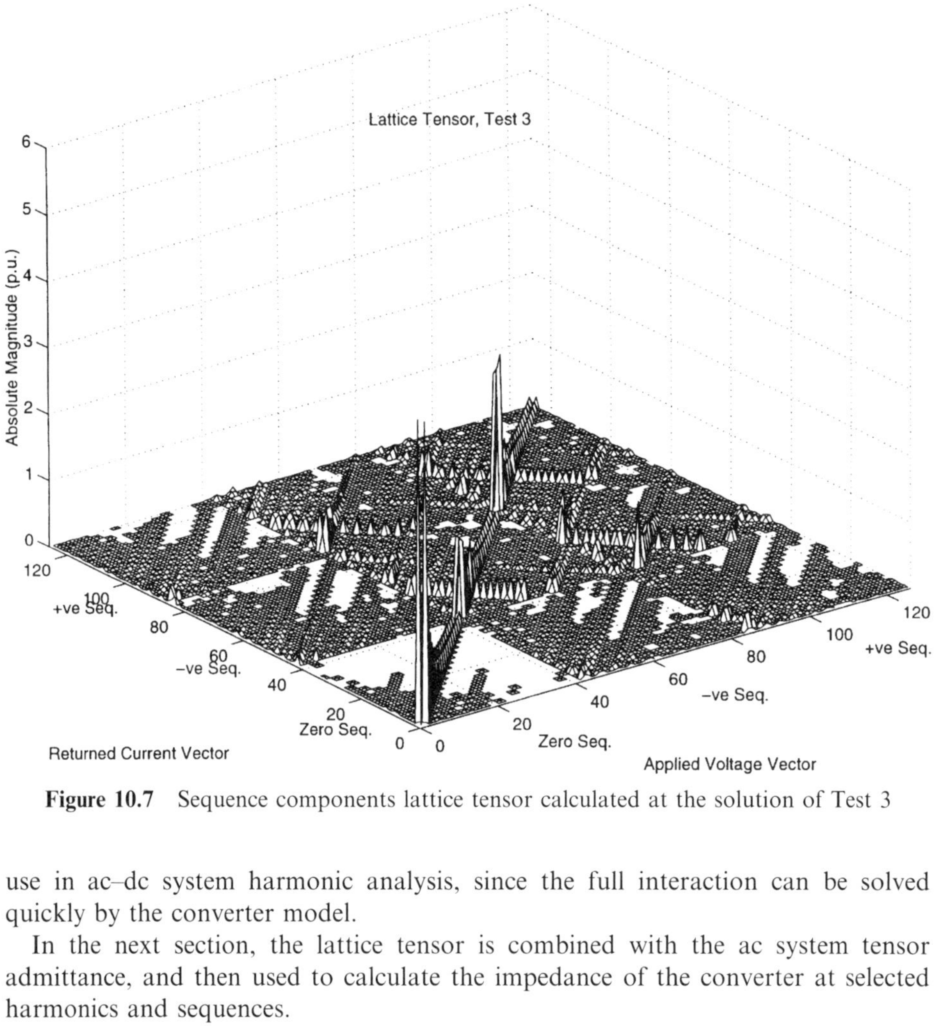

Figure 10.7 Sequence components lattice tensor calculated at the solution of Test 3

use in ac–dc system harmonic analysis, since the full interaction can be solved quickly by the converter model.

In the next section, the lattice tensor is combined with the ac system tensor admittance, and then used to calculate the impedance of the converter at selected harmonics and sequences.

10.2.3 Derivation of the Converter Impedance by Kron Reduction

Calculation of equivalent, or driving point impedances is widely used in system analysis to derive Norton or Thevenin equivalents. For example, the nodal analysis method is used in harmonic penetration and load flow programs, with each node corresponding to a physical node in phase components. When a Kron reduction is applied to a complex admittance matrix, the result is a matrix of reduced size, fully representing the interaction between sources connected to the remaining nodes.

In the harmonic domain, every harmonic of a particular sequence is a node, and provided there are no phase conjugating current injections, a positive frequency complex admittance matrix can be formed to represent the system. Thus an 'off the shelf' harmonic penetration program can analyse a network in which there are cross

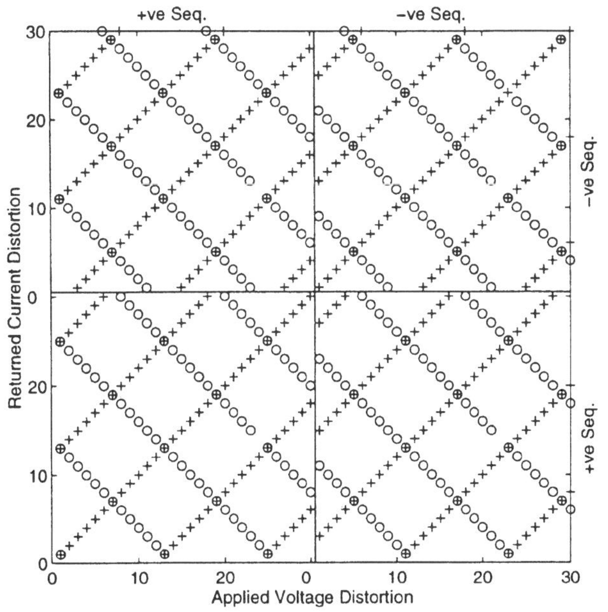

Figure 10.8 Lattice diagram calculated at the solution of Test 1

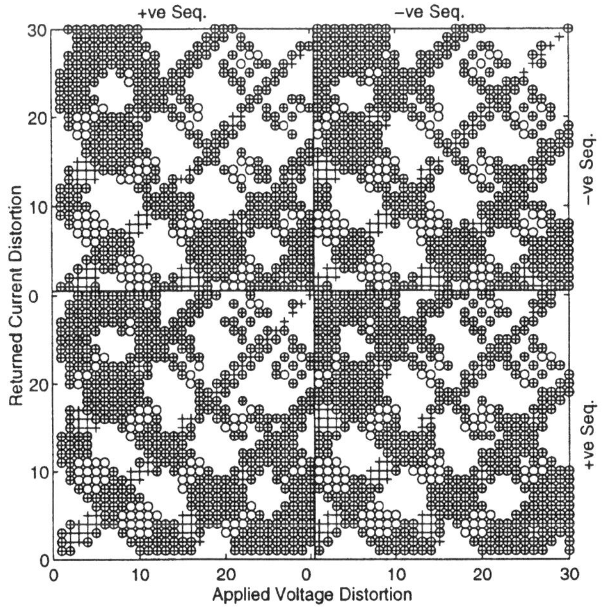

Figure 10.9 Lattice diagram calculated at the solution of Test 2

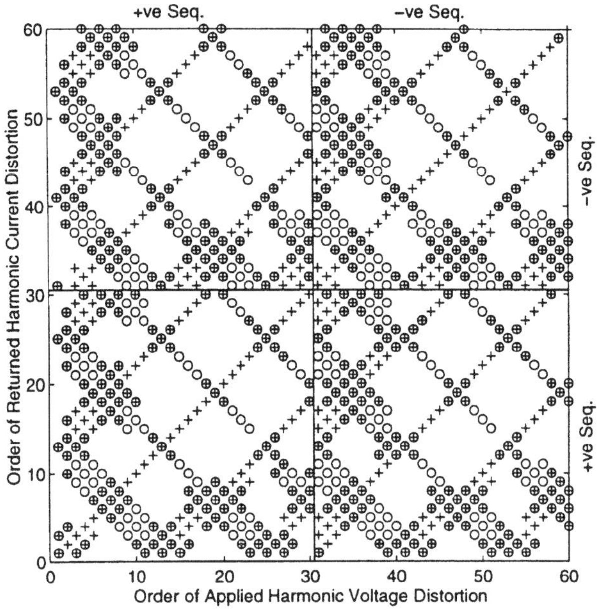

Figure 10.10 Lattice diagram calculated at a solution with 0.85% negative sequence fundamental voltage distortion at the converter terminal

harmonic couplings, simply by duplicating every bus at every harmonic. If there are phase conjugating admittances in the network, it would be necessary to create additional buses representing conjugate current injections and voltages.

Alternatively, creating vectors of the real and imaginary parts of voltage and current harmonic phasors, the system admittance matrix becomes a real valued tensor. The system tensor can include the linearization of any device in the system, including voltage controlled buses, converters, and load flow buses. In general then, the system tensor, when Kron reduced to a single three phase bus, will be a matrix $2 \times 3 \times n_h$ square. In phase components, the system tensor derived from a conventional equivalent complex admittance, will be block diagonal, with each 6×6 block being the three phase tensor corresponding to each three phase harmonic admittance. Assuming that there is no cross coupling between harmonics in the ac system, the connection of the converter to the ac system may be visualized as in Figure 10.11.

In this figure a harmonic current is injected into the three phase harmonic k node. In general currents will flow at every harmonic order into the ac system, causing harmonic voltages that influence the harmonic voltage at the kth node. The ac system has been disconnected from the converter at the kth harmonic, since we want to calculate the converter impedance only. The nodal equation corresponding to this scenario is:

$$\begin{bmatrix} [0] \\ \Delta I_k \end{bmatrix} = \begin{bmatrix} A + \mathrm{diag}(\bar{Y}_{sys}) & B \\ C & D \end{bmatrix} \begin{bmatrix} \Delta V_j \\ \Delta V_k \end{bmatrix} \tag{10.15}$$

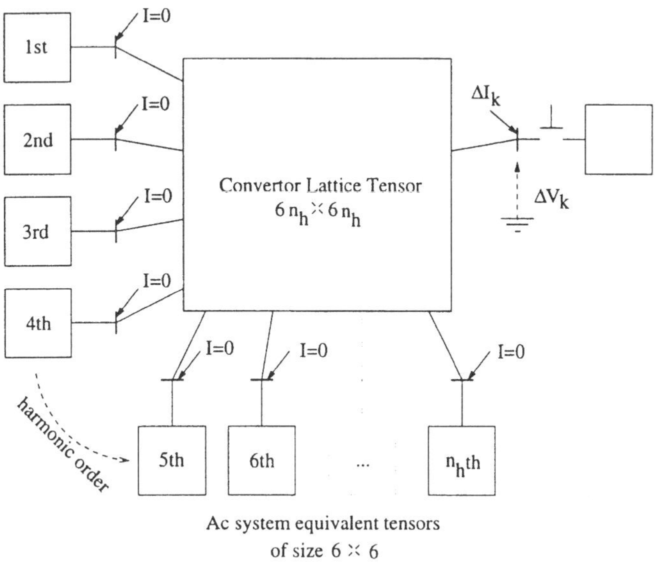

Figure 10.11 Linearized connection of the converter to an ac system

The matrices A, B, C, and D are obtained by moving the six rows and columns of the lattice tensor associated with harmonic order k, to the end of the tensor, and then partitioning off these last six rows and columns. The matrix A is therefore of size $6(n_h - 1) \times 6(n_h - 1)$. A current injected at any of the three phase harmonic nodes marked in Figure 10.11 will see the system and converter tensor admittances in parallel, consequently the system 6×6 equivalent admittance tensors are added as a block diagonal matrix to partition A of the converter lattice tensor. The system admittance at the kth harmonic has been disconnected from the converter, and so is not added to the D partition.

Applying the Kron reduction to Equation 10.15, yields the equivalent 6×6, kth harmonic three phase tensor admittance at the converter ac terminal:

$$\Delta I_k = (D - C(A + \text{diag}(\bar{Y}_{sys}))^{-1}B)\Delta V_k \tag{10.16}$$

$$\stackrel{\text{def}}{=} Y_{ck}\Delta V_k \tag{10.17}$$

This tensor admittance is a linearization of the converter response to applied kth harmonic three phase voltage perturbations, and includes the effect of control, switching angle variation, and coupling to the ac and dc systems at other harmonics. Once again, the large matrix $A + \text{diag}(\bar{Y}_{sys})$ is not actually inverted, but rather bifactorized to LU form, and then columns of $(A + \text{diag}(\bar{Y}_{sys}))^{-1}B$ are obtained by solving the linear system

$$(A + \text{diag}(\bar{Y}_{sys}))y_i = B_i, \tag{10.18}$$

where the y_i are the columns of $(A + \text{diag}(\bar{Y}_{sys}))^{-1}B$. Sequence transforming Y_{ck}, it is possible to apply another Kron reduction to obtain a single sequence equivalent. The situation here is more complicated than in Figure 10.11, as in general the ac system admittance at a particular harmonic will contain intersequence coupling. The situation is illustrated in Figure 10.12, where we wish to find the equivalent positive sequence 7th harmonic admittance of the converter, under the assumption that both the converter and ac system converter transform between sequences at this harmonic. The following equations describe the nodal voltages and currents of Figure 10.12:

$$\begin{bmatrix} \Delta I_7^0 \\ \Delta I_7^- \end{bmatrix} = A \begin{bmatrix} \Delta V_7^0 \\ \Delta V_7^- \end{bmatrix} + B \Delta V_7^+ \tag{10.19}$$

$$-\begin{bmatrix} \Delta I_7^0 \\ \Delta I_7^- \end{bmatrix} = a \begin{bmatrix} \Delta V_7^0 \\ \Delta V_7^- \end{bmatrix} + b \Delta v_7^+ \tag{10.20}$$

$$\Delta I_7^+ = C \begin{bmatrix} \Delta V_7^0 \\ \Delta V_7^- \end{bmatrix} + D \Delta V_7^+ \tag{10.21}$$

$$0 = c \begin{bmatrix} \Delta V_7^0 \\ \Delta V_7^- \end{bmatrix} + d \Delta v_7^+ \tag{10.22}$$

$\Delta V_7^0, \Delta V_7^-$ and Δv_7^+ are readily eliminated to yield

$$\Delta I_7^+ = \left(D - C[A + a - bd^{-1}c]^{-1}B\right)\Delta V_7^+ \tag{10.23}$$

$$\stackrel{\text{def}}{=} Y_{c7}^+ \Delta V_7^+ \tag{10.24}$$

Equation 10.23 indicates that the positive sequence 7th harmonic admittance of the converter is obtained by reducing the ac system admittance to a zero and negative sequence equivalent, adding this to the converter three sequence equivalent, and Kron reducing to the positive sequence. Inverting Y_{c7}^+ and multiplying by the impedance base yields the impedance tensor in ohms. For the rectifier end of the CIGRE benchmark (Appendix VI), at the Test 1 solution,

$$Z_{c7}^+ = \begin{bmatrix} 465.39 & -306.22 \\ 719.96 & 93.47 \end{bmatrix} \Omega.$$

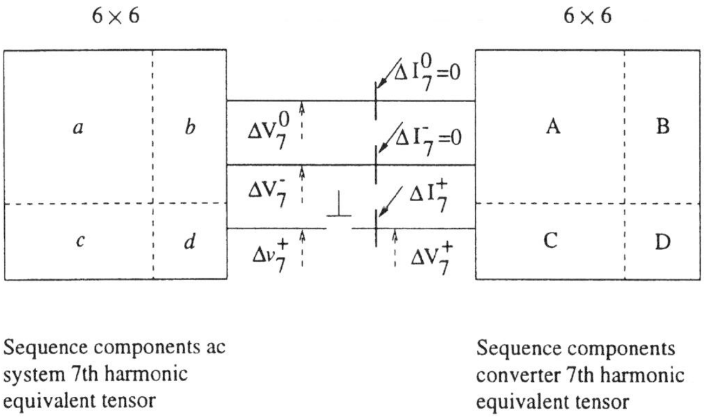

Figure 10.12 Linearized connection of the converter to an ac system at the 7th harmonic

Repeating these calculations for every harmonic from 2 to 37, in positive and negative sequence, the resulting circular impedance loci have been plotted in Figure 10.13. Data points obtained by the perturbation method have also been plotted on this diagram.

The method developed here for calculating the converter impedance is somewhat circuitous, as it is not necessary to first calculate the lattice tensor, and then reduce to a single harmonic equivalent. For example to calculate the 7th harmonic impedance from the Jacobian, it would be possible to retain voltage mismatches for every

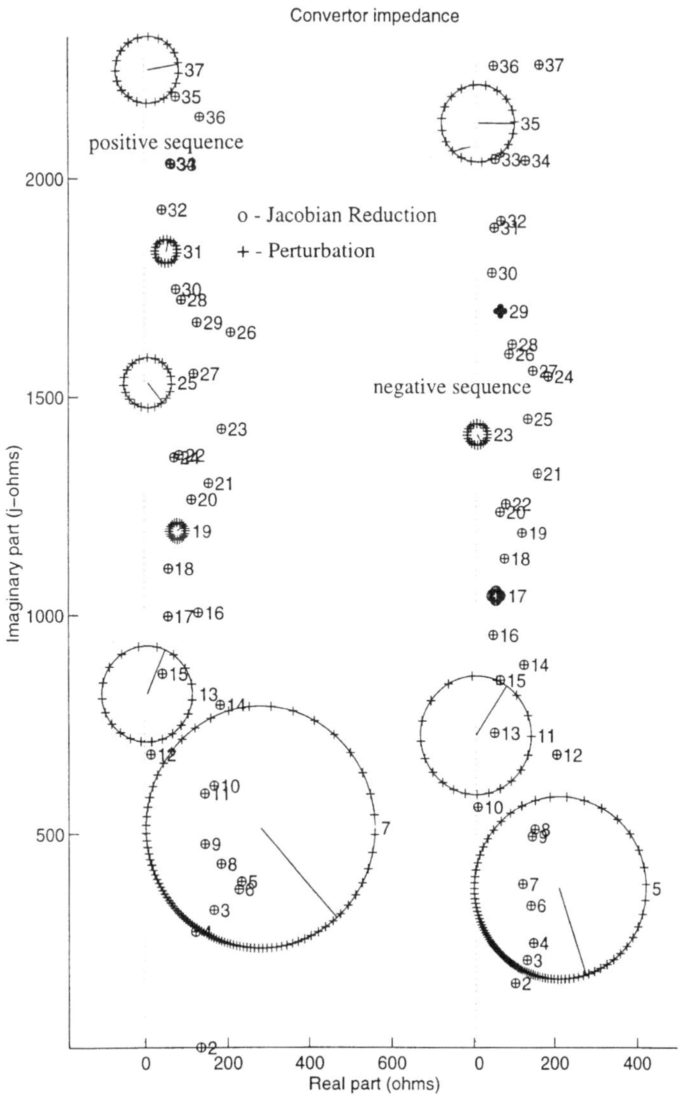

Figure 10.13 Intervalidation of the analytic and perturbation methods of calculating the converter impedance

harmonic except the 7th, and then Kron reduce the Jacobian to the 7th harmonic partition. This method would be faster for calculating a single harmonic impedance, but slower for calculating many. Another method would be to repeatedly Kron reduce to small lattice tensors covering a limited range of frequencies. Typically we are interested in harmonics below the 11th, and could therefore retain voltage mismatches in the Jacobian above the 10th harmonic, reduce to a ten harmonic lattice tensor, and then sequentially reduce this tensor to harmonic equivalents. This method would offer a $(49/9)^3 = 161$-fold improvement in speed for the calculation of each harmonic, after the initial calculation of the lattice tensor, but would give the same results. Nevertheless, by obtaining exact agreement with the perturbation method, the lattice tensor itself, and the nodal analysis utilizing that tensor, have been validated.

10.2.4 Sparse Implementation of the Kron Reduction

The motivation for developing a sparse Kron reduction, is the assumption that small elements in the Jacobian will have little, or no effect on the calculated admittance. Having calculated the full Jacobian matrix at the converter solution, elements larger than a preset tolerance are copied into sparse storage arrays. The Kron reduction to a single harmonic can then be performed using the same sparse routines as were employed in the Newton solution. Since this is a feasibility study only, the method will be implemented for the dc side impedance, as it is single phase.

The starting point is to calculate the full and unmodified Jacobian matrix at the converter solution. Next, elements larger in absolute size than η, are retained and copied into sparse storage. The full Jacobian is stored in a two dimensional array, and $\eta = 0.001$ initially. The sparse Jacobian is represented by a list of elements, row indices, and column indices. In order to perform a Kron reduction, rows and columns associated with harmonic k of the dc current must be identified and removed from the list. This is illustrated in Figure 10.14, where only parts ‘a’, ‘b’, ‘c’

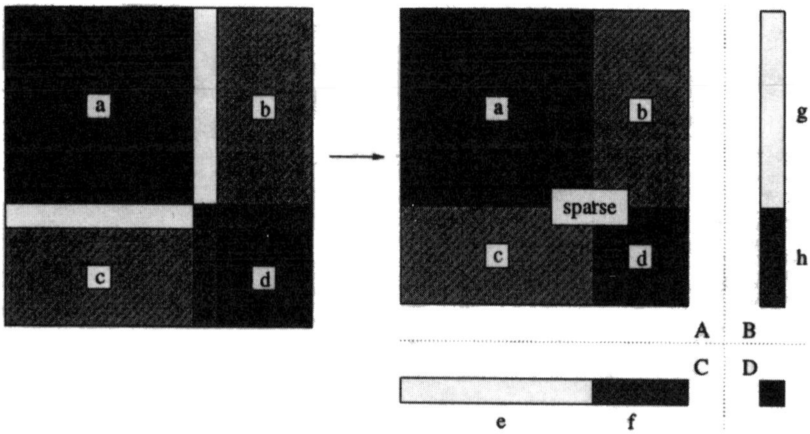

Figure 10.14 Elimination of rows and columns from the sparse Jacobian associated with dc harmonic k

and 'd' of the Jacobian are retained and stored in a new sparse matrix 'A'. This process requires that the row indices of elements in partitions 'c' and 'd' be decremented by two, and similarly for the column indices of elements in partitions 'b' and 'd'. For a dc side harmonic of order k, the rows and column indices to be removed are $6n_h + 2k - 1$, and $6n_h + 2k$. The matrices 'B', 'C', and 'D' are copied directly from the full Jacobian, and so are themselves full.

The Kron reduction consists in calculating $D - CA^{-1}B$, with A sparse in this case. For a dc side harmonic impedance, B contains only two columns, and letting b_i represent sequentially the first and second columns of B, the solution to the linear equation

$$Ax_i = b_i, \tag{10.25}$$

will be the corresponding columns of $A^{-1}B$. This sparse linear equation is readily solved by first calculating the sparse LU decomposition of A, and then solving with the two right hand sides b_i. The remaining calculation in the sparse Kron reduction is a straightforward matrix multiplication. The Kron reduction yields the matrix $D - CA^{-1}B$, which is a reduction of the Jacobian to a 2×2 matrix consisting of the partial derivatives of the dc current mismatches of order k with respect to dc ripple of order k. The dc mismatch is the calculated dc voltage, applied to the dc system to yield the dc current, and then subtracted from the estimated dc current. The dc side impedance is therefore:

$$\bar{Z}_{dk} = -[D - CA^{-1}B - I]\bar{Y}_{dk}^{-1} \tag{10.26}$$

where the dc side admittance, Y_{dk}, has been written as a tensor. The multiplication by $\bar{Y}_{dk}^{-1}$, and subtraction of I, reverse the Jacobian calculations whereby the dc mismatch partial derivatives are written in terms of the calculated dc voltage partial derivatives.

The dc side impedance calculated by the sparse method may contain errors due to the large number of small Jacobian elements that are neglected. The effect of the small terms on the impedance can be determined, without sacrificing sparsity, by means of the iterative refinement method. When the linear system described by equation 10.25 is solved, the error due to the Jacobian terms less than η is determined by multiplying the solution x_i by the full matrix J_A:

$$r = J_A x_i - b_i \tag{10.27}$$

where r is the residual, and J_A is the full partition of J corresponding to A. Since A has already been bifactorized, it is easy to calculate a correction vector, Δx_i, by solving a sparse linear system with the residual on the RHS:

$$A\Delta x_i = r, \tag{10.28}$$

so that $x_i' = x_i - \Delta x_i$ will be a much better approximation to the solution of $J_A x_i = b_i$. This process could, if desired, be repeated for several iterations, however a single iteration of the refinement method is more than adequate. The main overhead associated with the refinement is the matrix multiplication of a vector, $J_A x_i$, which

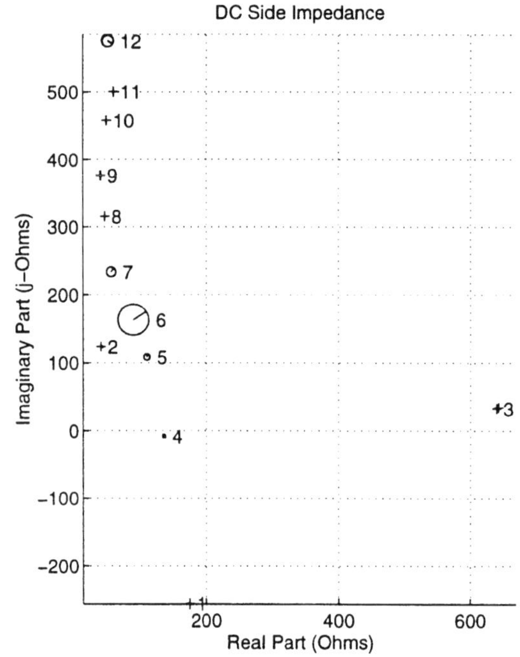

Figure 10.15 Calculated dc side impedances of the CIGRE rectifier using the sparse Kron reduction technique, with and without refinement

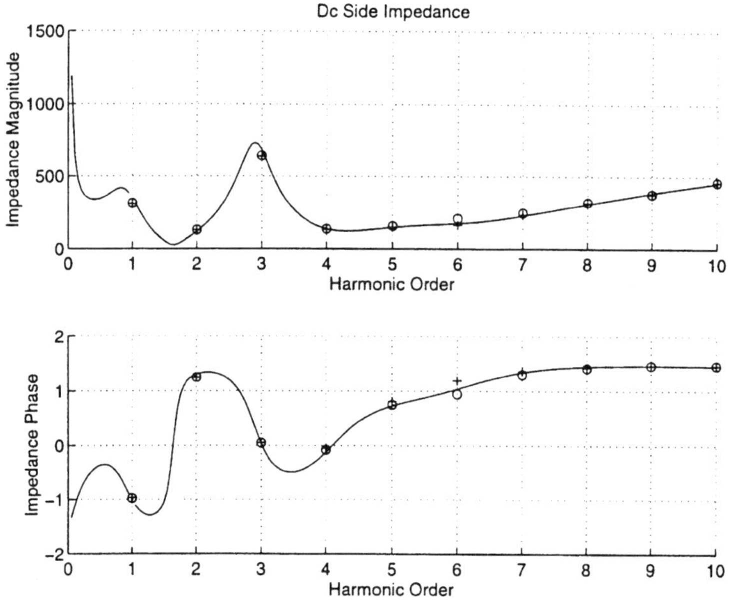

Figure 10.16 Comparison of harmonic and frequency domain solutions to the CIGRE rectifier dc side impedance. '+' and '○' mark the range of the complex locus at each harmonic

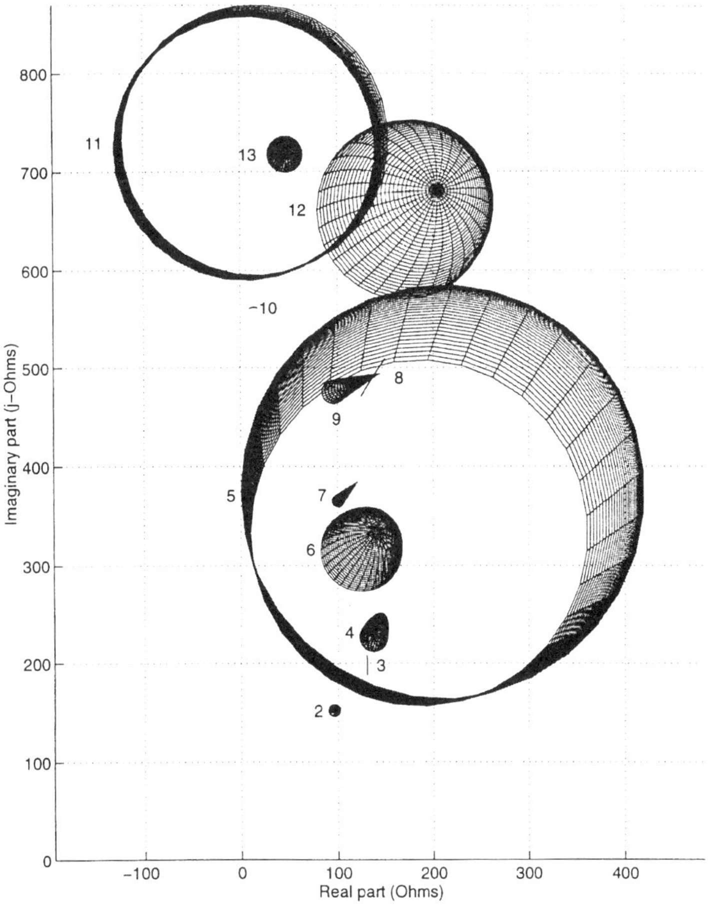

Figure 10.17 Variation in the negative sequence impedance of the CIGRE rectifier as the third harmonic positive sequence terminal voltage distortion is increased from 0 to 0.12 p.u

requires n^2 flops. The solution of the linear system is less than n^2, since A has already been reduced.

The calculated dc side impedances for the CIGRE rectifier are plotted as loci in Figure 10.15 for harmonics 1 to 12. Figure 10.15 indicates that in this case the neglected Jacobian terms have had virtually no effect on the dc side impedance, as the loci have been plotted twice, first with no refinement, and then with one iteration of the refinement method.

The sparse calculation of the dc side impedance is validated against a direct frequency domain calculation in Figure 10.16 for harmonics up to the 10th. The frequency domain calculation of converter impedances is described in detail in Chapter 5, and was there validated against time domain calculations of the converter dc side impedance using the perturbation method.

10.3 Variation of the Converter Impedance

As indicated by Figures 10.5 to 10.7, the converter lattice tensor is affected by ac terminal distortions, consequently the equivalent impedance of the converter at a particular harmonic and sequence is likely to vary also. This effect is evident in Figure 10.17, where the negative sequence impedances at low order harmonics have been plotted in the presence of third harmonic positive sequence terminal voltage distortion. The terminal voltage distortion was obtained by adding a third harmonic voltage component to the ac system Thevenin equivalent, and varying the magnitude from 0 to 0.2 p.u., which resulted in a maximum of 0.12 p.u. distortion at the converter terminal. Figure 10.18 is a similar plot of the positive sequence impedances, with a negative sequence fifth harmonic terminal voltage distortion of up to 0.1 p.u. Clearly, the main effect of an existing distortion is to introduce phase dependence into most of the converter impedances. If the effect of harmonic

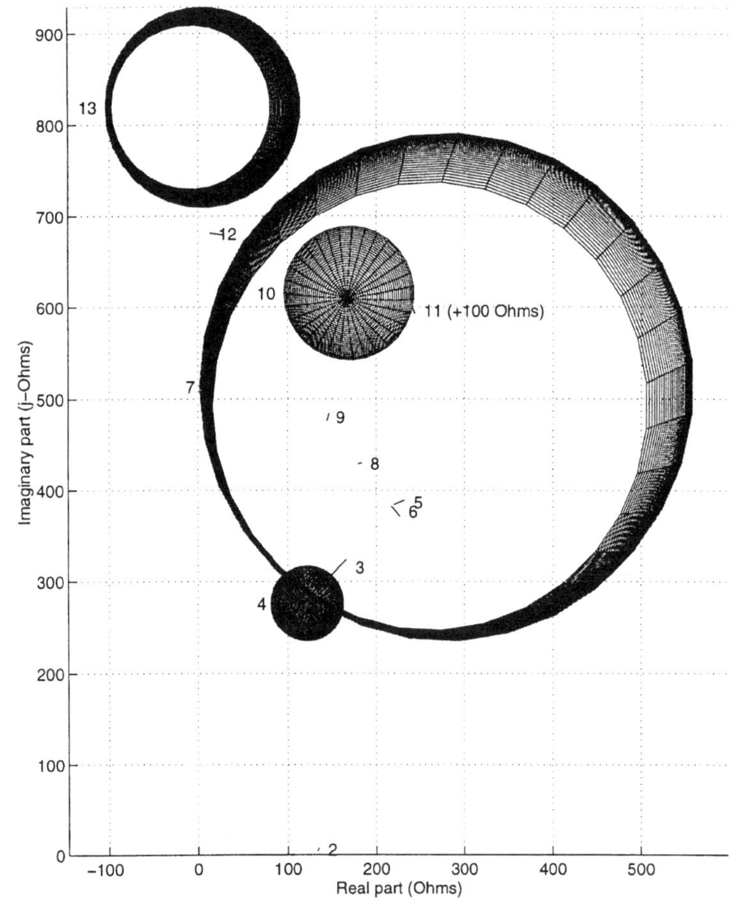

Figure 10.18 Variation in the positive sequence impedance of the CIGRE rectifier as the fifth harmonic negative sequence terminal voltage distortion is increased from 0 to 0.12 p.u

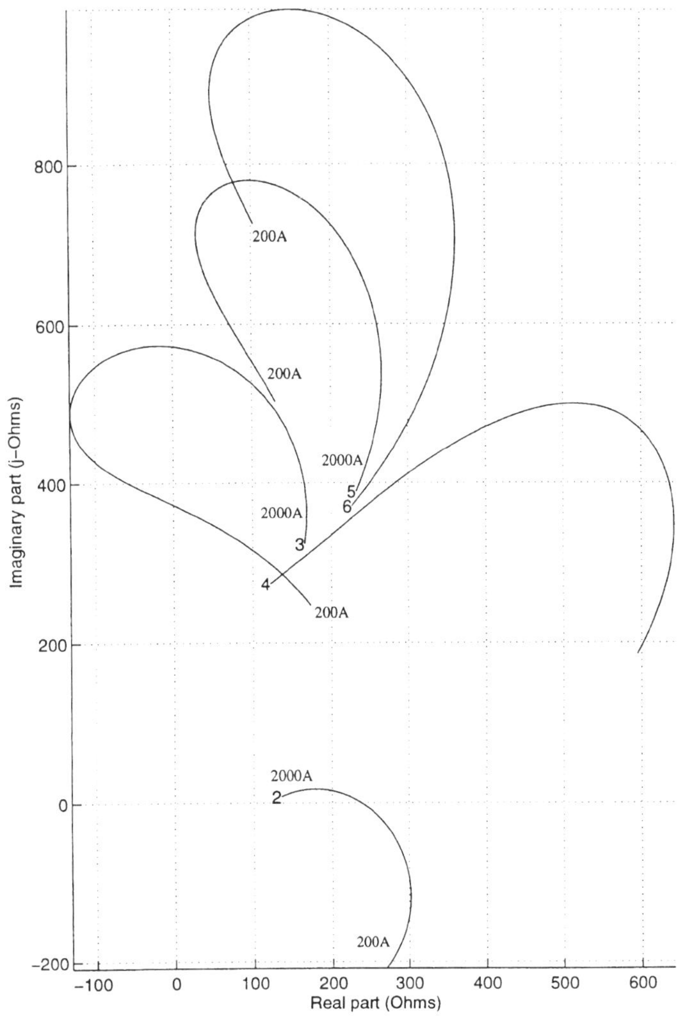

Figure 10.19 Variation in the positive sequence impedance of the CIGRE rectifier as the current order is decreased from 2000A to 200A. Harmonics 2 to 6

distortion at the converter terminal is ignored, the calculated impedance at other harmonics will be incorrect, but not greatly so.

The converter impedance is affected to a much larger extent by the operating point of the converter. In a real dc link, as the current order is reduced, the firing order is held approximately constant by tap change control, and reactive power compensation banks are switched out. Both of these activities will alter the ac side impedance of the converter.

At lower dc current orders, the commutation periods will be shorter, and an applied voltage distortion will affect the commutation less. The dc current modulated back onto the ac side is also smaller. It can therefore be expected that the converter impedance will increase as the current order is reduced. In fact, the

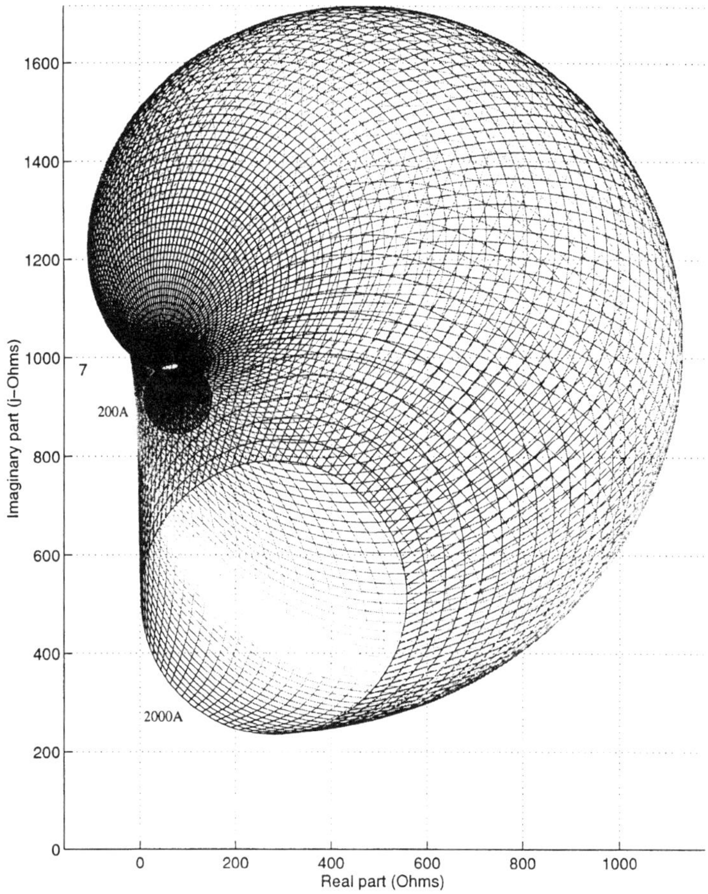

Figure 10.20 Variation in the positive sequence impedance of the CIGRE rectifier as the current order is decreased from 2000 A to 200 A. Harmonic 7

analysis of Chapter 5 predicts a sinc function envelope for the converter impedance magnitude as a function of commutation angle. Figures 10.19 to 10.22 plot the variation in the converter positive sequence impedance loci as a function of the current order. The current order was reduced in one hundred steps from 2000 A to 200 A, which resulted in a variation of firing angle from 14° to 50°, and commutation duration from 23° to 0.7°. The large variation in firing angle was due to a combination of factors; there was no tap change control, the inverter was represented by a constant dc voltage source, and since there was no reduction in reactive power compensation, the fundamental component of the terminal voltage increased. Clearly, a comprehensive study of the converter impedance variation would require additional steady state control equations to be integrated with the Jacobian matrix. Nevertheless, the procedure for calculating the converter impedance would remain the same, and the results obtained demonstrate the converter impedance is a strong function of the operating state.

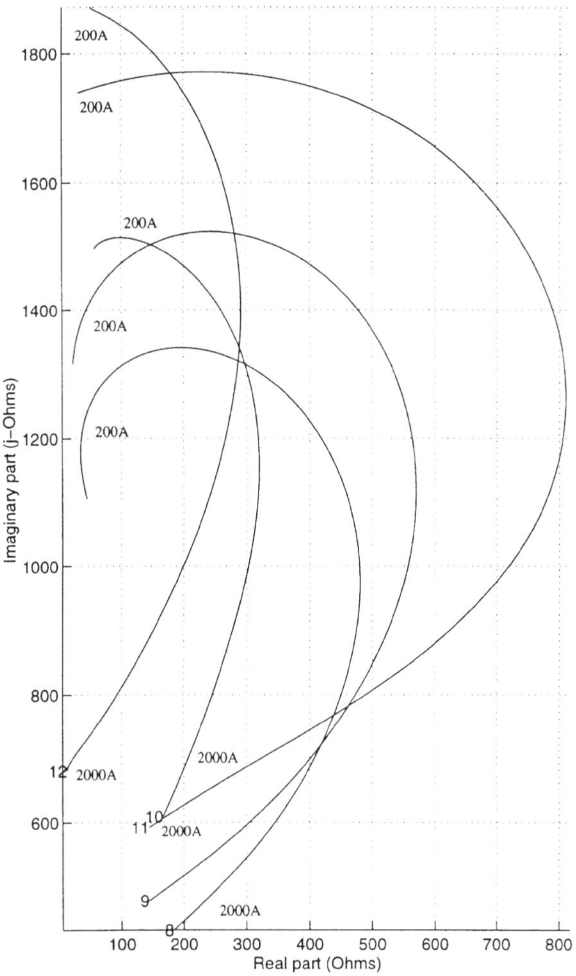

Figure 10.21 Variation in the positive sequence impedance of the CIGRE rectifier as the current order is decreased from 2000 A to 200 A. Harmonics 8 to 12

10.4 Summary

The representation of phase dependent admittances by second rank tensors has been described, and interpreted geometrically as a circular admittance locus on the complex plane. The Jacobian of the converter model is used to directly calculate tensor admittances for the converter on the ac side. A nodal analysis of the cross coupling converter lattice tensor attached to an ac system is applied to obtain equivalent impedances at the converter bus. The impedances thus obtained are verified against impedances calculated by a perturbation study at the converter bus. The dc side impedances have also been obtained and verified against those obtained from a frequency domain converter model.

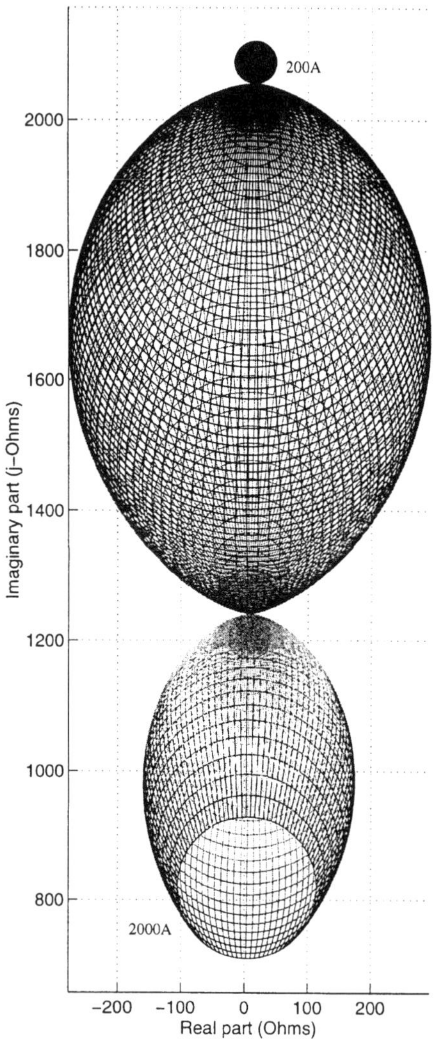

Figure 10.22 Variation in the positive sequence impedance of the CIGRE rectifier as the current order is decreased from 2000 A to 200 A. Harmonic 13

The characteristic harmonic impedances are strongly phase dependent, however none of the converter impedances are magnitude dependent, indicating that the converter is quite linear in the harmonic domain. The converter impedance is found to be a strong function of the converter operating point.

The utility of this method will increase as the generality of the Newton solution is expanded. For example, a unified Newton solution of several ac systems connected by dc links, with load flow and transformer saturation represented, would enable the calculation of any desired linear relationship from the Jacobian.

10.5 References

1. Bahrman, MP, *et al.*, (1986). Dc system resonance analysis, *IEEE Transactions on Power Delivery*, **PWRD-2**(1), 156–164.
2. Dickmander, DL, *et al.*, (1994). Ac/dc harmonic interactions in the presence of gic for the Quebec-New England phase II hvdc transmission, *IEEE Transactions on Power Delivery*, **9**(1), 68–78.
3. Semlyen, A, Acha, E and Arrillaga, J, (1988). Newton-type algorithms for the harmonic phasor analysis of non-linear power circuits in periodical steady state with special reference to magnetic non-linearities, *IEEE Transactions on Power Delivery*, **3**(3), 1090–1097.
4. Persson, EV, (1970). Calculation of transfer functions in grid controlled convertor systems, *IEE Proceedings*, **177**(5), 989–997.
5. Wood, AR, Smith, BC and Arrillaga, J, (1995). The harmonic impedance of an hvdc converter, *6th European Conference on Power Electronics and Applications (EPE 95), Seville*, 1039–44.
6. Larson, EV, Baker, DH and McIver, JC, (1989). Low order harmonic interaction on ac/dc systems, *IEEE Transactions on Power Delivery*, **4**(1), 493–501.
7. Wood, AR and Arrillaga, J, (1995). Composite resonance; a circuit approach to the waveform distortion dynamics of an hvdc converter, *IEEE Transactions on Power Delivery*, **10**(4), 1882–1888.

Appendix I

EFFICIENT DERIVATION OF IMPEDANCE LOCI

I.1 Adaptive Sampling Scheme

For a given system configuration and operating conditions, the impedance locus, described in Section 4.3.5, provides complete information of the system harmonic impedances, both in phase and magnitude.

The required information at the harmonic frequencies can be derived from tests or from the program HARMZ. The discrete information obtained is then subjected to a cubic spline interpolation to the real and imaginary parts separately and then plotted together to produce smooth curves.

As an example, the New Zealand south island network operating at 100% load, and seen from the Benmore converter station, is used as a test system. In order to plot the smooth curve of Figure I.1, the system impedances were calculated at frequency intervals of 1 Hz and this curve is, therefore, taken to represent the modelled impedances to high accuracy throughout the interval $50 \geqslant f \geqslant 1250$ Hz.

It should be clear from inspection of the locus that it is wasteful to use such a small frequency increment between impedance evaluations over the whole range. The locus can be summarized much more efficiently by using small frequency increments where there are tight loops and much larger increments elsewhere. Because the loops correspond to resonances it is important to be able to display them accurately and the danger in using a larger frequency increment is that some loops may be missed completely. Of course, the real difficulty is that the positions of the loops are not known until after the curve has been plotted.

I.2 Winding Angle Criterion

What is required is an adaptive scheme, capable of automatically adjusting the frequency increment according to the local geometry of the curve. When the curvature is large the impedance should be sampled at small frequency intervals and when the curvature is small it should be sufficient to sample at intervals which are large, provided that the interval is not so large that complete loops are *smoothed*

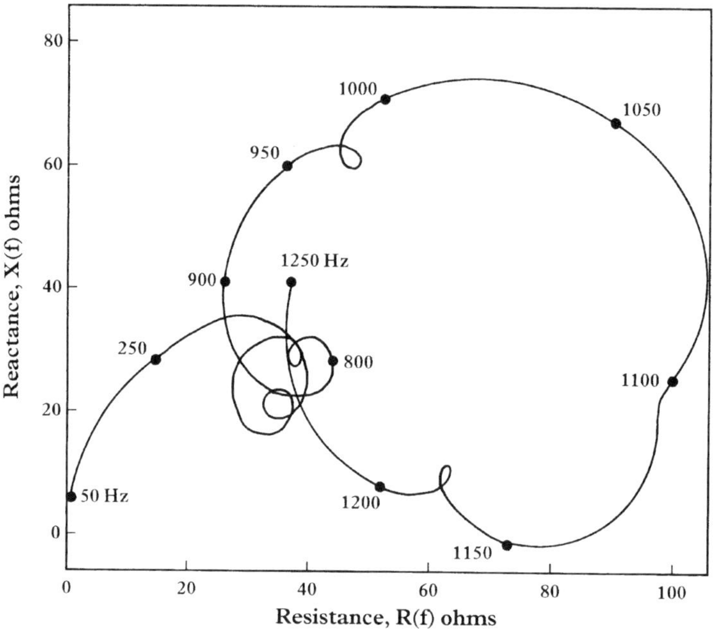

Figure I.1 Impedance of a 220 kV network at 100% (Benmore, South Island, New Zealand)

away. Ideally then, it might be argued that the impedances should be calculated at frequencies $\{f_k, k = 1, 2, \ldots\}$ such that $\psi_{k+1} - \psi_k = \alpha$, where $\psi_k = \psi(f_k)$ is the angle between the tangent to the locus at the frequency f_k and the R-axis (or some other fixed line) and α represents a constant angle. This approach would guarantee that small loops are given as much attention as large loops. For example, if the true locus were a circle and if α were fixed at 45°, then the locus would be summarized by eight equally spaced points around the circumference, *irrespective* of the radius of the circle.

The change in tangent angle, $\Delta\psi_k = \psi_{k+1} - \psi_k$, as the frequency changes from f_k to f_{k+1} is referred to as the winding angle. If the estimated $\Delta\psi_k$ is too large the impedance should be evaluated again at some intermediate point, say $f = \frac{1}{2}(f_k + f_{k+1})$.

The diagram in Figure I.2 represents a magnified view of a small portion of the impedance locus. Suppose that the impedance $Z(f) = R(f) + jX(f)$ has already been calculated at each of two frequencies f_k, f_{k+1}. Then the angle through which the tangent to the curve winds as the locus is traversed from $Z_k = Z(f)$ to $Z_{k+1} = Z(f_{k+1})$ can be estimated by evaluating the impedance $Z(\tau)$ at an intermediate point τ in the interval $[f_k, f_{k+1}]$. If θ denotes the angle between the chords $Z(\tau) - Z_k$ and $Z_{k+1} - Z(\tau)$ and if the locus has constant curvature over the interval $f_k \leqslant f \leqslant f_{k+1}$ (i.e. it is a circular arc), then some simple geometry shows that the tangent winds through an angle $\Delta\psi = 2\theta$. Thus, it can be hoped that the winding angle is estimated well by 2θ, if the curvature does not vary too much over

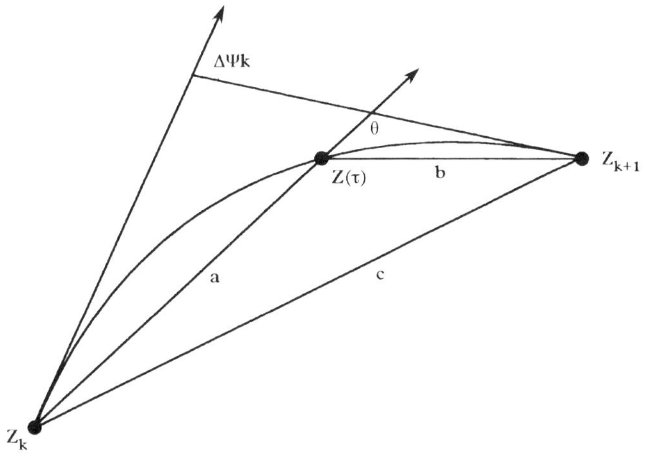

Figure I.2 Estimating the winding angle, $\Delta\psi_k$, using the chord angle θ

the interval in question. Moreover, for a smooth locus (at least differentiable) the mean value theorem can be invoked twice to show that the winding angle must be at least θ on an interval that contains τ and which is a strict subset of the interval $[f_k, f_{k+1}]$. In any event, if the angle θ is not small then clearly more points are required if the locus is to be summarized well.

The diagram in Figure I.2 is appropriate only if $\theta < 90^\circ$. This situation is easily detected by calculating the chord lengths a, b and c. If $a^2 + b^2 < c^2$ then the angle θ is acute and $\Delta\psi_k$ can be estimated by 2θ; the smaller this angle the more accurate is the approximation. If $a^2 + b^2 \geqslant c^2$ then it is not even worth calculating the angle θ, clearly the points Z_k, $Z(\tau)$ and Z_{k+1} are spaced too far apart. In this case, each of the sub-intervals $[f_k, \tau]$ and $[\tau, f_{k+1}]$ can be processed separately by evaluating the impedance again at an intermediate frequency and estimating the winding angles on each new sub-interval.

The calculation of the chord lengths is not wasted when $\theta < 90^\circ$, because the lengths a, b and c can be used to get estimates of the winding angles $\Delta\psi_{k+1}$ and $\Delta\psi_{k+2}$ on each of the sub-intervals. Again, some simple geometry provides the estimates:

$$\Delta\psi_{k+1} = \arcsin\left(\frac{a}{c}\sin\theta\right), \qquad \Delta\psi_{k+2} = \arcsin\left(\frac{b}{c}\sin\theta\right),$$

and it is straightforward to verify that these estimates satisfy $\Delta\psi_{k+1} + \Delta\psi_{k+2} = 2\theta$, as expected. In fact, the angle θ can also be calculated from the chord lengths since it satisfies the equation.

$$\cos|\theta| = \frac{c^2 - a^2 - b^2}{2ab}$$

but it is preferable to calculate θ in terms of the arctan function in order to preserve the orientation of the winding angle. The advantage in obtaining these estimates is that they can be used to get error estimates whenever the interval is subdivided and the winding angle subsequently recalculated. If two successive estimates agree within a pre-specified tolerance and if the best estimate indicates that the winding angle is sufficiently small (again to within a pre-specified tolerance), then there should be no need to subdivide further. This is the basis of the adaptive scheme.

The impedance is evaluated initially at the two end points of the interval in question before the first call. The length of the chord between these two initial points is also required together with an estimate of the winding angle over the entire length of the locus. Any value exceeding 180° will suffice initially since this value is re-estimated by the routine. A tolerance parameter is then required to be set before entry to the subroutine. Each subsequent call of the subroutine causes the impedance to be evaluated at the midpoint of the frequency interval. This effectively produces two sub-intervals. The chord lengths corresponding to each of these intervals are calculated and if the angle between them is not acute or if this angle is positive (corresponding to an anti-clockwise rotation), then the subroutine is called again immediately on each of the sub-intervals. The subroutine is also called again if either the initial estimate of the winding angle (passed in the parameter list) or the new estimate of the winding angle calculated y the subroutine exceeds the given tolerance. To prevent a possibly infinite recursion when the true locus is not differentiable the subroutine is not called if the range of a frequency interval is too small. Moreover, as an extra safeguard, the subroutine is also called automatically if the range is considered to be too large.

In practice, the user provides values for minimum and maximum frequency increments and the algorithm effectively varies the actual increment between these bounds in a way that ultimately causes the winding angle between adjacent points to be approximately constant.

The algorithm was applied to the power system locus of Figure I.1, and the results are displayed in Figure I.3. It can be seen that the adaptive scheme has worked quite well on this curve. In the region near 250 Hz the curve has been sampled at the minimum increment because the orientation is anti-clockwise and the algorithm tries hard to discover an extra loop that is not there. The algorithm has also started to subdivide more finely in the region near 1100 Hz but the given tolerances have prevented the algorithm from trying too hard here. A parametric cubic spline fitted through the points is displayed in Figure I.4 which cannot be distinguished by eye from the true curve of Figure I.1; moreover, the number of samples used is an order of magnitude lower than in the original plot.

Although the adaptive scheme provides a perfect match, the number of sample points required is still unnecessarily high. Further savings can be made by relaxing the equal curvature criterion to reduce the number of samples in regions of high curvature. This can be achieved by fixing the winding angle to suit the region of large frequencies and increasing its value in inverse proportion to the cord between successive samples.

As the discrete harmonic impedances are always required, these points provide the initialisation of the adaptive scheme. The magnitudes of the outer radius and largest cord between successive harmonics can be used to decide the value of the winding angle to be used as a reference. The magnitude of the cords between samples can then be used to decide the winding angle.

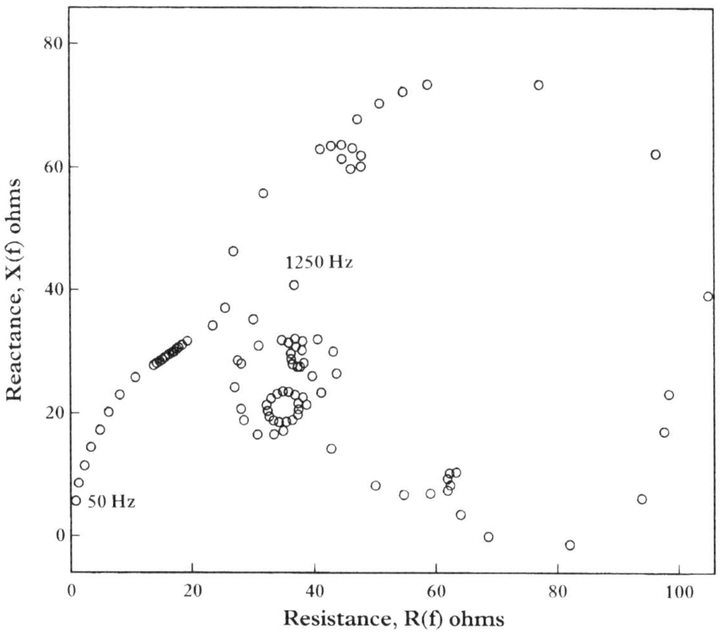

Figure I.3 113 adaptively chosen points

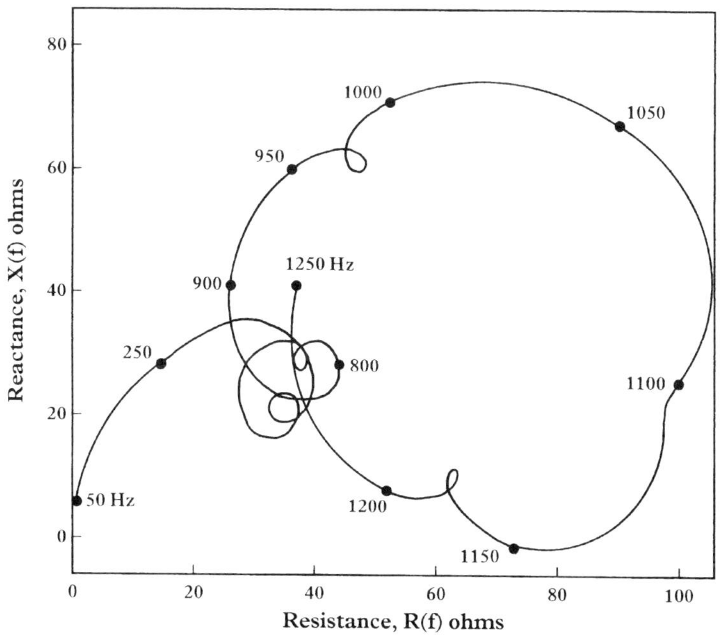

Figure I.4 A parametric cubic spline on 113 points

A more rigorous approach, still untried in practical systems, is the impedance contour concept [1] which combines the main features of the adaptive sampling and discrete impedance scatter techniques. The following steps are required:

1 For each selected configuration two discrete samples are initially derived for each harmonic of interest, one at the maximum expected frequency (i.e. $f_{max} . h$) and the other at the minimum frequency ($f_{min} . h$).

2 The interval between $f_{max} . h$ and $f_{min} . h$ is then subjected to the modified adaptive sampling criterion and a cubic spline curve fitted to the resulting points (taking advantage of the previously derived harmonic contour); it is expected that in most cases an extra intermediate point will provide sufficient information.

I.3 Reference

1. Dominguez, M, *et al.*, (1994). An adaptive scheme for the derivation of harmonic impedance contours, *IEEE Transactions on Power Delivery*, **9**(2), 879–886.

Appendix II

PULSE POSITION MODULATION ANALYSIS

II.1 The PPM Spectrum

In 1966 Schwarz [1] published an analysis of the frequency spectrum of a Pulse Position Modulated (PPM) waveform. He related the spectrum directly to the parameters in a purpose built PPM modulator. In this section the spectrum is rewritten in a form suitable for use in HVdc converter control analysis. In Schwarz's analysis the PPM waveform is generated by adding the modulating signal to a steady sawtooth sweep voltage. When this sweep is rising, and crosses a preset threshold, a pulse is generated. The unmodulated sweep waveform, threshold, and resulting pulses are shown in Figure II.1.

Letting the sweep be as shown in Figure II.1, and the modulating signal be $Q\cos(qt)$, and allowing each pulse to be a delta function $\delta(t)$ of area 1, Schwarz calculates the resulting frequency spectrum

$$E(t) = \frac{p}{2\pi} - \frac{qQ}{2P}\sin(qt) + \sum_{m=1}^{\infty}\sum_{n=-\infty}^{\infty}\frac{mp+nq}{m\pi}J_n\left(\frac{m\pi Q}{P}\right)\cos\left[(mp+nq)t+\frac{n\pi}{2}\right] \tag{II.1}$$

where p is the angular sweep frequency.

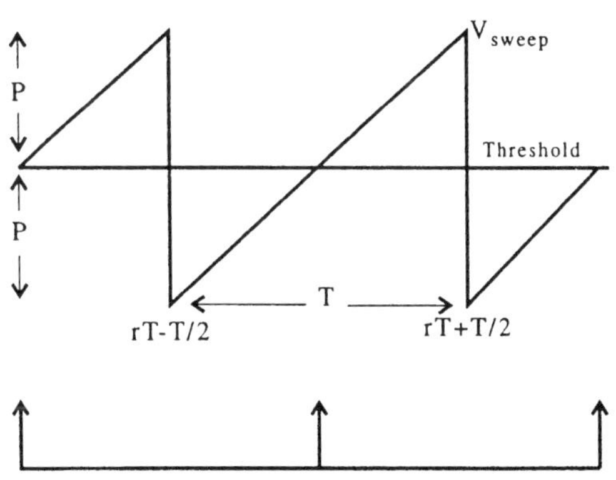

Figure II.1 PPM generation waveforms

Redefining the PPM in terms of the pulse angular position, in radians, as $b\cos(k\omega_0 t + \delta_k)$, results in the substitution

$$Q = \frac{-Pb}{\pi} \tag{II.2}$$

and

$$q = k\omega_0 \tag{II.3}$$

Letting p be the fundamental frequency ω_0, the frequency spectrum becomes

$$\begin{aligned} E(t) = {} & \frac{\omega_0}{2\pi} + \frac{bk\omega_0}{2\pi}\sin(k\omega_0 t + \delta_k) \\ & + \sum_{m=1}^{\infty}\sum_{n=-\infty}^{\infty} \frac{(m+nk)\omega_0}{\pi}\frac{J_n(mb)}{m}\cos\left[(m+nk)\omega_0 t + n\delta_k - \frac{n\pi}{2}\right] \end{aligned} \tag{II.4}$$

This statement of the PPM spectrum forms the basis of the analysis of the non-ideal converter transfer function spectra.

II.2 Contribution of Commutation Duration to DC Voltage

The waveform of the transfer function that, when added to the ideal 6 pulse converter transfer function, results in the non-ideal transfer function is shown in Figure II.2. It can be defined as the sum of four different waveforms, each of which has one pulse per fundamental cycle.

Each of the four waveforms is exactly a PPM waveform as described by Schwarz, but with the shape of each pulse being rectangular rather than a delta function. Schwarz's PPM spectrum is summed for the four waveforms, and then the Fourier transform of the rectangular pulse is calculated and applied as a frequency response function to the solution.

If the four waveforms are called F_i, where $i = 1, 2, 3, 4$, then each has a different pulse height, and each a different time reference. The relevant values are $h_1 = +1$,

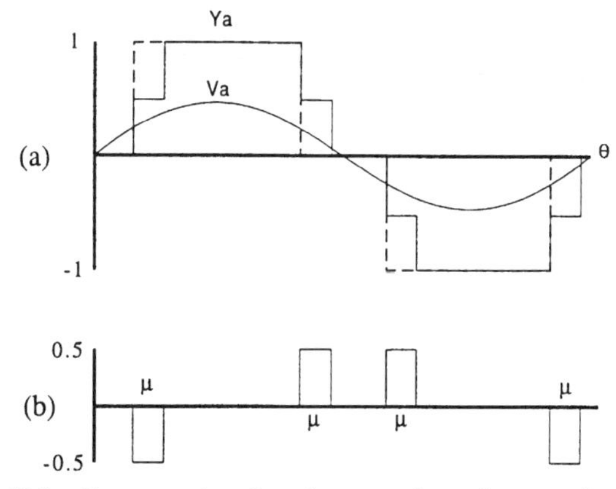

Figure II.2 Commutation function waveform for transfer to dc voltage

$h_2 = +1$, $h_3 = -1$, $h_4 = -1$, and $T_1 = (\pi/3 + \alpha_0)/\omega_0$, $T_2 = (2\pi/3 + \alpha_0)/\omega_0$, $T_3 = (-2\pi/3 + \alpha_0)/\omega_0$, and $T_4 = (-\pi/3 + \alpha_0)/\omega_0$. These can be incorporated into Schwarz's spectrum as follows

$$F_i(t) = h_i \frac{\omega_0}{2\pi} + h_i \frac{bk\omega_0}{2\pi} \sin(k\omega_0 t + \delta_k) + h_i \sum_{m=1}^{\infty} \sum_{n=-\infty}^{\infty} \frac{(m+nk)\omega_0}{\pi} \frac{J_n(mb)}{m} \cos\left[m(\omega_0 t - T_i) + nk\omega_0 t + n\delta_k - \frac{n\pi}{2}\right] \quad \text{(II.5)}$$

Summing over i results in the following spectrum

$$F(t) = \frac{2\sqrt{3}}{\pi} \sum_{m} \sum_{n=-\infty}^{\infty} (\pm)(m+nk)\omega_0 \frac{J_n(mb)}{m} \cdot \cos\left[(m+nk)\omega_0 t - m\alpha_0 + n\delta_k - \frac{n\pi}{2} - \frac{\pi}{2}\right] \quad \text{(II.6)}$$

for $m = 1, 5, 7, 11$ etc.

The next step is to calculate the Fourier transform of the rectangular pulse. The Fourier transform of a rectangular pulse of duration T and height A, centred around time $t = T/2$, is

$$G(f) = AT \frac{\sin(\pi f T)}{\pi f T} \underline{/-\pi f T} \quad \text{(II.7)}$$

Applying this for a commutation period of μ_0 radians at fundamental frequency ω_0, with a pulse height of 0.5 allows the transform to be rewritten

$$G(\omega) = \frac{\sin(\mu_0 \omega / 2\omega_0)}{\omega} \underline{/\frac{-\mu_0 \omega}{2\omega_0}} \quad \text{(II.8)}$$

Applying this to the calculated spectrum, and extending to three phases, yields

$$F_\psi(t) = \frac{2\sqrt{3}}{\pi} \sum_{m} \sum_{n=-\infty}^{\infty} (\pm) \sin\left[\frac{(m+nk)\mu_0}{2}\right] \frac{J_n(mb)}{m} \times \cos\left[(m+nk)\omega_0 t - m\left(\alpha_0 + \frac{\mu_0}{2}\right) + n\left(\delta_k - \frac{k\mu_0}{2} - \frac{\pi}{2}\right) - \frac{\pi}{2} - m\psi\right] \quad \text{(II.9)}$$

for $m = 1, 5, 7, 11$ etc. If desired, this can be expanded out to

$$F_\psi(t) = \frac{\sqrt{3}}{\pi} \sum_{m} \sum_{n=-\infty}^{\infty} (\pm) \frac{J_n(mb)}{m} \cos\left[(m+nk)\omega_0 t - m(\alpha_0 + \mu_0) + n\left(\delta_k - k\mu_0 - \frac{\pi}{2}\right) - m\psi\right] - \frac{\sqrt{3}}{\pi} \sum_{m} \sum_{n=-\infty}^{\infty} (\pm) \frac{J_n(mb)}{m} \cos\left[(m+nk)\omega_0 t - m\alpha_0 + n\left(\delta_k - \frac{\pi}{2}\right) - m\psi\right] \quad \text{(II.10)}$$

which can be further expanded to

$$
\begin{aligned}
F_\psi(t) = & \frac{2\sqrt{3}}{\pi}\sum_m (\pm)\sin\left(\frac{m\mu_0}{2}\right)\frac{J_0(mb)}{m}\cos\left[m\omega_0 t - m\left(\alpha_0 + \frac{\mu_0}{2}\right) - \frac{\pi}{2} - m\psi\right] \\
& + \frac{\sqrt{3}}{\pi}\sum_m\sum_{n=1}^{\infty}(\pm)\frac{J_n(mb)}{m}\cos\left[(m+nk)\omega_0 t - m(\alpha_0+\mu_0) + n\left(\delta_k - k\mu_0 - \frac{\pi}{2}\right) - m\psi\right] \\
& + \frac{\sqrt{3}}{\pi}\sum_m\sum_{n=1}^{\infty}(\pm)\frac{J_n(mb)}{m}\cos\left[(m-nk)\omega_0 t - m(\alpha_0+\mu_0) - n\left(\delta_k - k\mu_0 + \frac{\pi}{2}\right) - m\psi\right] \\
& - \frac{\sqrt{3}}{\pi}\sum_m\sum_{n=1}^{\infty}(\pm)\frac{J_n(mb)}{m}\cos\left[(m+nk)\omega_0 t - m\alpha_0 + n\left(\delta_k - \frac{\pi}{2}\right) - m\psi\right] \\
& - \frac{\sqrt{3}}{\pi}\sum_m\sum_{n=1}^{\infty}(\pm)\frac{J_n(mb)}{m}\cos\left[(m-nk)\omega_0 t - m\alpha_0 - n\left(\delta_k + \frac{\pi}{2}\right) - m\psi\right] \qquad \text{(II.11)}
\end{aligned}
$$

This is the contribution of the commutation period to the converter transfer function frequency spectrum.

II.3 Contribution of Commutation Duration to AC Current

The waveform of the transfer function that, when added to the ideal transfer function, results in the non-ideal transfer function is shown in Figure II.3. It can also be defined as the sum of four different waveforms, each of which has one pulse per fundamental cycle.

Each of the four waveforms is exactly a PPM waveform as described by Schwarz, but with the shape of each pulse being triangular rather than a delta function. Schwarz's PPM spectrum is summed for the four waveforms, and then the Fourier transform of the actual pulse is calculated and applied as a frequency response function to the solution.

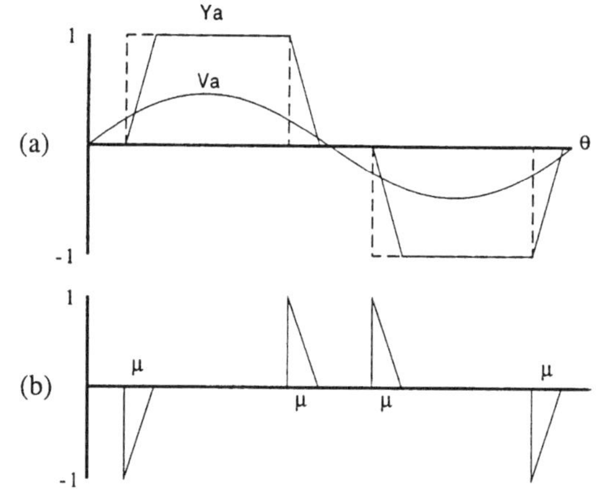

Figure II.3 Commutation function waveform for transfer to ac current

The sum of the four spectra, approximating each pulse with a delta function, is described in Equation II.6.

The Fourier transform for the triangular pulse of duration T and height A is

$$G(\omega) = \int_0^T A(1 - t/T)\mathrm{e}^{-\mathrm{j}\omega t}\,\mathrm{d}t \tag{II.12}$$

which is, in polar notation,

$$G(\omega) = \frac{A}{\omega}\left[1\angle -\frac{\pi}{2} + \frac{2}{\omega T}\sin\left(\frac{\omega T}{2}\right)\angle \frac{\pi}{2} - \frac{\omega T}{2}\right] \tag{II.13}$$

Applying this for a commutation period of μ_1 radians at fundamental frequeny ω_0 with a pulse height of 1 allows the transform to be rewritten

$$G(\omega) = \frac{1}{\omega}\left[1\angle -\frac{\pi}{2} + \frac{2\omega_0}{\mu_1\omega}\sin\left(\frac{\mu_1\omega}{2\omega_0}\right)\angle \frac{\pi}{2} - \frac{\mu_1\omega}{2\omega_0}\right] \tag{II.14}$$

Applying this to Equation II.6 and extending to three phases yields the spectrum of the commutation function.

$$\begin{aligned}
F_\psi(t) = &-\frac{2\sqrt{3}}{\pi}\sum_m \sum_{n=-\infty}^{\infty}(\pm)\frac{J_n(mb)}{m}\cos\left[(m+nk)\omega_0 t - m\alpha_0 + n\left(\delta_k - \frac{\pi}{2}\right) - m\psi\right] \\
&+\frac{2\sqrt{3}}{\pi}\sum_m \sum_{n=-\infty}^{\infty}(\pm)\frac{J_n(mb)}{m}\frac{2\sin\left[\dfrac{(m+nk)\mu_1}{2}\right]}{(m+nk)\mu_1}\times \\
&\cos\left[(m+nk)\omega_0 t - m\left(\alpha_0 + \frac{\mu_1}{2}\right) + n\left(\delta_k - \frac{\pi}{2} - \frac{k\mu_1}{2}\right) - m\psi\right]
\end{aligned} \tag{II.15}$$

for $m = 1, 5, 7, 11$ etc. If desired, this can be expanded out to

$$\begin{aligned}
F_\psi(t) = &-\frac{2\sqrt{3}}{\pi}\sum_m (\pm)\frac{J_0(mb)}{m}\cos[m\omega_0 t - m\alpha_0 - m\psi] \\
&-\frac{2\sqrt{3}}{\pi}\sum_m \sum_{n=1}^{\infty}(\pm)\frac{J_n(mb)}{m}\cos\left[(m+nk)\omega_0 t - m\alpha_0 + n\left(\delta_k - \frac{\pi}{2}\right) - m\psi\right] \\
&-\frac{2\sqrt{3}}{\pi}\sum_m \sum_{n=1}^{\infty}(\pm)\frac{J_n(mb)}{m}\cos\left[(m-nk)\omega_0 t - m\alpha_0 - n\left(\delta_k + \frac{\pi}{2}\right) - m\psi\right] \\
&+\frac{2\sqrt{3}}{\pi}\sum_m (\pm)\frac{J_0(mb)}{m}\frac{2\sin(m\mu_1/2)}{m\mu_1}\cos\left[m\omega_0 t - m\left(\alpha_0 + \frac{\mu_1}{2}\right) - m\psi\right]
\end{aligned}$$

(\over)

$$+\frac{2\sqrt{3}}{\pi}\sum_m\sum_{n=1}^{\infty}(\pm)\frac{J_n(mb)}{m}\frac{2\sin\left[\frac{(m+nk)\mu_1}{2}\right]}{(m+nk)\mu_1}\times$$
$$\cos\left[(m+nk)\omega_0 t - m\left(\alpha_0+\frac{\mu_1}{2}\right)+n\left(\delta_k-\frac{\pi}{2}-\frac{k\mu_1}{2}\right)-m\psi\right]$$
$$+\frac{2\sqrt{3}}{\pi}\sum_m\sum_{n=1}^{\infty}(\pm)\frac{J_n(mb)}{m}\frac{2\sin\left[\frac{(m-nk)\mu_1}{2}\right]}{(m-nk)\mu_1}\times$$
$$\cos\left[(m-nk)\omega_0 t - m\left(\alpha_0+\frac{\mu_1}{2}\right)-n\left(\delta_k+\frac{\pi}{2}-\frac{k\mu_1}{2}\right)-m\psi\right] \quad \text{(II.16)}$$

This is the contribution of the commutation period to the converter transfer function frequency spectrum.

II.4 Contribution of Commutation Period Variation to AC Current

The waveform of the transfer function that, when added to the non-ideal steady commutation period transfer function, results in the non-ideal variable commutation period transfer function is shown in Figure II.4, for one commutation period only. It can also be defined as the sum of four different waveforms, each of which has one pulse per fundamental cycle.

The waveform to be added to the non-ideal steady commutation period transfer function has three properties, namely area, position, and shape. The geometry of the

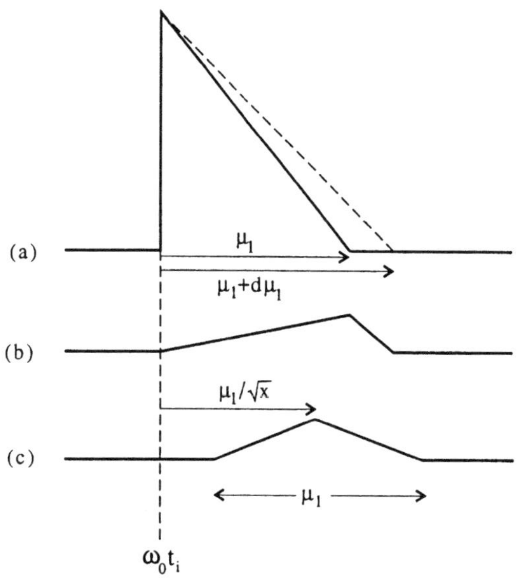

Figure II.4 Commutation period variation function, for transfer to ac current

waveform yields a time area of

$$Area = \frac{\Delta\mu_1}{2\omega_0} \tag{II.17}$$

Its centralized angular position depends on how the waveform area is distributed. For the rectifier, it will be somewhat before the end of the effective commutation period. Although the true commutation period ends earlier than the effective commutation period, the area change tends to be concentrated near the end. On this basis, the centralized angular position of the correction pulse is taken to be the same as that indicated by the linear transfer model. This is

$$\theta = \alpha_0 + \Delta\alpha + \sqrt{\frac{\mu_1(\mu_1 + \Delta\mu_1)}{2}} \tag{II.18}$$

The position of the pulse is dependent strongly on the firing angle variation $\Delta\alpha$, and only weakly on the variation in commutation period duration $\Delta\mu_1$. If $\Delta\mu_1$ is assumed to be small, equation II.18 can be reduced to

$$\theta = \alpha_0 + \Delta\alpha + \frac{\mu_1}{\sqrt{2}} \tag{II.19}$$

At the inverter, the true commutation period ends somewhat later than the effective commutation period. The centralized angular position of the correction pulse will be before the end of the true commutation period, and as an approximation the end of the effective commutation period is chosen. The relevant equation can be written

$$\theta = \alpha_0 + \Delta\alpha + \mu_1 \tag{II.20}$$

A variable x is set, such that $x = 1$ at the inverter, and $x = 2$ at the rectifier.

The duration of the waveform is fairly close to one commutation period. It is modelled initially as a dirac delta function, which is later converted to a symmetrical triangular pulse of duration μ_1.

From Equations II.17 and II.18, it can be seen that the area of the pulse is dependent only on the modulation of the end of the commutation period relative to the beginning of the commutation period.

If $\Delta\mu_1$ is held constant such that the area of the associated pulse is 1, and firing angle modulation such that $\Delta\alpha = b\cos(k\omega_0 t + \delta_k)$ is applied, the resulting impulse train has the PPM spectrum, derived from Equation II.6 as follows

$$F_\psi(t) = \frac{2\sqrt{3}}{\pi}\sum_m \sum_{n=-\infty}^{\infty} (\pm)(m+nk)\omega_0 \frac{J_n(mb)}{m} \times$$
$$\cos\left[(m+nk)\omega_0 t - m\left(\alpha_0 + \frac{\mu_1}{\sqrt{x}}\right) + n\left(\delta_k - \frac{k\mu_1}{\sqrt{x}} - \frac{\pi}{2}\right) - \frac{\pi}{2} - m\psi\right] \tag{II.21}$$

for $m = 1, 5, 7, 11$ etc. The impulses can be converted to symmetrical triangular pulses of duration $2T$ and area AT^2 by applying the Fourier transform.

$$G(f) = AT^2\left(\frac{\sin(\pi fT)}{\pi fT}\right)^2 \tag{II.22}$$

Choosing the appropriate dimensions for a triangular pulse of duration μ_1 yields the Fourier transform

$$G(\omega) = \frac{\Delta\mu_1}{2\omega_0}\left(\frac{\sin\left(\frac{\mu_1\omega}{4\omega_0}\right)}{\frac{\mu_1\omega}{4\omega_0}}\right)^2 \qquad \text{(II.23)}$$

Applying the Fourier transform of Equation II.23 to convert to a symmetrical triangular pulse yields

$$F_\psi(t) = \frac{2\sqrt{3}}{\pi}\frac{\Delta\mu_1}{2}\sum_m \sum_{n=-\infty}^{\infty}(\pm)(m+nk)\left\{\frac{\sin\left[\frac{(m+nk)\mu_1}{4}\right]}{\frac{(m+nk)\mu_1}{4}}\right\}^2 \frac{J_n(mb)}{m} \times$$

$$\cos\left[(m+nk)\omega_0 t - m\left(\alpha_0 + \frac{\mu_1}{\sqrt{x}}\right) + n\left(\delta_k - \frac{k\mu_1}{\sqrt{x}} - \frac{\pi}{2}\right) - \frac{\pi}{2} - m\psi\right] \qquad \text{(II.24)}$$

If $\Delta\mu_1$ is now allowed to vary as per $\Delta\mu_1 = b_1\cos(k\omega_0 t + \delta_{k1} - k\mu_1/\sqrt{x})$, which is the commutation period modulation referred to the centre of the correction pulse, this can be substituted directly into Equation II.24 to yield

$$F_\psi(t) = \frac{\sqrt{3}}{\pi}\frac{b_1}{2}\sum_m \sum_{n=-\infty}^{\infty}(\pm)(m+nk)\left\{\frac{\sin\left[\frac{(m+nk)\mu_1}{4}\right]}{\frac{(m+nk)\mu_1}{4}}\right\}^2 \frac{J_n(mb)}{m} \times$$

$$\left\{\cos\left[(m+nk+k_1)\omega_0 t - m\left(\alpha_0 + \frac{\mu_1}{\sqrt{x}}\right) + n\left(\delta_k - \frac{k\mu_1}{\sqrt{x}} - \frac{\pi}{2}\right) + \delta_{k1} - \frac{k\mu_1}{\sqrt{x}} - \frac{\pi}{2} - m\psi\right]\right.$$

$$\left. + \cos\left[(m+nk-k_1)\omega_0 t - m\left(\alpha_0 + \frac{\mu_1}{\sqrt{x}}\right) + n\left(\delta_k - \frac{k\mu_1}{\sqrt{x}} - \frac{\pi}{2}\right) - \delta_{k1} + \frac{k\mu_1}{\sqrt{x}} - \frac{\pi}{2} - m\psi\right]\right\}$$

(II.25)

for $m = 1, 5, 7, 11$ etc.

The largest terms of this series are those for $n = 0$. and they are expanded out here

$$F_\psi(t) = \frac{\sqrt{3}}{\pi}\frac{b_1}{2}\sum_m(\pm)\left\{\frac{\sin(m\mu_1/4)}{m\mu_1/4}\right\}^2 J_0(mb)\cos\left[(m+k_1)\omega_0 t - m\left(\alpha_0 + \frac{\mu_1}{\sqrt{x}}\right)\right.$$

$$\left. + \delta_{k1} - \frac{k\mu_l}{\sqrt{x}} - \frac{\pi}{2} - m\psi\right] + \frac{\sqrt{3}}{\pi}\frac{b_l}{2}\sum_m(\pm)\left\{\frac{\sin(m\mu_l/4)}{m\mu_l/4}\right\}^2 \times$$

$$J_0(mb)\cos\left[(m-k_1)\omega_0 t - m\left(\alpha_0 + \frac{\mu_1}{\sqrt{x}}\right) - \delta_{kl} + \frac{k\mu_1}{\sqrt{x}} - \frac{\pi}{2} - m\psi\right] \qquad \text{(II.26)}$$

for $m = 1, 5, 7, 11$ etc. These frequencies are dependent strongly on the amplitude modulation of the correction function, and only relatively weakly on its position modulation.

The remaining terms expand out to the following

$$F_{\psi}(t) = \frac{\sqrt{3}\,b_1}{\pi}\frac{}{2}\sum_{m}\sum_{n=1}^{\infty}(\pm)(m+nk)\left\{\frac{\sin\left[\dfrac{(m+nk)\mu_1}{4}\right]}{\dfrac{(m+nk)\mu_1}{4}}\right\}^2 \frac{J_n(mb)}{m}\times$$
$$\left\{\cos\left[(m+nk+k_1)\omega_0 t - m\left(\alpha_0+\frac{\mu_1}{\sqrt{x}}\right)+n\left(\delta_k-\frac{k\mu_1}{\sqrt{x}}-\frac{\pi}{2}\right)+\delta_{k1}-\frac{k\mu_1}{\sqrt{x}}-\frac{\pi}{2}-m\psi\right]\right.$$
$$\left.+\cos\left[(m+nk-k_1)\omega_0 t - m\left(\alpha_0+\frac{\mu_1}{\sqrt{x}}\right)+n\left(\delta_k-\frac{k\mu_1}{\sqrt{x}}-\frac{\pi}{2}\right)-\delta_{k1}+\frac{k\mu_1}{\sqrt{x}}-\frac{\pi}{2}-m\psi\right]\right\}$$
$$+\frac{\sqrt{3}\,b_1}{\pi}\frac{}{2}\sum_{m}\sum_{n=1}^{\infty}(\pm)(m-nk)\left\{\frac{\sin\left[\dfrac{(m-nk)\mu_1}{4}\right]}{\dfrac{(m-nk)\mu_1}{4}}\right\}^2 \frac{J_n(mb)}{m}\times$$
$$\left\{\cos\left[(m-nk+k_1)\omega_0 t - m\left(\alpha_0+\frac{\mu_1}{\sqrt{x}}\right)-n\left(\delta_k-\frac{k\mu_1}{\sqrt{x}}+\frac{\pi}{2}\right)+\delta_{k1}-\frac{k\mu_1}{\sqrt{x}}-\frac{\pi}{2}-m\psi\right]\right.$$
$$\left.+\cos\left[(m-nk-k_1)\omega_0 t - m\left(\alpha_0+\frac{\mu_1}{\sqrt{x}}\right)-n\left(\delta_k-\frac{k\mu_1}{\sqrt{x}}+\frac{\pi}{2}\right)-\delta_{k1}+\frac{k\mu_1}{\sqrt{x}}-\frac{\pi}{2}-m\psi\right]\right\} \tag{II.27}$$

for $m = 1, 5, 7, 11$ etc. The most significant terms of this series relate to $n = 1$ and are at the frequencies $(m+k+k_1)\omega_0$, $(m+k-k_1)\omega_0$, $(m-k+k_1)\omega_0$, and $(m-k-k_1)\omega_0$. These frequencies are at low levels.

Only the terms listed in Equation II.26 are carried on in the main text.

II.5 Reference

1. Schwarz, M, Bennett, WR and Stein, S, (1966). *Communication Systems and Techniques*. McGraw-Hill.

Appendix III

PULSE DURATION MODULATION ANALYSIS

III.1 The PDM spectrum

In 1966 Schwarz [1], after analysing the frequency spectrum of a Pulse Position Modulated (PPM) waveform, applied a similar analysis to a Pulse Duration Modulated (PDM) waveform. The spectrum is rewritten here in a form suitable for use in HVdc converter control analysis. In Schwarz's analysis the PDM waveform is generated by adding the modulating signal to a steady sawtooth sweep voltage. The pulse is initiated at the beginning of each voltage sweep, and is terminated when the sweep crosses a preset threshold. The unmodulated sweep waveform, threshold, and resulting pulses are shown in figure III.1.

Letting the sweep be as shown in Figure III.1, the modulating signal be $Q\cos(qt)$, and the pulse height h, Schwarz calculates the resulting frequency spectrum

$$E(t) = \frac{h}{2} - \frac{hQ}{2P}\cos(qt) + \frac{h}{\pi}\sum_{m=1}^{\infty}\left[(-1)^m - J_0\left(\frac{m\pi Q}{P}\right)\right]\frac{\sin(mpt)}{m}$$
$$-\frac{h}{\pi}\sum_{m=1}^{\infty}\sum_{n=1}^{\infty}\frac{J_n(m\pi Q/P)}{m}\sin\left[(mp+nq)t+\frac{n\pi}{2}\right]$$
$$-\frac{h}{\pi}\sum_{m=1}^{\infty}\sum_{n=1}^{\infty}\frac{J_n(m\pi Q/P)}{m}\sin\left[(mp-nq)t+\frac{n\pi}{2}\right] \qquad \text{(III.1)}$$

where p is the angular sweep frequency, which will now be known as ω_0.

Redefining the PDM in terms of the pulse angular position (radians) as $b\cos(k\omega_0 t + \delta_k)$, results in the substitution

$$Q = \frac{-Pb}{\pi} \qquad \text{(III.2)}$$

and

$$q = k\omega_0 \qquad \text{(III.3)}$$

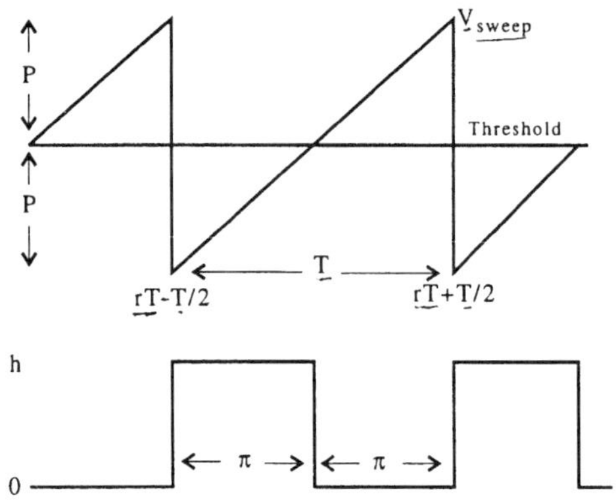

Figure III.1 PDM generation waveforms

The frequency spectrum becomes

$$E(t) = \frac{h}{2} + \frac{hb}{2\pi}\cos(k\omega_0 t + \delta_k) + \frac{h}{\pi}\sum_{m=1}^{\infty}[(-1)^m - J_0(mb)]\frac{\sin(m\omega_0 t)}{m}$$
$$-\frac{h}{\pi}\sum_{m=1}^{\infty}\sum_{n=1}^{\infty}\frac{J_n(mb)}{m}\sin\left[(m+nk)\omega_0 t + n\delta_k - \frac{n\pi}{2}\right]$$
$$-\frac{h}{\pi}\sum_{m=1}^{\infty}\sum_{n=1}^{\infty}\frac{J_n(mb)}{m}\sin\left[(m-nk)\omega_0 t - n\delta_k - \frac{n\pi}{2}\right] \quad \text{(III.4)}$$

In the following applications, only the change in spectrum that results from an existing wave being modulated needs to be considered. The spectrum of the unmodulated pulse train should be subtracted from that of the modulated pulse train to define the harmonic contribution of the modulation.

The unmodulated pulse train has the spectrum

$$E(t) = \frac{h}{2} + \frac{h}{\pi}\sum_{m=1}^{\infty}[(-1)^m - 1]\frac{\sin(m\omega_0 t)}{m} \quad \text{(III.5)}$$

Thus the contribution of modulation to the harmonic spectrum is

$$E(t) = \frac{hb}{2\pi}\cos(k\omega_0 t + \delta_k) + \frac{h}{\pi}\sum_{m=1}^{\infty}[1 - J_0(mb)]\frac{\sin(m\omega_0 t)}{m}$$
$$-\frac{h}{\pi}\sum_{m=1}^{\infty}\sum_{n=1}^{\infty}\frac{J_n(mb)}{m}\sin\left[(m+nk)\omega_0 t + n\delta_k - \frac{n\pi}{2}\right]$$
$$-\frac{h}{\pi}\sum_{m=1}^{\infty}\sum_{n=1}^{\infty}\frac{J_n(mb)}{m}\sin\left[(m-nk)\omega_0 t - n\delta_k - \frac{n\pi}{2}\right] \quad \text{(III.6)}$$

This statement of the PDM spectrum forms the basis of the analysis of firing angle modulation applied to the ideal model, and of natural modulation to the commutation period.

III.2 Firing Angle Modulation Applied to the Ideal Transfer Function

The waveform of the transfer function that, when added to the ideal transfer function, results in the firing angle modulated transfer function is shown in figure III.2. It can be defined as the sum of four different waveforms, each of which has one pulse per fundamental cycle.

Each of the four waveforms is exactly the waveform described by equation III.13, with different pulse heights and different time references.

If the four waveforms are called F_i, where $i = 1, 2, 3, 4$, then each has a different pulse height, each a different time reference, and each a different $\Delta\alpha$. The relevant values are $h_1 = +1$, $h_2 = +1$, $h_3 = -1$, $h_4 = -1$, and $T_1 = (\pi/3 + \alpha_0)/\omega_0$, $T_2 = (2\pi/3 + \alpha_0)/\omega_0$, $T_3 = (-2\pi/3 + \alpha_0)/\omega_0$, and $T_4 = (-\pi/3 + \alpha_0)/\omega_0$. These can be incorporated into equation III.6 as follows

$$
\begin{aligned}
F_i(t) = {} & \frac{h_i b}{2\pi}\cos(k\omega_0 t + \delta_k) + \frac{h_i}{\pi}\sum_{m=1}^{\infty}[1 - J_0(mb)]\frac{\sin(m\omega_0(t - T_i))}{m} \\
& - \frac{h_i}{\pi}\sum_{m=1}^{\infty}\sum_{n=1}^{\infty}\frac{J_n(mb)}{m}\sin\left[m\omega_0(t - T_i) + nk\omega_0 t + n\delta_k - \frac{n\pi}{2}\right] \\
& - \frac{h_i}{\pi}\sum_{m=1}^{\infty}\sum_{n=1}^{\infty}\frac{J_n(mb)}{m}\sin\left[m\omega_0(t - T_i) - nk\omega_0 t - n\delta_k - \frac{n\pi}{2}\right]
\end{aligned} \qquad \text{(III.7)}
$$

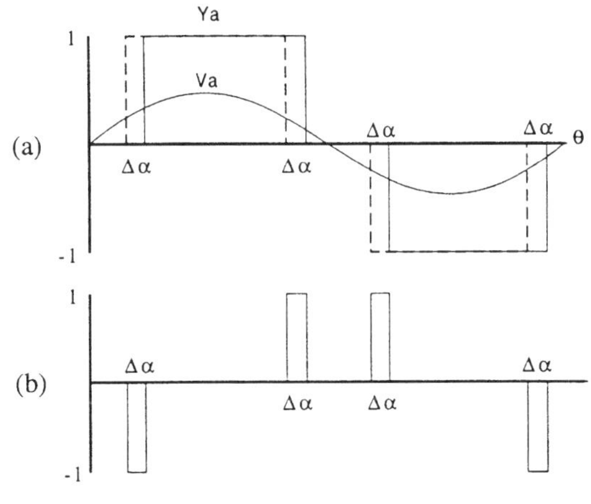

Figure III.2 Transfer function for firing angle modulation

Summing over i and extending to three phases results in the following spectrum

$$F_{\psi}(t) = \frac{\sqrt{3}}{\pi}\sum_{m}(\pm)\frac{J_0(mb)-1}{m}\cos[m(\omega_0 t-\alpha_0-\psi)]$$
$$+\frac{2\sqrt{3}}{\pi}\sum_{m}\sum_{n=1}^{\infty}(\pm)\frac{J_n(mb)}{m}\cos\left[(m+nk)\omega_0 t - m\alpha_0 + n\left(\delta_k-\frac{\pi}{2}\right)-m\psi\right]$$
$$+\frac{2\sqrt{3}}{\pi}\sum_{m}\sum_{n=1}^{\infty}(\pm)\frac{J_n(mb)}{m}\cos\left[(m-nk)\omega_0 t - m\alpha_0 - n\left(\delta_k+\frac{\pi}{2}\right)-m\psi\right] \tag{III.8}$$

for $m = 1, 5, 7 \ldots$.

This is the spectrum of frequencies in the six-pulse converter transfer function that result from applying firing angle modulation of $\alpha = \alpha_0 + b\cos(k\omega_0 t + \delta_k)$.

III.3 Reference

1. Schwarz, M, Bennett, WR and Stein, S, (1966). *Communication Systems and Techniques*. McGraw-Hill.

Appendix IV

DERIVATION OF THE JACOBIAN

In this Appendix, the full Jacobian matrix is derived for a six-pulse rectifier attached to the ac system via a star-g/star transformer. The elements of the Jacobian are the partial derivatives of the mismatch functions with respect to the variables that are being solved for by Newton's method. An important distinction is made in Newtons method between functions and variables. A variable is never a function of something else. A function is only a function of the variables being solved for, and never another mismatch function. The variables that are being solved for are defined to be V_k^a, V_k^b, V_k^c, I_{d_k}, ϕ_i, θ_i, α_0. Thus I_{d_k}, the dc current harmonics, are defined not to be a function of V_k, the terminal voltage harmonics. In the analysis to follow, any other quantity is either a function of these variables or a constant.

In finding the partial derivative of a mismatch function with respect to one of these variables, all other variables are held constant. Thus the chain rule is never applied to give a derivative of one variable with respect to another. For example, in the dc current mismatch equations, the partial derivative with respect to the average delay angle is always zero. Intuitively this is incorrect, as we expect a change in delay angle to change all the dc harmonics. However, in terms of the converter mismatch equations, this requires several applications of the chain rule to quantities that have been defined as variables, not functions. The overall network of interdependency between all the converter quantities is only fully represented in the inversion of the Jacobian matrix, and in the definition of the mismatch functions.

All of the partial derivatives obtained below have been confirmed by comparison with the numerical partial derivative obtained from the corresponding mismatch equation. That is;

$$\frac{\partial F}{\partial x} \simeq \frac{F(x + \Delta x) - F(x)}{\Delta x} \qquad \text{(IV.1)}$$

IV.1 Voltage Mismatch Partial Derivatives

The voltage mismatch partial derivatives are obtained by first deriving the phase current partial derivatives, and then applying the differential current to the three-phase system impedance. The phase current partial derivatives are derived in turn by considering the partial derivatives of the sampled commutation and dc currents of which they are composed.

IV.1.1 With Respect to AC Phase Voltage Variation

First, the partial derivative of the terminal voltage mismatch with respect to terminal voltage variation is obtained. In this analysis all other variables are held constant. The commutation current, sampled dc current, and phase currents are all functions, and require an application of the chain rule. The phase voltage mismatch equation, which represents the interaction of the calculated phase currents with the ac system impedance is

$$F_{V_k^\alpha} = V_k^\alpha - [Z_k^{\alpha\alpha}(I_k^\alpha + [Y_{tc}V_{th}]_k^\alpha) + Z_k^{\alpha\beta}(I_k^\beta + [Y_{tc}V_{th}]_k^\beta) + Z_k^{\alpha\gamma}(I_k^\gamma + [Y_{tc}V_{th}]_k^\gamma)] \quad \text{(IV.2)}$$

In order to generalize over all phases, $[\alpha\ \beta\ \gamma]$ is a permutation of $[a\ b\ c]$. $Z = Y_{cc}^{-1}$ is the ac system impedance, and I^α etc are the phase currents. Note that $I(V_k^a, V_k^b, V_k^c, I_{d_k}, \phi_i, \theta_i)$ is a function of the converter variables. Since the terminal voltages are represented as complex phasors, it is necessary to differentiate with respect to the real and imaginary parts separately. Differentiating Equation IV.2 partially with respect to a variation in one of the phase voltages:

$$\frac{\partial\mathcal{R}\{F_{V_k^\alpha}\}}{\partial\mathcal{R}\{V_m^\delta\}} = \mathcal{R}\begin{cases} -Z_k^{\alpha\alpha}\dfrac{\partial I_k^\alpha}{\partial\mathcal{R}\{V_m^\delta\}} - Z_k^{\alpha\beta}\dfrac{\partial I_k^\beta}{\partial\mathcal{R}\{V_m^\delta\}} - Z_k^{\alpha\gamma}\dfrac{\partial I_k^\gamma}{\partial\mathcal{R}\{V_m^\delta\}}, & \alpha\neq\delta \text{ or } k\neq m \\ 1 - Z_k^{\alpha\alpha}\dfrac{\partial I_k^\alpha}{\partial\mathcal{R}\{V_m^\delta\}} - Z_k^{\alpha\beta}\dfrac{\partial I_k^\beta}{\partial\mathcal{R}\{V_m^\delta\}} - Z_k^{\alpha\gamma}\dfrac{\partial I_k^\gamma}{\partial\mathcal{R}\{V_m^\delta\}}, & \alpha=\delta \text{ and } k=m \end{cases}, \quad \text{(IV.3)}$$

$$\frac{\partial\mathcal{R}\{F_{V_k^\alpha}\}}{\partial\mathcal{I}\{V_m^\delta\}} = \mathcal{I}\left\{-Z_k^{\alpha\alpha}\frac{\partial I_k^\alpha}{\partial\mathcal{I}\{V_m^\delta\}} - Z_k^{\alpha\beta}\frac{\partial I_k^\beta}{\partial\mathcal{I}\{V_m^\delta\}} - Z_k^{\alpha\gamma}\frac{\partial I_k^\gamma}{\partial\mathcal{I}\{V_m^\delta\}}\right\}, \quad \text{(IV.4)}$$

$$\frac{\partial\mathcal{I}\{F_{V_k^\alpha}\}}{\partial\mathcal{I}\{V_m^\delta\}} = \mathcal{I}\begin{cases} -Z_k^{\alpha\alpha}\dfrac{\partial I_k^\alpha}{\partial\mathcal{I}\{V_m^\delta\}} - Z_k^{\alpha\beta}\dfrac{\partial I_k^\beta}{\partial\mathcal{I}\{V_m^\delta\}} - Z_k^{\alpha\gamma}\dfrac{\partial I_k^\gamma}{\partial\mathcal{I}\{V_m^\delta\}}, & \alpha\neq\delta \text{ or } k\neq m \\ j - Z_k^{\alpha\alpha}\dfrac{\partial I_k^\alpha}{\partial\mathcal{I}\{V_m^\delta\}} - Z_k^{\alpha\beta}\dfrac{\partial I_k^\beta}{\partial\mathcal{I}\{V_m^\delta\}} - Z_k^{\alpha\gamma}\dfrac{\partial I_k^\gamma}{\partial\mathcal{I}\{V_m^\delta\}}, & \alpha=\delta \text{ and } k=m \end{cases}, \quad \text{(IV.5)}$$

$$\frac{\partial\mathcal{I}\{F_{V_k^\alpha}\}}{\partial\mathcal{R}\{V_m^\delta\}} = \mathcal{I}\left\{-Z_k^{\alpha\alpha}\frac{\partial I_k^\alpha}{\partial\mathcal{R}\{V_m^\delta\}} - Z_k^{\alpha\beta}\frac{\partial I_k^\beta}{\partial\mathcal{R}\{V_m^\delta\}} - Z_k^{\alpha\gamma}\frac{\partial I_k^\gamma}{\partial\mathcal{R}\{V_m^\delta\}}\right\}, \quad \text{(IV.6)}$$

Since the ac phase currents are calculated by sampling the dc and commutation currents, the effect of phase voltage variation on phase current is determined from the effect of phase voltage variation on the sampled commutation currents alone.

This is because the dc current and switching angles are variables which are held constant, while the commutation currents are functions of the terminal voltage. Four commutations contribute to one cycle of ac current, consequently

$$\frac{\partial I_k^\alpha}{\partial \mathcal{R}\{V_m^\delta\}} = \sum_{i=1}^{6} C_{\alpha\delta_i} \frac{\partial I_{ci}}{\partial \mathcal{R}\{V_m^\delta\}}, \tag{IV.7}$$

and

$$\frac{\partial I_k^\alpha}{\partial \mathcal{I}\{V_m^\delta\}} = \sum_{i=1}^{6} C_{\alpha\delta_i} \frac{\partial I_{ci_k}^s}{\partial \mathcal{I}\{V_m^\delta\}}, \tag{IV.8}$$

where I_{ci}^s is the ith commutation current, I_{ci}, sampled over the commutation period. $C_{\alpha\delta_i} \in [-1, 0, 1]$ is a coefficient matrix that defines whether for the ith commutation the phase α current is a function of the phase β terminal voltage, and if so, in which sense. C is defined in Table IV.1 where i indexes each set of data:

The commutation current is given by

$$I_{ci}(t) = D + \sum_{k=1}^{n_h} I_{ci_k} e^{jk\omega t}, \tag{IV.9}$$

where

$$D = -\mathcal{I}\left\{\sum_{k=1}^{n_h} I_{ci_k} e^{jk\theta_i}\right\}, \tag{IV.10}$$

$$I_{ci_k} = \frac{jk\omega L_e I_{d_k} - V_{eb_k}}{jk\omega L_{eb}}. \tag{IV.11}$$

In Equation IV.11, e and b are subscript functions of i, and refer to the phases ending and beginning conduction respectively. Sampling the commutation current requires a convolution with the relevant sampling function:

$$I_{ci}^s = \left[D + \sum_{k=1}^{n_h} I_{c_k}\right] \otimes \Psi_{(2i-1)}. \tag{IV.12}$$

Table IV.1. The coefficient matrix $C_{\alpha\delta_i}$ which specifies the dependence between commutation current i, terminal voltage phase, and ac current phase

		δ		
		a	b	c
α	a	$\{-1, 0, -1, -1, 0, -1\}$	$\{0, 0, 1, 0, 0, 1\}$	$\{1, 0, 0, 1, 0, 0\}$
	b	$\{0, 0, 1, 0, 0, 1\}$	$\{0, -1, -1, 0, -1, -1\}$	$\{0, 1, 0, 0, 1, 0\}$
	c	$\{1, 0, 0, 1, 0, 0\}$	$\{0, 1, 0, 0, 1, 0\}$	$\{-1, -1, 0, -1, -1, 0\}$

Setting $I_{ci_0} = D$, expanding the convolution, and taking the kth component yields

$$I^s_{ci_k} = \frac{1}{2}\mathrm{j}\left\{\sum_{l=k}^{n_h}(I_{ci_l}\Psi^*_{(2i-1)_{l-k}}) + \sum_{l=0}^{n_h}(I_{ci_l}\Psi^*_{(2i-1)_{l+k}})^* - \sum_{l=0}^{k}(I_{ci_l}\Psi_{(2i-1)_{k-l}})\right\}, \quad k > 0. \tag{IV.13}$$

Since the commutation circuit is linear,

$$\frac{\partial I_{ci_l}}{\partial \mathcal{R}\{V^\delta_m\}} = 0 \quad \forall\, l, m \text{ such that } l \neq m \text{ and } l > 0.$$

This allows the following partial derivative to be written:

$$\frac{\partial I^s_{ci_k}}{\partial \mathcal{R}\{V^\delta_m\}} = \frac{1}{2}\mathrm{j}\left[-\frac{\partial I_{ci_0}}{\partial \mathcal{R}\{V^\delta_m\}}\Psi_{(2i-1)_k} + \left(\frac{\partial I_{ci_0}}{\partial \mathcal{R}\{V^\delta_m\}}\Psi^*_{(2i-1)_k}\right)^*\right.$$
$$\left. + \left(\frac{\partial I_{ci_m}}{\partial \mathcal{R}\{V^\delta_m\}}\Psi^*_{(2i-1)_{m+k}}\right)^* + \begin{cases} -\dfrac{\partial I_{ci_m}}{\partial \mathcal{R}\{V^\delta_m\}}\Psi_{(2i-1)_{k-m}}, & m \leqslant k \\ \dfrac{\partial I_{ci_m}}{\partial \mathcal{R}\{V^\delta_m\}}\Psi^*_{(2i-1)_{m-k}}, & m \geqslant k \end{cases}\right]. \tag{IV.14}$$

Similarly,

$$\frac{\partial I^s_{ci_k}}{\partial \mathcal{I}\{V^\delta_m\}} = \frac{1}{2}\mathrm{j}\left[-\frac{\partial I_{ci_0}}{\partial \mathcal{I}\{V^\delta_m\}}\Psi_{(2i-1)_k} + \left(\frac{\partial I_{ci_0}}{\partial \mathcal{I}\{V^\delta_m\}}\Psi^*_{(2i-1)_k}\right)^*\right.$$
$$\left. + \left(\frac{\partial I_{ci_m}}{\partial \mathcal{I}\{V^\delta_m\}}\Psi^*_{(2i-1)_{m+k}}\right)^* + \begin{cases} -\dfrac{\partial I_{ci_m}}{\partial \mathcal{I}\{V^\delta_m\}}\Psi_{(2i-1)_{k-m}}, & m \leqslant k \\ \dfrac{\partial I_{ci_m}}{\partial \mathcal{I}\{V^\delta_m\}}\Psi^*_{(2i-1)_{m-k}}, & m \geqslant k \end{cases}\right]. \tag{IV.15}$$

The calculations required to evaluate Equations IV.14 and IV.15 can be approximately halved by using the Cauchy Reimann equations for the partial derivatives of complex functions. For an analytic function

$$\frac{\partial F(z)}{\partial \mathcal{I}\{z\}} = \mathrm{j}\frac{\partial F(z)}{\partial \mathcal{R}\{z\}}. \tag{IV.16}$$

For the complex conjugate of an analytic function

$$\frac{\partial F(z)^*}{\partial \mathcal{I}\{z\}} = -\mathrm{j}\frac{\partial F(z)^*}{\partial \mathcal{R}\{z\}}. \tag{IV.17}$$

Thus, as each term of equation IV.14 is evaluated, the corresponding term in Equation IV.15 is obtained from a simple rearrangement of the real and imaginary parts. The remaining partial derivatives, $\partial I_{ci_0}/\partial\mathcal{R}\{V_m^\delta\}$ and $\partial I_{ci_m}/\partial\mathcal{R}\{V_m^\delta\}$, are obtained from equations IV.10 and IV.11;

$$\frac{\partial I_{ci_0}}{\partial\mathcal{R}\{V_m^\delta\}} = -\mathrm{j}\mathcal{I}\left\{\frac{e^{\mathrm{j}k\theta_i}}{\mathrm{j}k\omega L_{eb}}\right\}, \tag{IV.18}$$

$$\frac{\partial I_{ci_0}}{\partial\mathcal{I}\{V_m^\delta\}} = -\mathrm{j}\mathcal{R}\left\{\frac{e^{\mathrm{j}k\theta_i}}{\mathrm{j}k\omega L_{eb}}\right\}, \tag{IV.19}$$

$$\frac{\partial I_{ci_m}}{\partial\mathcal{R}\{V_m^\delta\}} = \frac{1}{\mathrm{j}k\omega L_{eb}}. \tag{IV.20}$$

It is assumed here that phase δ is beginning conduction; phase δ ending conduction is accounted for in the C matrix. This completes the derivation of the partial derivative of terminal voltage mismatch with respect to terminal voltage.

IV.1.2 With Respect to DC Ripple Current Variation

A variation in the dc ripple at a particular harmonic is sampled by the compound dc current sampling function, and so affects all phase current harmonics. In addition, the commutation currents are functions of the dc ripple, and so the resulting variation in commutation currents are sampled on the ac side by the commutation interval sampling functions. These two effects, when added together, are sufficient to give the variation in phase currents, and by injecting the phase current variation into the ac system, the required mismatch partial derivative for the Jacobian matrix is obtained. The voltage mismatch partial derivatives are therefore

$$\frac{\partial\mathcal{R}\{F_{V_k^\alpha}\}}{\partial\mathcal{R}\{I_{d_m}\}} = \mathcal{R}\left\{-Z_k^{\alpha\alpha}\frac{\partial I_k^\alpha}{\partial\mathcal{R}\{I_{d_m}\}} - Z_k^{\alpha\beta}\frac{\partial I_k^\beta}{\partial\mathcal{R}\{I_{d_m}\}} - Z_k^{\alpha\gamma}\frac{\partial I_k^\gamma}{\partial\mathcal{R}\{I_{d_m}\}}\right\}, \tag{IV.21}$$

$$\frac{\partial\mathcal{R}\{F_{V_k^\alpha}\}}{\partial\mathcal{I}\{I_{d_m}\}} = \mathcal{R}\left\{-Z_k^{\alpha\alpha}\frac{\partial I_k^\alpha}{\partial\mathcal{I}\{I_{d_m}\}} - Z_k^{\alpha\beta}\frac{\partial I_k^\beta}{\partial\mathcal{I}\{I_{d_m}\}} - Z_k^{\alpha\gamma}\frac{\partial I_k^\gamma}{\partial\mathcal{I}\{I_{d_m}\}}\right\}, \tag{IV.22}$$

$$\frac{\partial\mathcal{I}\{F_{V_k^\alpha}\}}{\partial\mathcal{R}\{I_{d_m}\}} = \mathcal{I}\left\{-Z_k^{\alpha\alpha}\frac{\partial I_k^\alpha}{\partial\mathcal{R}\{I_{d_m}\}} - Z_k^{\alpha\beta}\frac{\partial I_k^\beta}{\partial\mathcal{R}\{I_{d_m}\}} - Z_k^{\alpha\gamma}\frac{\partial I_k^\gamma}{\partial\mathcal{R}\{I_{d_m}\}}\right\}, \tag{IV.23}$$

$$\frac{\partial\mathcal{I}\{F_{V_k^\alpha}\}}{\partial\mathcal{I}\{I_{d_m}\}} = \mathcal{I}\left\{-Z_k^{\alpha\alpha}\frac{\partial I_k^\alpha}{\partial\mathcal{I}\{I_{d_m}\}} - Z_k^{\alpha\beta}\frac{\partial I_k^\beta}{\partial\mathcal{I}\{I_{d_m}\}} - Z_k^{\alpha\gamma}\frac{\partial I_k^\gamma}{\partial\mathcal{I}\{I_{d_m}\}}\right\}. \tag{IV.24}$$

Table IV.2. Coefficient matrix E_i^{α} defining the contribution of the commutation currents to each phase current. i is the commutation number

		α		
		a	b	c
i	1	1	0	−1
	2	0	−1	1
	3	−1	1	0
	4	1	0	−1
	5	0	−1	1
	6	−1	1	0

The phase currents are composed of commutation currents and the dc current sampled on the ac side:

$$I^{\alpha} = \sum_{i=1}^{6} [E_i^{\alpha} I_{ci} \otimes \Psi_{2i-1}] + I_d \otimes \Psi_{\alpha}. \tag{IV.25}$$

Here E_i^{α} is a coefficient matrix that specifies how the commutation currents contribute to each of the phase currents, and Ψ_{α} is the phase α dc current compound sampling function. E is listed in Table IV.2. Differentiating the kth component of equation IV.25 with respect to $\mathcal{R}\{I_{d_m}\}$ yields

$$\frac{\partial I_k^{\alpha}}{\partial \mathcal{R}\{I_{d_m}\}} = \frac{\partial \{\sum_i [E_i^{\alpha} I_{ci} \otimes \Psi_{2i-1}]\}_k}{\partial \mathcal{R}\{I_{d_m}\}} + \frac{\partial \{I_d \otimes \Psi_{\alpha}\}_k}{\partial \mathcal{R}\{I_{d_m}\}}. \tag{IV.26}$$

The first term is similar to the differential with respect to voltage already derived above, as a variation in I_{d_m} affects only I_{ci_m} and I_{ci_0}.Expanding the convolution in the first term, selecting the kth component, and differentiating yields

$$\frac{\partial \{\sum_i [E_i^{\alpha} I_{ci} \otimes \Psi_{2i-1}]\}_k}{\partial \mathcal{R}\{I_{d_m}\}} = \frac{1}{2} \mathrm{j} \sum_i E_i^{\alpha} \left[-\frac{\partial I_{ci_0}}{\partial \mathcal{R}\{I_{d_m}\}} \Psi_{(2i-1)_k} + \left(\frac{\partial I_{ci_0}}{\partial \mathcal{R}\{I_{d_m}\}} \Psi^*_{(2i-1)_k} \right)^* \right.$$
$$\left. + \left(\frac{\partial I_{ci_m}}{\partial \mathcal{R}\{I_{d_m}\}} \Psi^*_{(2i-1)_{m+k}} \right)^* + \begin{cases} -\dfrac{\partial I_{ci_m}}{\partial \mathcal{R}\{I_{d_m}\}} \Psi_{(2i-1)_{k-m}}, & m \leqslant k \\ \dfrac{\partial I_{ci_m}}{\partial \mathcal{R}\{I_{d_m}\}} \Psi^*_{(2i-1)_{m-k}}, & m \geqslant k \end{cases} \right]. \tag{IV.27}$$

Expanding the convolution in the second term of Equation IV.26 gives the kth component of the sampled dc current:

$$I_{d_k}^{s\alpha} = \frac{1}{2} \mathrm{j} \left\{ \sum_{l=k}^{n_h} (I_{d_l} \Psi^*_{\alpha_{l-k}}) + \sum_{l=0}^{n_h} (I_{d_l} \Psi^*_{\alpha_{l+k}})^* - \sum_{l=0}^{k} (I_{d_l} \Psi_{\alpha_{k-l}}) \right\}, \quad k > 0, \tag{IV.28}$$

where $I_{d_k}^{s\alpha}$ is the phase α sample of the dc current. This equation can be differentiated to yield:

$$\frac{\partial\{I_d \otimes \Psi_\alpha\}_k}{\partial\mathcal{R}\{I_{d_m}\}} = \frac{1}{2}\mathrm{j}\Psi_{\alpha_{k+m}} + \frac{1}{2}\mathrm{j}\begin{cases} -\Psi_{\alpha_{k-m}}, & m \leqslant k \\ \Psi^*_{\alpha_{m-k}}, & m \geqslant k \end{cases}. \tag{IV.29}$$

Applying the above analysis for variations in the imaginary part of I_{d_m} gives

$$\frac{\partial\{\sum_i [E_i^\alpha I_{ci} \otimes \Psi_{2i-1}]\}_k}{\partial\mathcal{I}\{I_{d_m}\}} = \frac{1}{2}\mathrm{j}\sum_i E_i^\alpha\left[-\frac{\partial I_{ci_0}}{\partial\mathcal{I}\{I_{d_m}\}}\Psi_{(2i-1)_k} + \left(\frac{\partial I_{ci_0}}{\partial\mathcal{I}\{I_{d_m}\}}\Psi^*_{(2i-1)_k}\right)^*\right.$$

$$\left. + \left(\frac{\partial I_{ci_m}}{\partial\mathcal{I}\{I_{d_m}\}}\Psi^*_{(2i-1)_{m+k}}\right)^* + \begin{cases} -\dfrac{\partial I_{ci_m}}{\partial\mathcal{I}\{I_{d_m}\}}\Psi_{(2i-1)_{k-m}}, & m \leqslant k \\ \dfrac{\partial I_{ci_m}}{\partial\mathcal{I}\{I_{d_m}\}}\Psi^*_{(2i-1)_{m-k}}, & m \geqslant k \end{cases}\right], \tag{IV.30}$$

and

$$\frac{\partial\{I_d \otimes \Psi_\alpha\}_k}{\partial\mathcal{I}\{I_{d_m}\}} = \frac{1}{2}\mathrm{j}(\mathrm{j}\Psi^*_{\alpha_{k+m}})^* + \frac{1}{2}\mathrm{j}\begin{cases} -\mathrm{j}\Psi_{\alpha_{k-m}}, & m \leqslant k \\ \mathrm{j}\Psi^*_{\alpha_{m-k}}, & m \geqslant k \end{cases}. \tag{IV.31}$$

These equations are then substituted into an equation analogous to Equation IV.26 and hence into Equations IV.22 and IV.24. The remaining partial derivatives, $\partial I_{ci_0}/\partial\mathcal{R}\{I_{d_m}\}$ and $\partial I_{ci_m}/\partial\mathcal{R}\{I_{d_m}\}$, are obtained from Equations IV.10 and IV.11:

$$\frac{\partial I_{ci_0}}{\partial\mathcal{R}\{I_{d_m}\}} = -\mathrm{j}\mathcal{I}\left\{\frac{L_e \mathrm{e}^{\mathrm{j}k\theta_i}}{L_{eb}}\right\}, \tag{IV.32}$$

$$\frac{\partial I_{ci_0}}{\partial\mathcal{I}\{I_{d_m}\}} = -\mathrm{j}\mathcal{R}\left\{\frac{L_e \mathrm{e}^{\mathrm{j}k\theta_i}}{L_{eb}}\right\}, \tag{IV.33}$$

$$\frac{\partial I_{ci_m}}{\partial\mathcal{R}\{I_{d_m}\}} = \frac{L_e}{L_{eb}}. \tag{IV.34}$$

IV.1.3 With Respect to End of Commutation Variation

A variation in the end of commutation angle, ϕ_h, affects the $(2h-1)$th and $2h$th sampling functions. This affects all harmonics of the ac side sampled commutation currents, and dc current. Combining these two effects into the phase current variation, and injecting into the ac system yields the variation in the terminal voltage mismatch;

$$\frac{\partial \mathcal{R}\{F_{V_k^\alpha}\}}{\partial \phi_h} = \mathcal{R}\left\{ -Z_k^{\alpha\alpha} \frac{\partial I_k^\alpha}{\partial \phi_h} - Z_k^{\alpha\beta} \frac{\partial I_k^\beta}{\partial \phi_h} - Z_k^{\alpha\gamma} \frac{\partial I_k^\gamma}{\partial \phi_h} \right\}, \tag{IV.35}$$

$$\frac{\partial \mathcal{I}\{F_{V_k^\alpha}\}}{\partial \phi_h} = \mathcal{I}\left\{ -Z_k^{\alpha\alpha} \frac{\partial I_k^\alpha}{\partial \phi_h} - Z_k^{\alpha\beta} \frac{\partial I_k^\beta}{\partial \phi_h} - Z_k^{\alpha\gamma} \frac{\partial I_k^\gamma}{\partial \phi_h} \right\}. \tag{IV.36}$$

Differentiating the kth component of Equation IV.25 with respect to ϕ_h,

$$\frac{\partial I_k^\alpha}{\partial \phi_h} = \frac{\partial \left\{ \sum_i \left[E_i^\alpha I_{ci} \otimes \Psi_{2i-1} \right] \right\}_k}{\partial \phi_h} + \frac{\partial \left\{ I_d \otimes \Psi_\alpha \right\}_k}{\partial \phi_h} \tag{IV.37}$$

A variation in ϕ_h affects only the sampling of the hth commutation current in the first term of Equation IV.37. Expanding only this convolution from the summation yields

$$I_{ch_k}^s = \frac{1}{2} \mathrm{j} E_h^\alpha \left\{ \sum_{l=k}^{n_h} (I_{ch_l} \Psi^*_{(2h-1)_{l-k}}) + \sum_{l=0}^{n_h} (I_{ch_l} \Psi^*_{(2h-1)_{l+k}})^* - \sum_{l=0}^{k} (I_{ch_l} \Psi_{(2h-1)_{k-l}}) \right\}, \quad k > 0. \tag{IV.38}$$

Differentiating this gives

$$\frac{\partial \left\{ \sum_i \left[E_i^\alpha I_{ci} \otimes \Psi_{2i-1} \right] \right\}_k}{\partial \phi_h} = \frac{1}{2} \mathrm{j} E_h^\alpha \left\{ \sum_{l=k}^{n_h} \left(I_{ch_l} \frac{\partial \Psi^*_{(2h-1)_{l-k}}}{\partial \phi_h} \right) \right.$$

$$\left. + \sum_{l=1}^{n_h} \left(I_{ch_l} \frac{\partial \Psi^*_{(2h-1)_{l+k}}}{\partial \phi_h} \right)^* - \sum_{l=0}^{k} \left(I_{ch_l} \frac{\partial \Psi_{(2h-1)_{k-l}}}{\partial \phi_h} \right) \right\}, \quad k > 0, \tag{IV.39}$$

where

$$\frac{\partial \Psi_{(2h-1)_m}}{\partial \phi_h} = -\frac{\partial \Psi_{2h_m}}{\partial \phi_h} = \frac{1}{\pi} \left[\sin(m\phi_h) + \mathrm{j}\cos(m\phi_h) \right]. \tag{IV.40}$$

The compound sampling function in the second term of Equation IV.37 is affected by variation in four of the six end of commutation angles. This is because the transfer of dc current to the ac side is defined by two conduction periods per cycle; the beginning and end of each conduction period corresponding to an end of commutation angle. The effect of a variation in ϕ_h therefore depends upon whether it corresponds to the beginning or end of a positive or negative conduction period for phase α. This information is already collated as the coefficient matrix $-E_{\alpha_i}$.

A similar analysis to the above for the second term of Equation IV.37 results in

$$\frac{\partial\{I_d \otimes \Psi_\alpha\}_k}{\partial\phi_h} = -\frac{1}{2}\mathrm{j}E_h^\alpha\left\{\sum_{l=k}^{n_h}\left(I_{d_l}\frac{\partial\Psi^*_{2h_{l-k}}}{\partial\phi_h}\right) + \sum_{l=1}^{n_h}\left(I_{d_l}\frac{\partial\Psi^*_{h_{l+k}}}{\partial\phi_h}\right)^*\right.$$

$$\left. - \sum_{l=0}^{k}\left(I_{d_l}\frac{\partial\Psi_{2h_{k-l}}}{\partial\phi_h}\right)\right\}, \quad k > 0. \tag{IV.41}$$

Since the compound sampling function is a sum of constituent sampling functions, it has been replaced in the partial derivative by the only term which is a function of ϕ_h. This is then substituted back into Equation IV.37 to give the required partial derivative.

IV.1.4 With Respect to Firing Angle Variation

The transfer of dc current to the ac side of the converter is defined entirely with reference to the end of commutation angles. The effect of a variation in the firing instants on the phase currents can therefore be obtained by analysing just the sampling of the commutation currents. The partial derivative of the ac voltage mismatch with respect to a firing angle variation is

$$\frac{\partial\mathcal{R}\{F_{V_k^\alpha}\}}{\partial\theta_h} = \mathcal{R}\left\{-Z_k^{\alpha\alpha}\frac{\partial I_k^\alpha}{\partial\theta_h} - Z_k^{\alpha\beta}\frac{\partial I_k^\beta}{\partial\theta_h} - Z_k^{\alpha\gamma}\frac{\partial I_k^\gamma}{\partial\theta_h}\right\}. \tag{IV.42}$$

Since the ac side sampled dc current is not a function of θ_h

$$-\frac{\partial I_k^\alpha}{\partial\theta_h} = \frac{\partial\left\{\sum_{i=1}^{6} E_i^\alpha I_{ci}^s\right\}_k}{\partial\theta_h} = \frac{\partial\left\{\sum_{i=1}^{6} E_i^\alpha I_{ci}\otimes\Psi_{(2i-1)}\right\}_k}{\partial\theta_h} = \frac{\partial\{E_h^\alpha I_{ch}\otimes\Psi_{(2h-1)}\}_k}{\partial\theta_h}. \tag{IV.43}$$

Noting that a change in θ_h affects I_{ch_0} and $\Psi_{(2h-1)}$, applying the product rule to the expansion of the convolution leads to

$$\frac{\partial I_k^\alpha}{\partial\theta_h} = -\frac{1}{2}\mathrm{j}E_h^\alpha\left[\sum_{l=k}^{n_h}\left(I_{ch_l}\frac{\partial\Psi^*_{(2h-1)_{l-k}}}{\partial\theta_h}\right) + \sum_{l=0}^{n_h}\left(I_{ch_l}\frac{\partial\Psi^*_{(2h-1)_{l+k}}}{\partial\theta_h}\right)^*\right.$$

$$\left. - \frac{\partial I_{ch_0}}{\partial\theta_h}\Psi_{(2h-1)_k} + \left(\frac{\partial I_{ch_0}}{\partial\theta_h}\Psi^*_{(2h-1)_k}\right)^* - \sum_{l=0}^{k}\left(I_{ch_l}\frac{\partial\Psi_{(2h-1)_{k-l}}}{\partial\theta_h}\right)\right]. \tag{IV.44}$$

This completes the analysis of the terminal voltage mismatch equation partial derivatives, as there is no dependence upon α_0, the only remaining variable.

IV.2 Direct Current Partial Derivatives

The direct current partial derivatives are obtained by a process analogous to that used in obtaining the terminal voltage partial derivatives. The dc voltage partial derivatives are obtained and then applied to the dc system admittance to calculate the direct current differential. The direct current mismatch equation is:

$$F_{I_k} = I_{d_k} - V_{d_k} Y_{d_k} = 0. \tag{IV.45}$$

Finding the partial derivatives of this equation is therefore mainly concerned with V_{d_k}, which is a function of all the converter variables except α_0.

IV.2.1 With Respect to AC Phase Voltage Variation

Differentiating Equation IV.45 with respect to terminal voltage yields

$$\frac{\partial\mathcal{R}\{F_{I_k}\}}{\partial\mathcal{R}\{V_m^\delta\}} = \mathcal{R}\left\{-Y_{d_k}\frac{\partial V_{d_k}}{\partial\mathcal{R}\{V_m^\delta\}}\right\}, \tag{IV.46}$$

$$\frac{\partial\mathcal{I}\{F_{I_k}\}}{\partial\mathcal{R}\{V_m^\delta\}} = \mathcal{I}\left\{-Y_{d_k}\frac{\partial V_{d_k}}{\partial\mathcal{R}\{V_m^\delta\}}\right\}, \tag{IV.47}$$

$$\frac{\partial\mathcal{R}\{F_{I_k}\}}{\partial\mathcal{I}\{V_m^\delta\}} = \mathcal{R}\left\{-Y_{d_k}\frac{\partial V_{d_k}}{\partial\mathcal{I}\{V_m^\delta\}}\right\}, \tag{IV.48}$$

$$\frac{\partial\mathcal{I}\{F_{I_k}\}}{\partial\mathcal{I}\{V_m^\delta\}} = \mathcal{I}\left\{-Y_{d_k}\frac{\partial V_{d_k}}{\partial\mathcal{I}\{V_m^\delta\}}\right\}. \tag{IV.49}$$

V_{d_k} is given by:

$$V_{d_k} = \frac{1}{2}\mathrm{j}\sum_{i=1}^{12}\left\{\sum_{l=k}^{n_h}(V_{di_l}\Psi_{i_{l-k}}{}^*) + \sum_{l=1}^{n_h}(V_{di_l}\Psi_{i_{l+k}}{}^*)^* - \sum_{l=0}^{k}(V_{di_l}\Psi_{i_{k-l}})\right\}, \quad k > 0. \tag{IV.50}$$

The only terms in this equation which are a function of V_m^δ are the twelve pre-convolved dc voltage phasors V_{di_l}. There are three equations which define the twelve pre-convolved dc voltage phasors;

$$V_{d,normal_l} = V_l^+ - V_l^- - \mathrm{j}\omega k(L^+ + L^-)I_{d_l} \tag{IV.51}$$

for a 'normal' conduction interval,

$$V_{d,commutation_l} = \frac{L_e V_l^b + L_b(V_l^e - \mathrm{j}k\omega L_e I_{d_l})}{L_e + L_b} - V_l^o - \mathrm{j}k\omega L_o I_{d_l} \tag{IV.52}$$

for a commutation on the positive dc rail, and

$$V_{d,commutation_l} = V_l^o - jk\omega L_o I_{d_l} - \frac{L_e V_l^b + L_b(V_l^e + jl\omega L_e I_{d_l})}{L_e + L_b} \tag{IV.53}$$

for a commutation on the negative rail. Differentiating these three equations with respect to V_m^δ gives 36 possible values for $\partial V_{di_l}/\partial V_m^\delta$ according to the sample number i, and the phase of δ. The required partial derivatives are summarized in Table IV.3 by reference to the partial derivatives of Equations IV.51, IV.52, and IV.53 listed below:

$$\begin{aligned} \frac{\partial V_{di_l}}{\partial \mathcal{R}\{V_l^+\}} &= 1 \\ \frac{\partial V_{di_l}}{\partial \mathcal{R}\{V_l^-\}} &= -1 \end{aligned} \tag{IV.54}$$

for a 'normal' conduction interval,

$$\begin{aligned} \frac{\partial V_{di_l}}{\partial \mathcal{R}\{V_l^e\}} &= \frac{L_b}{L_b + L_e} \\ \frac{\partial V_{di_l}}{\partial \mathcal{R}\{V_l^b\}} &= \frac{L_e}{L_b + L_e} \\ \frac{\partial V_{di_l}}{\partial \mathcal{R}\{V_l^o\}} &= -1 \end{aligned} \tag{IV.55}$$

for a commutation on the positive dc rail, and

$$\begin{aligned} \frac{\partial V_{di_l}}{\partial \mathcal{R}\{V_l^e\}} &= -\frac{L_b}{L_b + L_e} \\ \frac{\partial V_{di_l}}{\partial \mathcal{R}\{V_l^b\}} &= -\frac{L_e}{L_b + L_e} \\ \frac{\partial V_{di_l}}{\partial \mathcal{R}\{V_l^o\}} &= 1 \end{aligned} \tag{IV.56}$$

for a commutation on the negative rail. The Cauchy Reimann equations can be used to give the partial derivative with respect to the imaginary part of the voltage variation as j times that listed above.

Table IV.3. Assembly of dc voltage partial derivatives

sample (i)	Dc Voltage derivative					
	e	b	o	+	−	equation
1	A	C	B	.	.	(IV.55)
2	.	.	.	A	B	(IV.54)
3	C	B	A	.	.	(IV.56)
4	.	.	.	A	C	(IV.54)
5	B	A	C	.	.	(IV.55)
6	.	.	.	B	C	(IV.54)
7	A	C	B	.	.	(IV.56)
8	.	.	.	B	A	(IV.54)
9	C	B	A	.	.	(IV.55)
10	.	.	.	C	A	(IV.54)
11	B	A	C	.	.	(IV.56)
12	.	.	.	C	B	(IV.54)

The partial derivatives in Equations IV.54, IV.55, and IV.56 are then substituted into the partial derivatives of Equation IV.50 which are given below:

$$\frac{\partial V_{d_k}}{\partial \mathcal{R}\{V_m^\delta\}} = \frac{1}{2}\mathrm{j}\sum_{i=1}^{12}\left[\left(\frac{\partial V_{di_m}}{\partial \mathcal{R}\{V_m^\delta\}}\Psi^*_{i_{m+k}}\right)^* + \begin{cases} -\dfrac{\partial V_{di_m}}{\partial \mathcal{R}\{V_m^\delta\}}\Psi_{i_{k-m}} & m \leqslant k \\ \dfrac{\partial V_{di_m}}{\partial \mathcal{R}\{V_m^\delta\}}\Psi^*_{i_{m-k}} & m \geqslant k \end{cases}\right], k > 0, \quad \text{(IV.57)}$$

$$\frac{\partial V_{d_k}}{\partial \mathcal{I}\{V_m^\delta\}} = \frac{1}{2}\mathrm{j}\sum_{i=1}^{12}\left[\left(\frac{\partial V_{di_m}}{\partial \mathcal{I}\{V_m^\delta\}}\Psi^*_{i_{m+k}}\right)^* + \begin{cases} -\dfrac{\partial V_{di_m}}{\partial \mathcal{I}\{V_m^\delta\}}\Psi_{i_{k-m}} & m \leqslant k \\ \dfrac{\partial V_{di_m}}{\partial \mathcal{I}\{V_m^\delta\}}\Psi^*_{i_{m-k}} & m \geqslant k \end{cases}\right], k > 0. \quad \text{(IV.58)}$$

This completes the derivation of the partial derivative of dc current mismatch equation with respect to terminal voltage.

IV.2.2 With Respect to Direct Current Ripple Variation

Apart from being the most significant term in the dc current mismatch equation, the dc ripple affects the commutation currents, and also causes a voltage drop through the commutating reactance. These last two effects mean that the dc voltage is a function of the dc current ripple. It is therefore necessary to obtain partial derivatives of the dc voltage harmonics in a similar manner to that undertaken already for the derivative with respect to terminal voltage. Differentiating the dc current mismatch equation;

$$\frac{\partial \mathcal{R}\{F_{I_k}\}}{\partial \mathcal{R}\{I_{d_m}\}} = \mathcal{R}\begin{cases} -Y_{d_k}\dfrac{\partial V_{d_k}}{\partial \mathcal{R}\{I_{d_m}\}} & k \neq m \\ 1 - Y_{d_k}\dfrac{\partial V_{d_k}}{\partial \mathcal{R}\{I_{d_m}\}} & k = m \end{cases}, \quad \text{(IV.59)}$$

$$\frac{\partial \mathcal{R}\{F_{I_k}\}}{\partial \mathcal{I}\{I_{d_m}\}} = \mathcal{R}\left\{-Y_{d_k}\frac{\partial V_{d_k}}{\partial \mathcal{I}\{I_{d_m}\}}\right\}, \quad \text{(IV.60)}$$

$$\frac{\partial \mathcal{I}\{F_{I_k}\}}{\partial \mathcal{I}\{I_{d_m}\}} = \mathcal{I}\begin{cases} -Y_{d_k}\dfrac{\partial V_{d_k}}{\partial \mathcal{I}\{I_{d_m}\}} & k \neq m \\ \mathrm{j} - Y_{d_k}\dfrac{\partial V_{d_k}}{\partial \mathcal{I}\{I_{d_m}\}} & k = m \end{cases}, \quad \text{(IV.61)}$$

$$\frac{\partial \mathcal{I}\{F_{I_k}\}}{\partial \mathcal{R}\{I_{d_m}\}} = \mathcal{I}\left\{-Y_{d_k}\frac{\partial V_{d_k}}{\partial \mathcal{R}\{I_{d_m}\}}\right\}. \quad \text{(IV.62)}$$

Differentiating Equation IV.50 with respect to $\mathcal{R}\{I_{d_m}\}$ yields

$$\frac{\partial V_{d_k}}{\partial \mathcal{R}\{I_{d_m}\}} = \frac{1}{2}\mathrm{j}\sum_{i=1}^{12}\left[\left(\frac{\partial V_{di_m}}{\partial \mathcal{R}\{I_{d_m}\}}\Psi^*_{i_{m+k}}\right)^* + \begin{cases} -\dfrac{\partial V_{di_m}}{\partial \mathcal{R}\{I_{d_m}\}}\Psi_{i_{k-m}} & m \leqslant k \\ \dfrac{\partial V_{di_m}}{\partial \mathcal{R}\{I_{d_m}\}}\Psi^*_{i_{m-k}} & m \geqslant k \end{cases}\right], \quad k > 0.$$

(IV.63)

Similarly,

$$\frac{\partial V_{d_k}}{\partial \mathcal{I}\{I_{d_m}\}} = \frac{1}{2}\mathrm{j}\sum_{i=1}^{12}\left[\left(\frac{\partial V_{di_m}}{\partial \mathcal{I}\{I_{d_m}\}}\Psi^*_{i_{m+k}}\right)^* + \begin{cases} -\dfrac{\partial V_{di_m}}{\partial \mathcal{I}\{I_{d_m}\}}\Psi_{i_{k-m}} & m \leqslant k \\ \dfrac{\partial V_{di_m}}{\partial \mathcal{I}\{I_{d_m}\}}\Psi^*_{i_{m-k}} & m \geqslant k \end{cases}\right], \quad k > 0.$$

(IV.64)

The partial derivatives $\partial V_{di_m}/\partial \mathcal{I}\{I_{d_m}\}$ are obtained from Equations IV.51, IV.52, and IV.53:

$$\frac{\partial V_{di_m}}{\partial \mathcal{R}\{I_{d_m}\}} = -\mathrm{j}k\omega(L_+ + L_-) \quad \text{(IV.65)}$$

during normal conduction, and

$$\frac{\partial V_{di_m}}{\partial \mathcal{R}\{I_{d_m}\}} = \mathrm{j}k\omega\left[L_0 + \frac{L_e L_b}{L_e + L_b}\right] \quad \text{(IV.66)}$$

during any commutation. Again, the imaginary partial derivatives are obtained by the Cauchy Reimann equations. The correct phase subscripts can be obtained from Table IV.3. This completes the linearized dependence of the dc mismatch upon dc ripple variation.

IV.2.3 With Respect to End of Commutation Variation

The effect of variation in the end of commutation angle ϕ_h is limited solely to the sampling of relevant dc voltage sections. This is best explained with reference to the following equation for the dc voltage:

$$V_d = \frac{1}{2}\mathrm{j}\sum_{i=1}^{12} V_{di} \otimes \Psi_i. \qquad \text{(IV.67)}$$

In this equation, only two of the twelve Ψ_i are functions of ϕ_h. Table IV.4 shows that Ψ_{2h} and Ψ_{2h-1} are functions of ϕ_h. Equation IV.50 can therefore be differentiated to yield:

$$\begin{aligned}\frac{\partial V_{d_k}}{\partial \phi_h} = \frac{1}{2}\mathrm{j}\Bigg(&\sum_{l=0}^{k} V_{d(2h-1)_l}\frac{\partial \Psi_{2h_{k-l}}}{\partial \phi_h} - \sum_{l=k}^{n_h} V_{d(2h-1)_l}\left(\frac{\partial \Psi_{2h_{l-k}}}{\partial \phi_h}\right)^* \\ &- \sum_{l=0}^{n_h}\left[V_{d(2h-1)_l}\left(\frac{\partial \Psi_{2h_{l-k}}}{\partial \phi_h}\right)^*\right]^* - \sum_{l=0}^{k} V_{d(2h)_l}\frac{\partial \Psi_{2h_{k-l}}}{\partial \phi_h} \\ &+ \sum_{l=k}^{n_h} V_{d(2h)_l}\left(\frac{\partial \Psi_{2h_{l-k}}}{\partial \phi_h}\right)^* + \sum_{l=0}^{n_h}\left[V_{d(2h)_l}\left(\frac{\partial \Psi_{2h_{l-k}}}{\partial \phi_h}\right)^*\right]^*\Bigg). \end{aligned} \qquad \text{(IV.68)}$$

Table IV.4. Limits of converter states for use in sampling functions

sample (i)	a_i	b_i
1	θ_1	ϕ_1
2	ϕ_1	θ_2
3	θ_2	ϕ_2
4	ϕ_2	θ_3
5	θ_3	ϕ_3
6	ϕ_3	θ_4
7	θ_4	ϕ_4
8	ϕ_4	θ_5
9	θ_5	ϕ_5
10	ϕ_5	θ_6
11	θ_6	ϕ_6
12	ϕ_6	θ_1

Differentiating the dc ripple mismatch Equation IV.45 yields:

$$\frac{\partial \mathcal{R}\{F_{I_k}\}}{\partial \phi_h} = \mathcal{R}\left\{ -Y_{d_k} \frac{\partial V_{d_k}}{\partial \phi_h} \right\}, \tag{IV.69}$$

and

$$\frac{\partial \mathcal{I}\{F_{I_k}\}}{\partial \phi_h} = \mathcal{I}\left\{ -Y_{d_k} \frac{\partial V_{d_k}}{\partial \phi_h} \right\}. \tag{IV.70}$$

Substituting Equation IV.68 into IV.69 and IV.70 gives the required partial derivative of dc ripple mismatch with respect to end of commutation.

IV.2.4 With Respect to Firing Angle Variation

The partial derivatives of dc ripple mismatch with respect to firing angle are obtained in an exactly similar manner. The result is:

$$\frac{\partial \mathcal{R}\{F_{I_k}\}}{\partial \theta_h} = \mathcal{R}\left\{ -Y_{d_k} \frac{\partial V_{d_k}}{\partial \theta_h} \right\}, \tag{IV.71}$$

and

$$\frac{\partial \mathcal{I}\{F_{I_k}\}}{\partial \theta_h} = \mathcal{I}\left\{ -Y_{d_k} \frac{\partial V_{d_k}}{\partial \theta_h} \right\}, \tag{IV.72}$$

where

$$\begin{aligned}\frac{\partial V_{d_k}}{\partial \theta_h} = \frac{1}{2}\mathrm{j}\Bigg(& \sum_{l=0}^{k} V_{d(2h-2)_l} \frac{\partial \Psi_{2h_{k-l}}}{\partial \theta_h} - \sum_{l=k}^{n_h} V_{d(2h-2)_l} \left(\frac{\partial \Psi_{2h_{l-k}}}{\partial \theta_h} \right)^* \\ & - \sum_{l=0}^{n_h} \left[V_{d(2h-2)_l} \left(\frac{\partial \Psi_{2h-1_{l-k}}}{\partial \theta_h} \right)^* \right]^* - \sum_{l=0}^{k} V_{d(2h-1)_l} \frac{\partial \Psi_{2h-1_{k-l}}}{\partial \theta_h} \\ & + \sum_{l=k}^{n_h} V_{d(2h-1)_l} \left(\frac{\partial \Psi_{2h-1_{l-k}}}{\partial \theta_h} \right)^* + \sum_{l=0}^{n_h} \left[V_{d(2h-1)_l} \left(\frac{\partial \Psi_{2h-1_{l-k}}}{\partial \theta_h} \right)^* \right]^* \Bigg).\end{aligned} \tag{IV.73}$$

IV.3 End of Commutation Mismatch Partial Derivatives

The end of commutation mismatch is the current in the phase that is commutating off, at the end of the commutation. This current should be zero, and is given by

$$F_{\phi_i} = \mathcal{I}\left\{ \mathrm{j}(-D + I_{d_0}) + \sum_{k=1}^{n_h} F_{\phi_{i_k}} \mathrm{e}^{\mathrm{j}k\phi_i} \right\} = 0 \tag{IV.74}$$

for a commutation on the positive rail, where

$$D = -\mathcal{I}\left\{\sum_{k=1}^{n_h} I_{ci_k} e^{jk\theta_i}\right\} \tag{IV.75}$$

$$F_{\phi_{i_k}} = I_{d_k} - I_{ci_k} \tag{IV.76}$$

$$I_{ci_k} = \frac{jk\omega L_e I_{d_k} - V_k^{eb}}{jk\omega L_{eb}}. \tag{IV.77}$$

The end of commutation mismatch equation is therefore a function of all the variables except the average delay angle α_0. A commutation on the negative rail is accounted for by substituting $-I_{d_m}$ in the above equations for I_{d_m}.

VI.3.1 With Respect to AC Phase Voltage Variation

Differentiating Equation IV.74 with respect to an arbitrary voltage phase and harmonic yields:

$$\frac{\partial F_{\phi_i}}{\partial \mathcal{R}\{V_m^\delta\}} = \mathcal{I}\left\{ -\frac{\partial D}{\partial \mathcal{R}\{V_m^\delta\}} - \frac{\partial I_{ci_m}}{\partial \mathcal{R}\{V_m^\delta\}} e^{jm\phi_i}\right\} \tag{IV.78}$$

and

$$\frac{\partial F_{\phi_i}}{\partial \mathcal{I}\{V_m^\delta\}} = \mathcal{R}\left\{ -\frac{\partial D}{\partial \mathcal{I}\{V_m^\delta\}} - \frac{\partial I_{ci_m}}{\partial \mathcal{I}\{V_m^\delta\}} e^{jm\phi_i}\right\}. \tag{IV.79}$$

The partial derivatives in these equations are obtained from Equations IV.75 to IV.77 as

$$\frac{\partial D}{\partial \mathcal{R}\{V_m^\delta\}} = \frac{1}{jm\omega(L_e + L_b)} e^{jk\theta_i}, \tag{IV.80}$$

$$\frac{\partial I_{ci_m}}{\partial \mathcal{R}\{V_m^\delta\}} = \frac{-1}{jm\omega(L_e + L_b)}, \tag{IV.81}$$

$$\frac{\partial D}{\partial \mathcal{I}\{V_m^\delta\}} = \frac{j}{jm\omega(L_e + L_b)} e^{jk\theta_i}, \tag{IV.82}$$

$$\frac{\partial I_{ci_m}}{\partial \mathcal{I}\{V_m^\delta\}} = \frac{-j}{jm\omega(L_e + L_b)}. \tag{IV.83}$$

It is assumed in the above analysis that V_m^δ corresponds to a phase ending conduction. Multiplying by -1 gives the required partial derivative for the case that V_m^δ corresponds to a phase beginning conduction.

IV.3.2 With Respect to Direct Current Ripple Variation

The analysis is similar to that for ac phase voltage variation. Differentiating Equation IV.74 with respect to dc ripple yields

$$\frac{\partial F_{\phi_i}}{\partial \mathcal{R}\{I_{d_m}\}} = \mathcal{I}\left\{ -\frac{\partial D}{\partial \mathcal{R}\{I_{d_m}\}} - \frac{\partial I_{ci_m}}{\partial \mathcal{R}\{I_{d_m}\}} e^{jm\phi_i} \right\}, \qquad \text{(IV.84)}$$

and

$$\frac{\partial F_{\phi_i}}{\partial \mathcal{I}\{I_{d_m}\}} = \mathcal{R}\left\{ -\frac{\partial D}{\partial \mathcal{I}\{I_{d_m}\}} - \frac{\partial I_{ci_m}}{\partial \mathcal{I}\{I_{d_m}\}} e^{jm\phi_i} \right\}. \qquad \text{(IV.85)}$$

The partial derivatives in these equations are obtained from Equations IV.75 to IV.77 as

$$\frac{\partial D}{\partial \mathcal{R}\{I_{d_m}\}} = \frac{L_e}{L_e + L_b} e^{jk\theta_i}, \qquad \text{(IV.86)}$$

$$\frac{\partial I_{ci_m}}{\partial \mathcal{R}\{I_{d_m}\}} = \frac{-L_e}{L_e + L_b}, \qquad \text{(IV.87)}$$

$$\frac{\partial D}{\partial \mathcal{I}\{I_{d_m}\}} = \frac{jL_e}{L_e + L_b} e^{jk\theta_i}, \qquad \text{(IV.88)}$$

$$\frac{\partial I_{ci_m}}{\partial \mathcal{I}\{I_{d_m}\}} = \frac{-jL_e}{L_e + L_b}. \qquad \text{(IV.89)}$$

This analysis assumes that the commutation is on the positive rail. A similar analysis holds for a commutation on the negative rail, but with $-I_{d_m}$ substituted into equation IV.74.

IV.3.3 With Respect to End of Commutation Variation

This partial derivative gives the effect on the 'residual' commutating-off current at the end of the commutation, if the end of commutation is moved. It is obtained simply by differentiating Equation IV.74 with respect to ϕ_i to yield

$$\frac{\partial F_{\phi_i}}{\partial \phi_i} = \mathcal{I}\left\{ \sum_{k=1}^{n_h} jkF_{\phi_{i_k}} e^{jk\phi_i} \right\} \qquad \text{(IV.90)}$$

IV.3.4 With Respect to Firing Instant Variation

The dc offset to the commutation, D, is a function of the firing instant, and so the only effect of θ_i on F_{ϕ_i} is through D. Differentiating the expression for D, Equation IV.75, gives the required partial derivative.

$$\frac{\partial F_{\phi_i}}{\partial \theta_i} = \mathcal{I}\left\{ -\sum_{k=1}^{n_h} \mathrm{j}kI_{c_k}\mathrm{e}^{\mathrm{j}k\theta_i} \right\} \tag{IV.91}$$

IV.4 Firing Instant Mismatch Equation Partial Derivatives

For the implemented constant current controller, the firing instant mismatch is not a function of the ac terminal voltage. This is because the firing order is obtained solely from monitoring the dc current. This also means that the firing mismatch is not a function of the end of commutation instants. The firing instant mismatch is a function only of the dc ripple, the firing angle, and the average firing order. The firing mismatch equation is

$$F_{\theta_i} = \mathcal{I}\left\{ \mathrm{j}(\beta_i + \alpha_0 - \theta_i) + \sum_{k=1}^{n_h} \alpha_{i_k}\mathrm{e}^{\mathrm{j}k\theta_i} \right\} = 0, \tag{IV.92}$$

where

$$\alpha_{i_k} = \frac{G}{1 + \mathrm{j}k\omega T}\left(P + \frac{1}{\mathrm{j}k\omega T_I}\right) I_{dk}. \tag{IV.93}$$

The partial derivatives are easily obtained as;

$$\frac{\partial F_{\theta_i}}{\partial \alpha_0} = 1, \tag{IV.94}$$

$$\frac{\partial F_{\theta_i}}{\partial \theta_i} = \mathcal{I}\left\{ -\mathrm{j} + \sum_{k=1}^{n_h} \mathrm{j}k\theta_i \alpha_{i_k}\mathrm{e}^{\mathrm{j}k\theta_i} \right\}, \tag{IV.95}$$

$$\frac{\partial F_{\theta_i}}{\partial \mathcal{R}\{I_{d_k}\}} = \mathcal{I}\left\{ \sum_{k=1}^{n_h} \frac{\partial \alpha_{i_k}}{\partial I_{d_k}} \mathrm{e}^{\mathrm{j}k\theta_i} \right\}, \tag{IV.96}$$

$$\frac{\partial F_{\theta_i}}{\partial \mathcal{I}\{I_{d_k}\}} = \mathcal{R}\left\{ \sum_{k=1}^{n_h} \frac{\partial \alpha_{i_k}}{\partial I_{d_k}} \mathrm{e}^{\mathrm{j}k\theta_i} \right\}, \tag{IV.97}$$

where

$$\frac{\partial \alpha_{i_k}}{\partial I_{d_k}} = \frac{G}{1 + \mathrm{j}k\omega T}\left(P + \frac{1}{\mathrm{j}k\omega T_I}\right). \tag{IV.98}$$

IV.5 Average Delay Angle Partial Derivatives

The average delay angle, α_0, is obtained by requiring that the average dc voltage, when applied to the dc system, should result in the current order. Thus α_0 is a control variable, required for the case of constant current control. The mismatch equation is

$$F_{\alpha_0} = \mathcal{I}\{I_{d_0} - (V_{d_0} - V_{dc})Y_{d_0}\}, \tag{IV.99}$$

where V_{dc} represents a dc voltage source, and

$$V_{d_0} = \frac{1}{2}\mathrm{j}\sum_{i=1}^{12}\left\{\sum_{l=0}^{n_h}(V_{di_l}\Psi_{i_l}{}^{*})\right\}. \tag{IV.100}$$

Thus F_{α_0} is a function of all the converter variables, with the exception of α_0 itself. Analysis of the partial derivatives of Equation IV.99 is similar to that for the partial derivatives of the dc ripple mismatch.

IV.5.1 With Respect to AC Phase Voltage Variation

Differentiating the mismatch Equation IV.99 with respect to an arbitrary phase and harmonic of ac voltage yields

$$\frac{\partial F_{\alpha_0}}{\partial \mathcal{R}\{V_m^\delta\}} = \mathcal{I}\left\{-Y_{d_0}\frac{\partial V_{d_0}}{\partial \mathcal{R}\{V_m^\delta\}}\right\}, \tag{IV.101}$$

and

$$\frac{\partial F_{\alpha_0}}{\partial \mathcal{I}\{V_m^\delta\}} = \mathcal{I}\left\{-Y_{d_0}\frac{\partial V_{d_0}}{\partial \mathcal{I}\{V_m^\delta\}}\right\}. \tag{IV.102}$$

Differentiating Equation IV.100 yields

$$\frac{\partial V_{d_0}}{\partial \mathcal{R}\{V_m^\delta\}} = \frac{1}{2}\mathrm{j}\sum_{i=1}^{12}\left(\frac{\partial V_{di_m}}{\partial \mathcal{R}\{V_m^\delta\}}\Psi_{i_m}{}^{*}\right), \tag{IV.103}$$

and

$$\frac{\partial V_{d_0}}{\partial \mathcal{I}\{V_m^\delta\}} = \frac{1}{2}\mathrm{j}\sum_{i=1}^{12}\left(\frac{\partial V_{di_m}}{\partial \mathcal{I}\{V_m^\delta\}}\Psi_{i_m}{}^{*}\right), \tag{IV.104}$$

which when substituted back into Equations IV.101 and IV.102 give the required partial derivatives. The remaining partial derivatives, $\partial V_{di_m}/\partial\mathcal{R}\{V_m^\delta\}$ etc., have already been obtained in Equations IV.51, IV.52, and IV.53, and by reference to Table IV.3.

IV.5.2 With Respect to DC Ripple Current Variation

Variation in the dc ripple affects the dc voltage samples in a similar manner to that of a variation in the terminal voltage above. Differentiating the mismatch Equation IV.99 with respect to an arbitrary phase and harmonic of dc current ripple yields

$$\frac{\partial F_{\alpha_0}}{\partial \mathcal{R}\{I_{d_m}\}} = \mathcal{I}\left\{ - Y_{d_0} \frac{\partial V_{d_0}}{\partial \mathcal{R}\{I_{d_m}\}} \right\}, \tag{IV.105}$$

and

$$\frac{\partial F_{\alpha_0}}{\partial \mathcal{I}\{I_{d_m}\}} = \mathcal{I}\left\{ - Y_{d_0} \frac{\partial V_{d_0}}{\partial \mathcal{I}\{I_{d_m}\}} \right\}. \tag{IV.106}$$

Differentiating Equation IV.100 yields

$$\frac{\partial V_{d_0}}{\partial \mathcal{R}\{I_{d_m}\}} = \frac{1}{2}\mathrm{j}\sum_{i=1}^{12} \left(\frac{\partial V_{di_m}}{\partial \mathcal{R}\{I_{d_m}\}} \Psi^*_{i_m} \right), \tag{IV.107}$$

and

$$\frac{\partial V_{d_0}}{\partial \mathcal{I}\{I_{d_m}\}} = \frac{1}{2}\mathrm{j}\sum_{i=1}^{12} \left(\frac{\partial V_{di_m}}{\partial \mathcal{I}\{I_{d_m}\}} \Psi^*_{i_m} \right), \tag{IV.108}$$

which when substituted back into Equations IV.105 and IV.106 give the required partial derivatives. The remaining partial derivatives, $\partial V_{di_m}/\partial \mathcal{R}\{I_{d_m}\}$ etc., have already been obtained in Equations IV.65, and IV.66.

IV.5.3 With Respect to End of Commutation Variation

The effect of a variation in the end of commutation is to modify the sampling of the dc voltage sections in Equation IV.100. Differentiating Equation IV.99 with respect to ϕ_h yields

$$\frac{\partial F_{\alpha_0}}{\partial \phi_h} = \mathcal{I}\left\{ - Y_{d_0} \frac{\partial V_{d_0}}{\partial \phi_h} \right\}. \tag{IV.109}$$

This requires the partial derivative $\partial V_{d_0}/\partial \phi_h$. Differentiating Equation IV.100 yields

$$\frac{\partial V_{d_0}}{\partial \phi_h} = \frac{1}{2}\mathrm{j}\left\{ \sum_{l=1}^{n_h} (V_{d(2h-1)_l} \frac{\partial \Psi^*_{(2h-1)_l}}{\partial \phi_h}) + \sum_{l=1}^{n_h} \left(V_{d(2h)_l} \frac{\partial \Psi^*_{2h_l}}{\partial \phi_h} \right) \right\}, \tag{IV.110}$$

where use has been made of Table IV.4 to determine the only two sampling functions that are affected by ϕ_h.

IV.5.4 With Respect to Firing Angle Variation

The effect of a change in the firing angle on the average delay angle mismatch equation is similar to that for a change in the end of commutation angle. The sampling of the dc voltage sections is modified, and this changes the average dc voltage. The analysis carried out above for the end of commutation variation is also valid in this case, with only the two affected sampling functions, as determined from Table IV.4 being different. The result is that

$$\frac{\partial F_{\alpha_0}}{\partial \theta_h} = \mathcal{I}\left\{ - Y_{d_0} \frac{\partial V_{d_0}}{\partial \theta_h} \right\}, \tag{IV.111}$$

where

$$\frac{\partial V_{d_0}}{\partial \theta_h} = \frac{1}{2}\mathrm{j}\left\{ \sum_{l=1}^{n_h} \left(V_{d(2h-2)_l} \frac{\partial \Psi^*_{(2h-2)_l}}{\partial \theta_h} \right) + \sum_{l=1}^{n_h} \left(V_{d(2h-1)_l} \frac{\partial \Psi^*_{(2h-1)_l}}{\partial \theta_h} \right) \right\}. \tag{IV.112}$$

This completes the derivation of the partial derivatives required for the Jacobian matrix, as the average delay angle mismatch equation is not a function of the average delay angle.

Appendix V

THE IMPEDANCE TENSOR

V.1 Impedance Derivation

The impedance tensor is a convenient framework for the linearization of the converter in the steady state. It is used in Chapter 10 to derive a linearized converter equivalent impedance for use in the analysis of ac–dc system interactions.

A power system component can be represented by a voltage controlled current source: $I = F(V)$, where in general I and V are vectors of harmonic phasors. The function F, is a complex vector function, and may be non-linear, and non-analytic. If F is linear, it may include linear cross-coupling between harmonics, and may be non-analytic, i.e. harmonic cross coupling and phase dependence do not imply non-linearity in the harmonic domain.

The linearized response of F to a single applied harmonic may be calculated by taking the first partial derivatives in rectangular coordinates:

$$\begin{bmatrix} \Delta I_R \\ \Delta I_I \end{bmatrix} = \begin{bmatrix} \dfrac{\partial F_R}{\partial V_R} & \dfrac{\partial F_R}{\partial V_I} \\ \dfrac{\partial F_I}{\partial V_R} & \dfrac{\partial F_I}{\partial V_I} \end{bmatrix} \begin{bmatrix} \Delta V_R \\ \Delta V_I \end{bmatrix}, \tag{V.1}$$

where F has been expanded into its component parts:

$$I_R + \mathrm{j}I_I = F_R(V_R, V_I) + \mathrm{j}F_I(V_R, V_I). \tag{V.2}$$

If the Cauchy-Rieman conditions

$$\frac{\partial F_R}{\partial V_R} = \frac{\partial F_I}{\partial V_I} \tag{V.3}$$

$$\frac{\partial F_I}{\partial V_R} = -\frac{\partial F_R}{\partial V_I} \tag{V.4}$$

hold, then the matrix of partial derivatives is of the form

$$\begin{bmatrix} a & -b \\ b & a \end{bmatrix},$$

which is a type of matrix isomorphic with the complex number field. The linearization then becomes

$$\Delta I = \frac{\partial F}{\partial V}\Delta V. \tag{V.5}$$

The complex admittance representation of a linearization can therefore only be applied to analytic current injections. Non-analytic injections are quite common, for example in the load flow, $I = S^*/V^*$ for a $P + jQ$ bus, which is also nonlinear. An example of a linear, but nonanalytic injection would be:

$$I = Y_1 V + Y_2 V^*. \tag{V.6}$$

This is linear, since if

$$I_1 = Y_1 V_1 + Y_2 V_1^* \tag{V.7}$$

$$I_2 = Y_1 V_2 + Y_2 V_2^* \tag{V.8}$$

$$V_3 = aV_1 + bV_2 \tag{V.9}$$

then

$$I_3 = Y_1 V_3 + Y_2 V_3^* \tag{V.10}$$

$$= Y_1(aV_1 + bV_2) + Y_2(aV_1 + bV_2)^* \tag{V.11}$$

$$= aY_1 V_1 + bY_1 V_2 + aY_2 V_1^* + bY_2 V_2^* \tag{V.12}$$

$$= aI_1 + bI_2. \tag{V.13}$$

For such a source, the complex admittance is;

$$Y = \frac{I}{V} \tag{V.14}$$

$$= \frac{Y_1 V + Y_2 V^*}{V} \tag{V.15}$$

$$= Y_1 + |Y_2| \angle(\angle Y_2 - 2\angle V) \tag{V.16}$$

The complex admittance twice traces a circular locus centered at Y_1 in the clockwise direction, as the angle of the applied voltage is varied through 2π. By the inverse mapping, $Z = 1/Y$, also traces a circular locus, but in the anti-clockwise direction. Next, it will be shown that any current injection, when linearized, can be written in the form

$$\Delta I = Y_1 \Delta V + Y_2 \Delta V^* \tag{V.17}$$

Expanding equation V.17 into components,

$$\begin{bmatrix} \Delta I_R \\ \Delta I_I \end{bmatrix} = \begin{bmatrix} Y_{1R} + Y_{2R} & Y_{2I} - Y_{1I} \\ Y_{1I} + Y_{2I} & Y_{1R} - Y_{2R} \end{bmatrix} \begin{bmatrix} \Delta V_R \\ \Delta V_I \end{bmatrix}, \tag{V.18}$$

Equating the matrices on the right hand sides of equations V.18 and V.1, the real and imaginary parts of Y_1 and Y_2 are readily expressed in terms of the partial derivatives:

$$\begin{aligned} Y_{1R} &= \frac{1}{2}\left(\frac{\partial F_R}{\partial V_R} + \frac{\partial F_I}{\partial V_I}\right) \\ Y_{1I} &= \frac{1}{2}\left(\frac{\partial F_I}{\partial V_R} - \frac{\partial F_R}{\partial V_I}\right) \\ Y_{2R} &= \frac{1}{2}\left(\frac{\partial F_R}{\partial V_R} - \frac{\partial F_I}{\partial V_I}\right) \\ Y_{2I} &= \frac{1}{2}\left(\frac{\partial F_R}{\partial V_I} + \frac{\partial F_I}{\partial V_R}\right) \end{aligned} \tag{V.19}$$

For nodal analysis, the conjugate operator is eliminated by treating conjugated voltages and currents as additional variables. The conjugation operator then introduces a harmonic cross coupling term between the conjugated variables, and the linearized admittance of any source can be written:

$$\begin{bmatrix} \Delta I \\ \Delta I^* \end{bmatrix} = \begin{bmatrix} Y_1 & Y_2 \\ Y_2^* & Y_1^* \end{bmatrix} \begin{bmatrix} \Delta V \\ \Delta V^* \end{bmatrix}, \tag{V.20}$$

This is the nodal analysis used in the transformer model of Chapter 7. If F is non-analytic, $Y_2 \neq 0$, and the complex admittance will be a circular locus. For positive frequency nodal analysis, the real matrix of partial derivatives is retained, and is henceforth called the admittance or impedance *tensor*:

$$\bar{Y} = \begin{bmatrix} y_{11} & y_{12} \\ y_{21} & y_{22} \end{bmatrix} = \begin{bmatrix} \dfrac{\partial F_R}{\partial V_R} & \dfrac{\partial F_R}{\partial V_I} \\ \dfrac{\partial F_I}{\partial V_R} & \dfrac{\partial F_I}{\partial V_I} \end{bmatrix}. \tag{V.21}$$

The nodal analysis is then performed in real components using a Cartesian vector representation of voltage and current phasors, and a second rank tensor representation of admittances and impedances. Tensors are widely used in physics to represent a relationship between vectors that is invariant under rotation of the coordinate axes. In this case, the coordinate axes are the real and imaginary components of the complex voltage phasor, which rotate under a shift in phase reference. The components of the voltage therefore transform like the elements of a vector under this rotation. Since the current is a vector function of the voltage, the matrix of partial derivatives can be written as the direct product of the gradient operator with the current vector, yielding a second rank tensor

$$\bar{Y} = \nabla \hat{I}^T, \tag{V.22}$$

Although the elements of the tensor admittance are modified by a change in phase reference, they transform in such a way that the relationship described by the tensor, dependence of current on voltage, is invariant. When the tensor admittance is extended to three phases and multiple harmonics, the dimension will be $6n_h$, but the rank of the tensor will still be two.

V.2 Phase Dependent Impedance

The circular complex locus of an impedance tensor is obtained in this section by injecting a 2π phase range of currents into the impedance tensor, calculating the resulting voltage, and then showing that the complex impedance defined by the ratio of the voltage to the current, corresponds to a circle in the complex plane. Given a current injected into an impedance tensor, the voltage developed is

$$\begin{bmatrix} V_R \\ V_I \end{bmatrix} = \begin{bmatrix} z_{11} & z_{12} \\ z_{21} & z_{22} \end{bmatrix} \begin{bmatrix} I_R \\ I_I \end{bmatrix} \tag{V.23}$$

which can be written in complex form:

$$V = [(z_{11}I_R + z_{12}I_I) + \mathrm{j}(z_{21}I_R + z_{22}I_I)]. \tag{V.24}$$

Dividing now to obtain the complex impedance:

$$Z = \frac{[(z_{11}I_R + z_{12}I_I) + \mathrm{j}(z_{21}I_R + z_{22}I_I)]}{I_R + \mathrm{j}I_R} \tag{V.25}$$

$$= \frac{[(z_{11}I_R + z_{12}I_I) + \mathrm{j}(z_{21}I_R + z_{22}I_I)](I_R - \mathrm{j}I_I)}{|I|^2} \tag{V.26}$$

Separating out the real and imaginary parts of the complex impedance:

$$Z_R = \frac{z_{11}I_R^2 + z_{12}I_II_R + z_{21}I_RI_I + z_{22}I_I^2}{|I|^2} \tag{V.27}$$

$$Z_I = \frac{z_{21}I_R^2 + z_{22}I_II_R - z_{11}I_RI_I - z_{12}I_I^2}{|I|^2} \tag{V.28}$$

Since the objective is to obtain a phase dependent locus, the polar transformation,

$$I_R = |I|\cos\theta \tag{V.29}$$

$$I_I = |I|\sin\theta \tag{V.30}$$

is applied, to yield:

$$Z_R = \frac{z_{11}|I|^2\cos^2\theta + (z_{12} + z_{21})|I|^2\sin\theta\cos\theta + z_{22}|I|^2\sin^2\theta}{|I|^2}$$

$$Z_I = \frac{z_{21}|I|^2\cos^2\theta + (z_{22} - z_{11})|I|^2\sin\theta\cos\theta - z_{12}|I|^2\sin^2\theta}{|I|^2} \tag{V.31}$$

The square magnitude of the current evidently cancels from equations V.31, resulting in a phase dependent impedance:

$$\begin{aligned} Z_R &= z_{11}c^2 + (z_{12} + z_{21})sc + z_{22}s^2 \\ Z_I &= z_{21}c^2 + (z_{22} - z_{11})sc - z_{12}s^2, \end{aligned} \tag{V.32}$$

where for the sake of brevity, the following notation has been defined:

$$s \overset{\text{def}}{=} \sin\theta \tag{V.33}$$

$$c \overset{\text{def}}{=} \cos\theta \tag{V.34}$$

$$\tag{V.35}$$

The following identities are therefore valid:

$$s^2 = 1 - c^2 \tag{V.36}$$

$$s^4 = 1 - 2c^2 + c^4 \tag{V.37}$$

$$s^3c = sc - sc^3 \tag{V.38}$$

$$s^2c^2 = c^2 - c^4 \tag{V.39}$$

To show that the phase dependent impedance corresponds to a circular locus, it is necessary only to show that the real and imaginary parts of the complex impedance satisfy the quadratic form for a circle in the plane:

$$(Z_R - a)^2 + (Z_I - b)^2 = r^2, \tag{V.40}$$

where a and b, are the coordinates of the circle center, and r is the radius. The problem is to determine the values of a, b and r, since Z_R and Z_I have been defined parametrically in terms of θ, the angle of the current injection. A lucky guess gave:

$$a = \frac{1}{2}(z_{11} + z_{22}) \tag{V.41}$$

$$b = \frac{1}{2}(-z_{12} + z_{21}) \tag{V.42}$$

$$r = \frac{1}{2}\sqrt{(z_{11} - z_{22})^2 + (z_{12} + z_{21})^2} \tag{V.43}$$

which is now verified by substitution into Equation V.40:

$$\begin{aligned} (Z_R - a)^2 + (Z_I - b)^2 =& (z_{11}c^2 + (z_{12} + z_{21})sc + z_{22}s^2 - \tfrac{1}{2}(z_{11} + z_{22}))^2 \\ &+ ((z_{21}c^2 + (z_{22} - z_{11})sc - z_{12}s^2 - \tfrac{1}{2}(-z_{12} + z_{21}))^2 \end{aligned} \tag{V.44}$$

expanding the RHS of equation V.44:

$$
\begin{aligned}
\text{RHS} = {} & \\
& z_{11}^2 c^4 + 2z_{11}(z_{12} + z_{21})sc^3 + 2z_{11}z_{22}s^2c^2 + 2z_{22}(z_{12} + z_{21})cs^3 \\
& + (z_{12} + z_{21})^2 s^2 c^2 + z_{22}^2 s^4 + \tfrac{1}{4}(z_{11} + z_{22})^2 - z_{11}(z_{11} + z_{22})c^2 \\
& - (z_{11} + z_{22})(z_{12} + z_{21})sc - z_{22}(z_{11} + z_{22})s^2 + z_{21}^2 c^4 \\
& + 2z_{21}(z_{22} - z_{11})sc^3 - 2z_{21}z_{12}s^2c^2 + (z_{22} - z_{11})^2 s^2 c^2 \\
& - 2z_{12}(z_{22} - z_{11})cs^3 + z_{12}^2 s^4 + \tfrac{1}{4}(-z_{12} + z_{21})^2 - z_{21}(-z_{12} + z_{21})c^2 \\
& - (z_{22} - z_{11})(-z_{12} + z_{21})sc + z_{12}(-z_{12} + z_{21})s^2
\end{aligned}
\tag{V.45}
$$

Equation V.45 is a trigonometric polynomial in powers of s and c. In order to simplify it, the identities of Equations V.36 to V.39 are substituted, and the coefficients of c^2, sc, sc^3, c^4 and the constant component are collected:

$$
\begin{aligned}
\text{RHS} = {} & \\
& [z_{11}^2 - 2z_{11}z_{22} - (z_{12} + z_{21})^2 + z_{22}^2 + z_{21}^2 + 2z_{21}z_{12} - (z_{22} - z_{11})^2 + z_{12}^2]c^4 \\
& + [2z_{11}(z_{12} + z_{21}) - 2z_{22}(z_{12} + z_{21}) + 2z_{21}(z_{22} - z_{11}) + 2z_{12}(z_{22} - z_{11})]sc^3 \\
& + [2z_{11}z_{22} + (z_{12} + z_{21})^2 - 2z_{22}^2 + (z_{22} - z_{11})(z_{11} + z_{22}) - 2z_{21}z_{12} \\
& + (z_{22} - z_{11})^2 - 2z_{12}^2 - (z_{12} + z_{21})(-z_{12} + z_{21})]c^2 + [2z_{22}(z_{12} + z_{21}) \\
& - (z_{11} + z_{22})(z_{12} + z_{21}) - 2z_{12}(z_{22} - z_{11}) - (z_{22} - z_{11})(-z_{12} + z_{21})]sc \\
& + [z_{22}^2 + \tfrac{1}{4}(z_{11} + z_{22})^2 - z_{22}(z_{11} + z_{22}) \\
& + z_{12}^2 + \tfrac{1}{4}(-z_{12} + z_{21})^2 + z_{12}(-z_{12} + z_{21})] \\
= {} & \tfrac{1}{4}\left[(z_{11} - z_{22})^2 + (z_{12} + z_{21})^2\right] \\
= {} & r^2
\end{aligned}
\tag{V.46}
$$

The trigonometric terms all cancel, leaving only a constant term, which shows that all points on the locus are the same distance, r, from the point $(a + jb)$. It remains only to show that for a full 2π range of applied current angles, all points on the circular locus are visited by the impedance. This is done by subtracting the locus position, $(a+jb)$, from the complex impedance locus. Considering first the real part:

$$\begin{aligned}
Z_R - a &= z_{11}\cos^2\theta + \tfrac{1}{2}(z_{12}+z_{21})\sin 2\theta + z_{22}\sin^2\theta - \tfrac{1}{2}(z_{11}+z_{22}) \\
&= z_{11}\cos^2\theta + \tfrac{1}{2}(z_{12}+z_{21})\sin 2\theta + z_{22} - z_{22}\cos^2\theta - \tfrac{1}{2}(z_{11}+z_{22}) \\
&= (z_{11}-z_{22})\cos^2\theta + \tfrac{1}{2}(z_{12}+z_{21})\sin 2\theta - \tfrac{1}{2}(z_{11}-z_{22}) \\
&= \tfrac{1}{2}(z_{11}-z_{22}) + \tfrac{1}{2}(z_{11}-z_{22})\cos 2\theta + \tfrac{1}{2}(z_{12}+z_{21})\sin 2\theta - \tfrac{1}{2}(z_{11}-z_{22}) \\
&= \tfrac{1}{2}\sqrt{(z_{11}-z_{22})^2+(z_{12}+z_{21})^2}\left[\frac{(z_{11}-z_{22})}{\sqrt{(z_{11}-z_{22})^2+(z_{12}+z_{21})^2}}\cos 2\theta \right. \\
&\quad \left. + \frac{(z_{12}+z_{21})}{\sqrt{(z_{11}-z_{22})^2+(z_{12}+z_{21})^2}}\sin 2\theta\right] \\
&= \tfrac{1}{2}\sqrt{(z_{11}-z_{22})^2+(z_{12}+z_{21})^2}\cos(2\theta+\gamma)
\end{aligned} \tag{V.47}$$

where $\gamma = -\tan^{-1}((z_{12}+z_{21})/(z_{11}-z_{22}))$. The analysis for the imaginary part is similar, yielding:

$$Z_I - b = \frac{1}{2}\sqrt{(z_{11}-z_{22})^2+(z_{12}+z_{21})^2}\sin(2\theta+\gamma) \tag{V.48}$$

Equations V.47 and V.48 indicate that as the angle of the applied current is increased from zero to 2π radians, the circular impedance locus is traversed twice in a counter clockwise direction, starting from a point on the locus that makes an angle γ radians to the real axis. This is illustrated in Figure V.1 below.

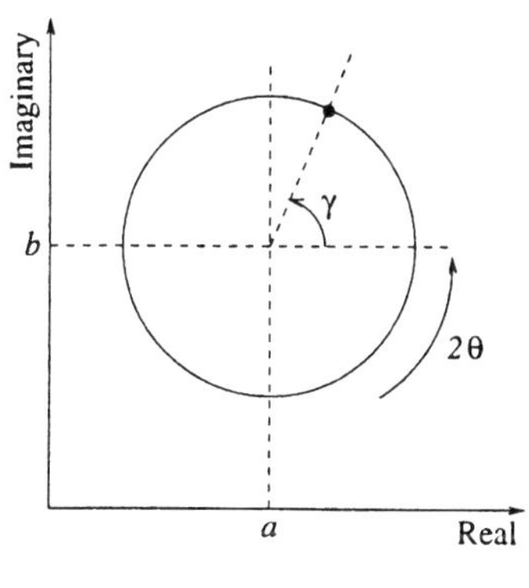

Figure V.1 Complex impedance locus for a tensor impedance. The locus point rotates counter clockwise twice, starting from the angle γ, as the current injection ranges in angle from 0 to 2π

Since the impedance tensor is a matrix, it may not be invertible, in which case the determinant is zero. However

$$
\begin{aligned}
a^2 + b^2 - r^2 &= \tfrac{1}{4}\left[(z_{11} + z_{22})^2 + (-z_{12} + z_{21})^2 - (z_{11} - z_{22})^2 - (z_{12} + z_{21})^2\right] \\
&= z_{11}z_{22} - z_{12}z_{21} \\
&= \det(\bar{Z}),
\end{aligned}
\tag{V.49}
$$

and the tensor is invertible if and only if the complex circular locus does not intersect the origin. The x-axis intercepts of the circular locus correspond to a resistive impedance, so that

$$\bar{Z}I = RI. \tag{V.50}$$

The eigenvalues of $\bar{Z}$ are therefore the values of resistance where the locus intersects the x-axis. The characteristic equation is

$$\begin{vmatrix} (z_{11} - R) & z_{12} \\ z_{21} & (z_{22} - R) \end{vmatrix} = R^2 - R(z_{11} + z_{22}) - z_{12}z_{21} + z_{11}z_{22} \tag{V.51}$$

$$= 0, \tag{V.52}$$

which is a quadratic in R, and it is easy to show that a real solution for R is only possible if $r^2 > b^2$, i.e. the circular locus is close enough to the x-axis. If the x-axis intercepts are considered as being a type of resonance, then the eigenvectors of $\bar{Z}$ are those currents that have the correct angle to excite the resonance.

Appendix VI

TEST SYSTEMS

VI.1 CIGRE Benchmark

The test systems are based on the rectifier end of the CIGRE benchmark model. The inverter side has been replaced by a constant dc voltage source, E, as illustrated in Figure VI.1.

The benchmark model consists of a weak ac system, parallel resonant at the second harmonic, coupled via the rectifier to a dc system that is series resonant at the fundamental frequency. These features are shown in the impedance plots of Figures VI.2 and VI.3. The system therefore displays a composite resonance between the ac and dc systems. The ac system is balanced, and is connected in grounded star, as are both converter transformers on the ac side. Additional parameters for the system are listed in Table VI.1.

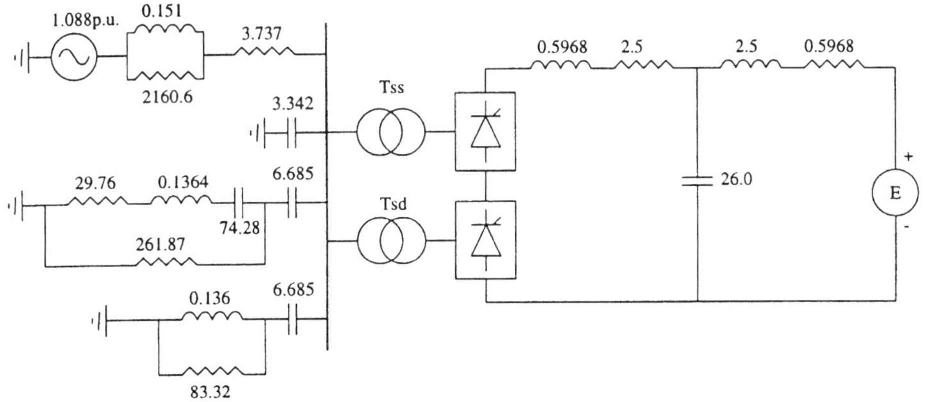

Figure VI.1 Rectifier end of the CIGRE benchmark model. Components values in Ω, H, and μF

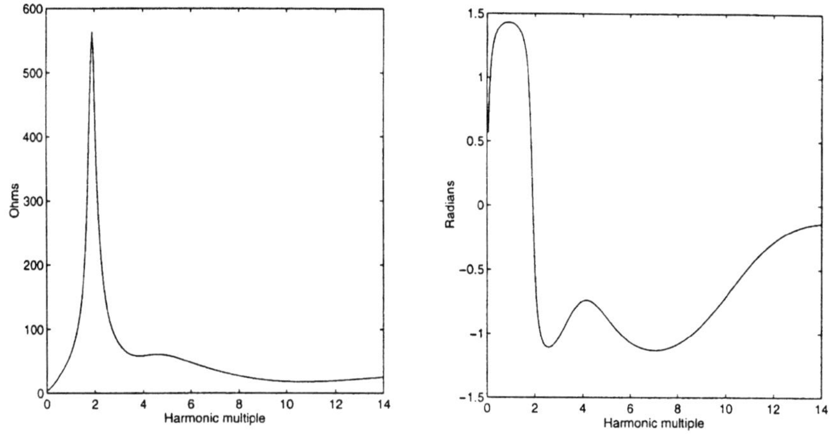

(a) Impedance magnitude (b) Impedance phase

Figure VI.2 Frequency scan of the CIGRE rectifier ac system impedance

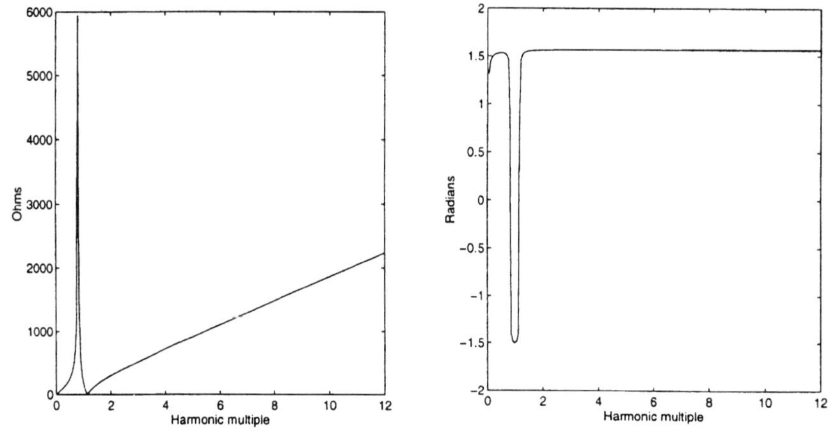

(a) Impedance magnitude (b) Impedance phase

Figure VI.3 Frequency scan of the CIGRE rectifier dc system impedance

Table VI.1 Parameters for the CIGRE benchmark rectifier

power base	603.73 MVA
primary voltage base	345 kV
secondary voltage base	213.4557 kV
nominal dc current	2000 A
nominal firing angle	15°
dc voltage source	4.179 p.u.
transformer leakage reactance	0.18 p.u.
transformer series resistance	0.01 p.u.
thyristor forward voltage drop	8.11E-6 p.u.
thyristor on resistance	0.001325 p.u.
dc current transducer time constant	0.001 s/rad
PI controller proportional gain	1.0989 rad/A(p.u.)
PI controller time constant	0.0091 s/rad

INDEX